how to
know the
freshwater
algae

The **Pictured Key Nature Series** has been published since 1944 by the Wm. C. Brown Company. The series was initiated in 1937 by the late Dr. H. E. Jaques, Professor Emeritus of Biology at Iowa Wesleyan University. Dr. Jaques' dedication to the interest of nature lovers in every walk of life has resulted in the prominent place this series fills for all who wonder **"How to Know."**

John F. Bamrick and Edward T. Cawley
Consulting Editors

The Pictured Key Nature Series

How to Know the
AQUATIC INSECTS, Lehmkuhl
AQUATIC PLANTS, Prescott
BEETLES, Arnett-Downie-Jaques, Second Edition
BUTTERFLIES, Ehrlich
FALL FLOWERS, Cuthbert
FERNS AND FERN ALLIES, Mickel
FRESHWATER ALGAE, Prescott, Third Edition
FRESHWATER FISHES, Eddy-Underhill, Third Edition
GILLED MUSHROOMS, Smith-Smith-Weber
GRASSES, Pohl, Third Edition
IMMATURE INSECTS, Chu
INSECTS, Bland-Jaques, Third Edition
LICHENS, Hale, Second Edition
LIVING THINGS, Winchester-Jaques, Second Edition
MAMMALS, Booth, Third Edition
MITES AND TICKS, McDaniel
MOSSES AND LIVERWORTS, Conard-Redfearn, Third Edition
NON-GILLED MUSHROOMS, Smith-Smith-Weber, Second Edition
PLANT FAMILIES, Jaques
POLLEN AND SPORES, Kapp
PROTOZOA, Jahn, Bovee, Jahn, Third Edition
SEAWEEDS, Abbott-Dawson, Second Edition
SEED PLANTS, Cronquist

SPIDERS, Kaston, Third Edition
SPRING FLOWERS, Cuthbert, Second Edition
TREES, Miller-Jaques, Third Edition
TRUE BUGS, Slater-Baranowski
TRUE SLIME MOLDS, Farr
WEEDS, Wilkinson-Jaques, Third Edition
WESTERN TREES, Baerg, Second Edition

how to
know the
freshwater
algae

Third Edition

G.W. Prescott
University of Montana

The Pictured Key Nature Series

Boston, Massachusetts   Burr Ridge, Illinois   Dubuque, Iowa
Madison, Wisconsin   New York, New York   San Francisco, California   St. Louis, Missouri

WCB/McGraw-Hill

A Division of The McGraw-Hill Companies

Copyright © 1954, 1964 by H.E. Jaques

Copyright © 1970, 1978 by Wm. C. Brown Company Publishers

Library of Congress Catalog Card Number: 77—91275

ISBN 0—697—04755—5 (Cloth)
ISBN 0—697—04754—7 (Paper)

Printed in the United States of America
22 23 24 25 26 27 28 29 30 DOC/DOC 0 9 8 7

Contents

Preface

This volume has been prepared to give the less experienced or non-professional student of the algae an opportunity to learn the scientific names (at least the genus name) of common freshwater forms. There are many fine manuals or handbooks that treat with freshwater algae, but none have an illustrated key to facilitate identification. It is hoped that once the key is understood, it will permit the identification of some 558 algal genera. (See p. 2 for definitions.)

The introductory section of the book gives information on where and how to collect algae, and how to preserve them. There is an introductory synopsis of the algal phyla to which freshwater algae belong, and in the Appendix there is a check list of algal genera according to families, orders and phyla so that the reader can learn how various algae are related or classified. A special Index has been prepared which includes a glossary in which many terms are illustrated.

Since the publication of previous editions of this illustrated key, many genera have been reported from the United States. The most common of these have been included and illustrated in the following key. This results in a rather lengthy key. Accordingly, the user will need to develop some patience.

The key is dichotomous, that is, the reader is presented with two descriptive choices and he must select the one which agrees with the alga in which he is interested—and then to follow the key according to numbers. Until some experience and judgment are developed, it may be necessary to follow out alternate dichotomies if one does not arrive at a satisfactory place in the key.

This book should be especially useful in view of the increased interest in algae as they relate to water quality, in addition to other economic importances possessed by these plants. They play an important role in the food chain for fish and sometimes in fish-kills. Some algae produce toxins which lead to the death of land animals and birds which make use of infested water. In addition to these reasons for becoming acquainted with algae, there are many interests involving purely scientific as well as aesthetic objectives.

Laboratory Procedure

Preserving

If samples are to be preserved an amount of 6-3-1 preservative equal to the volume of the specimens (and its water medium) may be added to the vial. This preservative is composed of 6 parts water, 3 parts 95% alcohol, and 1 part commercial formalin. If 5 cc of glycerin are added to each 100 cc of the preservative a medium is produced which protects the specimen against total loss should the preservative evaporate. Cork-stoppered vials, as a rule, are much more serviceable than screw-cap vials which permit a greater amount of evaporation because the tops loosen eespecially with some types of screw-caps) upon standing for a time.

Formalin-acetic acid-alcohol (FAA) makes an excellent preservative as well as a killing agent if material is later to be prepared for staining or cytological work. To 50 cc of 95% alcohol add 5 cc of glacial acetic acid, 10 cc of commercial formalin, and 35 cc of water. Proprionic acid may be substituted for the glacial acetic. This preservative is useless for Dinoflagellates with wall plates, however, because the acid causes plates to dissociate.

For general and incidental preserving, ordinary 3% formalin may be used if the above ingredients are not available. (Add 3 cc of commercial formalin to 37 cc of water to prepare 3% formalin.)

Preparing Mounts

For a study of most freshwater algae a compound microscope which has a 10X ocular and 10X and 43X objectives is needed. A 20X ocular is necessary if camera lucida (see glossary) drawings are to be made. Larger forms of algae such as the Characeae are best studied with a binocular dissecting microscope. Best illumination for the microscope is obtained from daylight because the algal pigments appear more naturally. In lieu of good daylight (light from a north window preferred), artificial light from a microscope lamp fitted with a daylight-blue bulb is used, or a lamp which has a blue filter. Naturally, all optical parts of the microscope should be kept free of dust, moisture, and fingerprints, using rice lens paper for cleaning. It is difficult enough to see microorganisms clearly when optical conditions are perfect. An eyepiece micrometer and a stage micrometer for microscope calibrations are essential if measurements are to be made.

In preparing mounts for the study of algae *small* amounts of material should be used, and spread out evenly in a thin layer. Dense clumps and opaque masses of algae in a microscope mount produce only disappointment, eye strain and headaches.

For the study of Diatoms it is necessary to have a clear and unobstructed view of the wall and its markings, and of the details of internal wall structures, free from chloroplasts and other cell contents. There are several methods for cleaning and clearing Diatom cells (frustules), some of which are rather complex, involving boiling in acid, *etc*. A simple procedure which is satisfactory for generic determination is to spread a bit of material in a generous drop of water on a microscope slide. The slide is held over a flame and the water brought to a boil. The smear is steamed thus for a few seconds, after which a drop of water or 5% glycerin can be added for microscopical examination.

For Diatom techniques see Burke (1937) and Fleming (1949). Also see Patrick and Reimer (1966) for an excellent description of collecting and clearing techniques. If semipermanent microscope mounts are desired, specimens may be placed on a slide, evenly spread out in a large drop of 5% glycerin. The slide should be set away under a dust-proof cover. Once or twice in an hour, and for several hours other drops of glycerin are added to the mount until, through evaporation of the water, approximately 100% glycerin remains about the specimens. To this a small drop of glycerin jelly is added and the coverslip put in place. Care should be used to add just enough jelly to fill out the area under the coverslip to be used so as not to allow leakage from beneath it. Excess glycerin and the jelly should be wiped away and then the coverslip may be ringed with a sealing material such as balsam, colorless

fingernail polish, Bismark Black, or Gold Size. (See catalogues of biological supply houses which list these and other mounting and sealing materials.)

A useful reagent to demonstrate the presence of starch ($C_6H_{10}O_5$) of the Chlorophyta is I-KI. Commonly employed solutions are: (1) Iodine, 2 gs, Potassium Iodide, 1 g Water, 200 cc; (2) Lugol's Iodine: Iodine 1 g, Potassium Iodide, 2 g, Water, 300 cc. A killing and fixing solution may be prepared by adding to 200 cc of water, 10 gs Iodine, 200 gs Potassium Iodide, and 20 cc Glacial Acetic Acid. Starch becomes dark, purplish-blue to black when stained with Iodine.

A depression slide may be used if specimens are to be examined over an extended period, for a study of motility, reproductive processes, *etc*. Rub a smear of vaseline around the margin of the depression. Place a drop of water or culture medium containing the specimen in the center of a cover glass. Then invert the cover glass with the drop suspended in the slide depression. Press down the margins of the cover slip in the vaseline film to seal the chamber. In such a chamber the supply of oxygen is limited of course. A similar and often more efficient type of mount is to seal a low, plastic collar to a slide (in a film of vaseline). Smear the top edge of the collar with vaseline and invert a hanging drop in the collar, pressing down the edges of the cover slip into the film of vaseline so as to seal the chamber. Such a mount can be used for low magnification study.

What Are Algae

Although most freshwater algae are microscopic, many kinds are gregarious and occur in such numbers as to form the well-known "water-blooms" or pond scums. A few genera are individually large enough to be seen readily without the aid of a microscope, *e.g.,* the stoneworts (Characeae) or some of the freshwater red algae such as *Batrachospermum* and *Lemanea,* or globular, colonial forms such as the blue-green alga *Nostoc.*

If it were possible for freshwater algae to grow as large as some other plants (mosses and ferns for example) and to live upon land, they would be considered highly attractive indeed and would be much cultivated as ornamentals. The symmetry of form and the patterns of external decorations possessed by many of them are not excelled in beauty by larger plants. The varied shapes of both marine and freshwater algae, coupled with their many colors and hues have made them the subject of observation and wonderment for a long time, especially since the invention of the microscope. Indeed, the microscopic size of most of the freshwater algae renders them all the more intriguing, and since the early days of the first microscopical club they have been used for pleasurable observation, contemplation and speculation.

It is not the aesthetic quality of freshwater algae alone which explains the amount of interest shown them. For small though they are, freshwater algae (like some of their microscopic kin in the oceans) have many economic importances and considerable biological significance. Their relationship to aquatic biological problems such as the food chain, their troublesome contamination of water supplies, and their use in general physiological research constitute just a few of the many aspects which lead to a study of them. Purely scientific problems, such as the role of algae in organic evolution, the biology of their reproduction and life histories, and their ecology are common subjects of investigation. Although much is still to be learned from them, the solution, or at least clarifications, of many problems in general biology and physiology have been obtained from studies of algae. At this time, for example, much attention is being given algae in culture for the study of highly important and practical problems in photosynthesis and the products of algal metabolism. Some genera of unicellular algae are being used for the assay and detection of biologicals (vitamins and growth-promoting or growth-inhibiting substances). Some cancer research involves studies in the physiology and reproduction of algal cells. The use of algae in sewage oxidation, and their role as indicators of water pollution are other well-known fields of study.

Whatever the interests in freshwater algae may be, the student who has access to a microscope can find many hours of fascination in a few drops of pond water.

The term "algae" derived from the Latin name for sea-wrack, has come to be applied to all relatively simple (thalloid) marine and freshwater vegetation. Actually, many different kinds of organisms are included among the plants which lie outside (or below) the realm of mosses (Bryophyta), the ferns (Pteridophyta), and the seed-bearing plants (Spermatophyta). Included under "algae" are the smallest and most simple of chlorophyll-bearing organisms, the entire plant being but a single cell. Some of these may be less than 1 micron in diameter. At the other extreme, some of the brown algae such as the kelps (Phaeophyta) are the longest plants in the world. *Macrocystis* of the Pacific Ocean, for example is commonly over 100 feet in length, and greater lengths up to 700 feet have been claimed.

The student soon learns that "algae" includes several divisions or phyla of the plant kingdom, and that there are incorporated even some groups of organisms which, strictly speaking, belong neither to the plant nor to the animal kingdom (Euglenophyta, Pyrrhophyta, and many of the yellow-green algae such as *Synura* and *Dinobryon*). These are forms which usually are treated as chlorophyll-bearing protozoans in a reference work dealing with one-celled animals. Several of the swimming, protozoanlike forms have definitely plantlike or even non-motile relatives, however, which more than justify their being given a place among the plantlike algae.

The reader who is not familiar with the classification of plants and animals, nor with the terminology used for the different categories, may wish to refer to the following definitions.

SPECIES

A particular kind of plant or animal population is called a species. For example, a certain kind of rose, or a particular alga such as a "pond silk," or a particular bird is known as a species and is given an identifying or specific name. Because there are so many (although slight) variations between individuals which are, in general, very much alike, the limitations or precise circumscriptions of a species of plant or animal are often difficult, and subject to different interpretations by specialists.

GENUS

All plants which obviously are roses, but not all the same kind, are grouped and constitute what is known as a genus (plural, genera). Thus, all different species of roses are placed in the genus *Rosa,* the Latin name for the genus. All species of "pond silk" are placed in the genus *Spirogyra.* The genus name, *Spirogyra,* and a species name (a particular kind) together constitute the *scientific* name. For example, *Spirogyra elongata* is the scientific name of a species which has long cells; *Rosa cinnamonea* is the cinnamon rose. This method of naming each kind of plant or animal with a double name identifies not only a particular kind of individual but also indicates to what group (genus) it belongs. From a taxonomic point of view a "species" is regarded as a *population* of similar organisms, genetically distinct and which, ordinarily, will not cross with members of another 'population'.

FAMILY

The genus *Rosa* has much in common with the strawberry genus *Fragaria,* and is much like the prune genus, *Prunus.* Similarly *Spirogyra* has much in common with another group of species which constitutes the genus *Mougeotia.* Therefore, *Rosa, Fragaria, Prunus,* and other genera that have characteristics much in common are grouped to form what is called a family. In this instance the Rosaceae or Rose Family. *Spiro-gyra, Mougeotia, Zygnema* and some other algal genera which have characteristics in common and which seem, therefore, to be related are grouped to form the family Zygnemataceae. All genera in a family and all species in a genus are assumed to have a common ancestor.

ORDER

In turn, families which are distinct from one another but which, nevertheless, have some few characteristics in common are grouped to form what is known as an *Order*. Thus we have the Rosales, the Zygnematales, *etc*.

DIVISION OR PHYLUM

Several related orders form a major category known as a Division or Phylum of the plant kingdom (or of the animal kingdom). Thus several orders of the green algae constitute the division (phylum) Chlorophyta. In many instances it may be convenient to subdivide the phylum into groups of orders called *Classes*. Hence in the Chlorophyta there are 2 classes (or Sub-phyla), Chlorophyceae and Charophyceae. In the key which follows only the genus names are given, with illustrations of 1 or 2 common species. In so far as possible, the most distinctive features of the genus are shown in the illustrations.

In some instances a genus may be monotypic, *i.e.,* contains only 1 species, or a family may have only 1 genus. For example the family Microsporaceae includes only the genus *Microspora* (Fig. 262). Likewise, it is possible for a phylum or organism to have but 1 order, all such compositions being subject to interpretations of specialists. Categories of classification are often referred to as *taxa* (singular, *taxon*).

Use of the Key

The task of the writer in describing freshwater algae is not made easier by their relatively small (mostly microscopic size). Hence it is necessary to employ special descriptive terms to differentiate these minute organisms, and to assign them properly to the families and phyla to which they belong. Many of the terms are defined in the Pictured Glossary.

In such a treatment as is presented here, only the more common and best-known genera can be given a place. Included are not only most of the genera reported from the United States, but also a number known from Europe which can be expected in North America. The reader should keep in mind that all genera are not considered when using the key. He should avail himself of other less abridged or specialized works if satisfactory identification of a plant in which he is interested does not appear possible by the use of the following key. But the almost world-wide distribution of a majority of freshwater genera gives such a key a usefulness beyond geographical limits. Such a key cannot be made as easy to use as are many keys to larger organisms. An attempt has been made to overcome some of the usual difficulties inherent in a key by leading to the same genus name at several different points; especially is this true for those genera which are so variable that selection of any one set of differentiating

criteria for them is impossible unless one should write out a full description of the genus.

A beginning student or one with limited familiarity with the algae must exercise patience until he has developed some degree of judgment and has become well-acquainted with the meaning of terms, and until he has discovered to what degree a plant may vary from the more usual character which is employed in making an identification. Many times the user will find it profitable, if not necessary to 'back track' in the key and to follow down both dichotomies of choice before arriving at a satisfactory determination. As mentioned before, in making use of the illustrations it must be remembered that only 1 or 2 species of a genus are pictured, and that the plant in question may not appear exactly like the forms which are figured. This is true for many of the genera in the Chlorococcales *(Scenedesmus, Oocystis, Tetraedon, etc.)* and of the Desmids *(Cosmarium, Euastrum, Micrasterias, e.g.)* also in the Chlorophyta, genera which have very many species and which show considerable variation.

One of the primary difficulties with which the inexperienced student is confronted when first using a general key to the algae is that of detecting and identifying colors, green, blue-green, yellow-green, *etc.* to which the key makes reference. Pigmentation in the different algal groups is a fundamental characteristic, in the main, and one which is very useful in making especially the primary identifications. But yellow-green algae at times appear decidedly grass-green, and the brown-pigmented algae may have a distinct tinge of green, especially when artificial light is used for the microscope. Hence, other characters, or a combination of characteristics excluding or in addition to color must be employed to make a choice in the key. Suggestions are given in appropriate places for making certain tests to help differentiate genera on the basis of color. Although it is a combination of characters which differentiate algae in the final analysis, the key can select these characters one by one only.

How and Where to Collect Freshwater Algae

Filamentous algae can be collected from mass growths by hand, and representative tufts placed in vials or collecting jars. Less conspicuous forms may be found as fuzzy films on submersed grasses, old rush culms, and sticks. Using the fingers these growths can be lifted away or pulled from their attachment, or short sections of stems and aquatic plants and grass leaves can be placed in vials and the algae removed with scraping tools in the laboratory. A dropping pipette and a pair of tweezers are useful for collecting minute forms. Also a 5 or 10 cc pipette fitted with a rubber bulb is appropriate for removing bottom sediments which contain algae.

Using the back of the thumb nail, or a dull-edged knife will serve, greenish coatings on rocks and submersed wood can be scraped away. Such an instrument is useful for removing samples of green or brown feltlike or mucilaginous growths from wet stones about waterfalls, from dripping cliffs and rocky outcrops.

Submersed glass, shells and bits of crockery in the water furnish substrates for many algae which occur as minute, green discs or tufts. Old, rotting wood may be perforated with algae which lie so far below the surface that they are scarcely visible, but wood that appears at all greenish from the exterior should be examined.

Feel under the rim of dams or along the edges of stones in flowing water. Many blue-green and also some of the more rare freshwater red algae occur in such habitats.

On and in damp soil are to be found numerous species of Cyanophyta and Diatoms. Sometimes algae occur in pure 'stands' and sheets or films of a single species and may be lifted or scraped from soil, wet boards, and from the face of moist cliffs. A dry desert does not seem a suitable habitat for algae, yet many species occur on desert soils where they may form 'rain crusts' or may be found by taking soil samples from beneath stones.

On beaches near the high water line, but back far enough where the sand lies unmolested most of the time, the upper dry layer of sand may be removed to disclose a densely green stratun of algae. The green sand can be scraped into a container and rinsed, and then when the water is poured off in the laboratory an interesting mixture of algae will be found, together with a variety of microscopic animals (protozoans, rotifers, copepods, *etc.*). This biotic cosmos is known as *psammon* and includes many organisms that normally occur in sandy beaches although not necessarily in the open water of a nearby lake or stream.

In *Nitella* (one of the larger green algae), in *Lemna trisulca* (one of the duck weeds), in *Ricciocarpus natans* and *Riccia fluitans* (floating liverworts) occur various green and blue-green, endophytic algae. Small portions of these aquatic plants, and others as well, may be allowed to age and to become discolored in laboratory dishes. The endophytes (and some epiphytes too) will then appear more clearly and can be dissected away for study.

In humid climates trunks of trees and surfaces of leaves may have epiphytic and endophytic (semi-parasitic) algae such as *Trentepohlia* and *Cephaleuros*. *Arisaema* (Indian Turnip, Jack-in-the-Pulpit) leaves invariably contain the parasitic alga *Phyllosiphon* which causes yellow or red spots in host tissue.

One interesting habitat is the back of snapping turtles where the coarse, wiry filamentous alga *Basicladia* is found invariably. Other algae may be associated with this genus on the 'mossy' backs of turtles whereas alligators are sometimes veritable algal gardens and offer a variety of species for the less timid collector. In the rain-forests of the tropics, Central America, *e.g.*, the three-toed sloth harbors among its hair scales a minute red alga, *Cyanoderma bradypodis,* and a filamentous green alga, *Trichophilus welcheri.*

In alpine and subalpine regions where there are banks of permanent snow, red streaks will be found in the snow fields, or footprints will turn red as the snow is compacted. The color is produced by unicellular green algae which exist as red cysts. *Chlamydomonas nivalis,* commonly found, contains a red carotenoid pigment (haematochrome) which develops in many species of organisms when exposed to intense illumination. Although such cells contain chlorophyll the green is masked by the more abundant haematochrome. Occasionally green and yellow snow banks are found. A small quantity of red or 'bloody' snow when allowed to melt in a jar may yield a surprising quantity of such genera as *Chloromonas (Scotiella), Ankistrodesmus, Raphidonema, Mycanthococcus* and certain dinoflagellates.

Specimens collected from open water (plankton organisms) are best taken with a cone-shaped, silk, bolting cloth net (No. 20 mesh). Plankton nets are obtainable from biological or equipment supply houses, or may be made up by securing a yard of the silk from an importer or from a flour mill. The American Limnological and Oceanographic Society publishes periodically a list of commercial houses and firms where various kinds of collecting equipment may be obtained. A light-weight, brass (preferred) or thin galvanized iron ring (stout wire), or band may be used for the mouth of the net. A convenient size is a ring about 6 or 8 inches in diameter. Using a pattern (See Welch, P.S. 1948. Limnological Methods, Blakiston Co., p. 234-235.) cut the silk so that when attached to the ring a cone about 14 inches long is formed. The silk should not be attached directly to the ring but sewed first to a band of stout muslin which then can be sewed over the ring or metal band. If a flat metal band is used for the mouth of the net the edges should be filed smooth and rounded to eliminate as much cutting and fraying of the muslin as possible. The net may be used as a closed cone in which condition it must be turned inside out and rinsed into a dish or jar after making a haul. More conveniently, the tip of the net may be cut off at a point about 1/2 inch or less from the end which will permit the insertion of a small homeopathic vial (4 to 6 dram capacity) which can be tied around its neck into the apex of the net. Thus the sample will become concentrated in the vial and when the net is reversed the collected material can be poured into a sample bottle, and the net then rinsed before another sample is taken. Of course the vial at the end of the net can be untied with each use and another inserted. Better still, a small metal (aluminum) band, threaded to receive a 6- or 8- dram screw cap vial can be sewed into the tip of the net. Then the vial can be unscrewed and a fresh one inserted conveniently after each use. Muslin should be used at the neck of the net in which the aluminum, threaded ring is inserted. When comparative habitat studies are made much care should be used to see that the net is well-rinsed if it is to be used to collect from more than one habitat for critical studies; or a separate net should be used for each habitat.

The net should have 3 leaders of equal

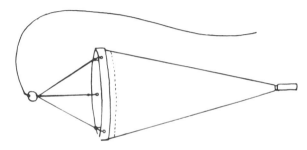

length attached to a small ring to which the tow cord is also attached. Use a heavy line such as a sash cord for the towing or casting line. The leaders may be wire or nylon cord. Braided copper wire is sometimes used for the leaders but these become worn quickly at points of attachment and snap.

Microscopic forms of algae may be obtained in great numbers from squeezings of *Sphagnum* (and other aquatic mosses), especially when the plants feel slippery or slimy. Overhanging grasses and sedges, and exposed plant roots in *Sphagnum* bog pools and seeps are usually coated with algae, especially desmids. Squeezings may be collected directly in a wide-mouth bottle, or vial, or it is often found desirable to collect the squeezings in a plankton net and so obtain a rich concentration of algae. *Utricularia* (bladderwort), especially when such plants occur in soft or acid lakes, is a veritable net itself. Water from handfuls of this plant can be squeezed into a plankton net with very fruitful results. The bladders of *Utricularia* sometimes contain an interesting assortment of small algae, especially desmids.

Specimens collected from the field should be put in receptacles with just enough water to cover them, leaving ample space for air, especially if the sample is to be stoppered for some time before arriving at the laboratory. Clots of larger, filamentous algae may have the excess water gently squeezed from them, rolled in wet and then dry paper (newspaper highly satisfactory) and so be kept in good condition. An efficient and inexpensive container for field collecting is a cellophane or plastic bag. A pack of sandwich bags can be taken into the field and used for individual collections, placing a small label or code number in the bag with a minimum amount of water.

Immediately upon returning from the field, vials or packets of material should be opened and poured into wide, shallow dishes so that they may be well-aerated using just enough water to cover the specimens. If the collection is not too crowded in a dish of water the plants may be kept alive and in good condition almost indefinitely, especially if the dishes are stored in a cool place with reduced illumination such as in a north-facing window. Some kinds of algae will remain in satisfactory condition for study (even though additional growth may not occur) when stored in a refrigerator kept at ordinary temperatures used in food storage.

Some collectors prefer to spread algae on cards or stiff paper to dry, and then make herbarium specimens of them. In working with such specimens later, a few drops of water placed on the dried plants will soak up the material well enough that it can be lifted away for mounting on a slide. Specimens so treated, however, are not satisfactory unless one has had a long experience in examining algae and is familiar with their appearance in the undried condition.

If it is desirable to keep a record of the location from which separate field collections are made, it is obviously necessary to give samples a code number or label at the time they are taken. One satisfactory way of doing this is to carry 3- × 5-inch cards, all but cut through into narrow strips which will fit into collecting vials. A number then can be written on a slip which is torn off from the card and inserted. Information bearing the same code number can be written into a field notebook for future reference. In the laboratory a permanent number can be assigned to the vials and written on the cork or label if the material is to be saved for subsequent studies.

The Phyla of Algae

The organisms which constitute what are commonly known as "algae" are extremely diverse in form, color, habit, and in their habitats. Actually there are as many as 8 separate phyla or divisions of the plant kingdom included under "algae," (9 if Cryptophyceae of uncertain position are given a phylum status). Hence, to write descriptively of algae, one is confronted with almost as great a task as if he were treating all the phyla of land plants, fungi, mosses, ferns and seed plants, plus 3 or 4 additional groups. To be sure, all the phyla of algae do not include as many families and genera as do some of the higher plants, but the green algae alone include some 20 or 25 thousand species, distributed among about 400 genera.

The 3 major phyla of algae (those which are the most common) are the Chlorophyta (green), Cyanophyta (blue-green) and Chrysophyta (yellow- or brown-green). It is suggested that to facilitate the differentiation of these 3 groups and to become acquainted with the colors that characterize them, a known green alga *(Spirogyra),* a blue-green *(Anabaena)* and a Diatom be mounted side by side on the same microscope slide. This will permit the ready comparison of the colors and also a comparison of the wall features, cell contents (chloroplast), *etc.* Then a series of illustrations depicting these 3 groups should be examined so that gross morphology can be associated with the respective pigmentations.

The phyla of freshwater algae herein recognized are as follows:

Cyanophyta (Blue-green, Prokaryotic Algae)

Plants unicellular, colonial, or in simple or branched (sometimes falsely branched) filaments; chloroplasts lacking, the pigments seemingly distributed throughout the entire protoplast (but contained in flat bodies known as thylakoids), often more dense in the peripheral region of the cell; pigments are chlorophyll-a, 3 carotenes, 2 or possibly as many as 15 xanthophylls (not necessarily present in all forms), phycoerythrin, phycocyanin; cell wall thin, a membrane which usually has a gelatinous outer sheath; cell contents often with false (pseudo-) vacuoles which refract the light and obscure the true color of the cells which may be green, blue-green, gray-green, violet, tan, brown or purple; definite nucleus lacking but nuclear material occurring as a cluster of chromatic granules or fibrils in the midregion (central body) of the cell; motile cells and sexual reproduction wanting; reproduction by cell division (fission) or by spores (endospores; akinetes); food storage questionably glycogen, possibly floridean starch; iodine test for starch negative.

Chlorophyta (Green, Eukaryotic Algae)

Plants unicellular, colonial, or filamentous; floating, swimming or attached and stationary; cells containing plastids (chloroplasts) in which chlorophyll (grass-green) is predominant, and in which there is usually a shiny, starch-storing body, the pyrenoid; pigments are chlorophyll-a, chlorophyll-b, 2 (possibly 3) carotenes, as many as 6, possibly 10 xanthophylls, and with red carotenoids (haematochrome) sometimes present; starch test with iodine positive (in almost every instance); nucleus definite (although often small and inconspicuous); cell wall (rarely lacking) composed of cellulose and pectose; swimming cells or motile reproductive elements furnishsd with 2 (usually), 4, or rarely as many as 8 flagella of equal length and attached in the anterior end; sexual reproduction by iso- aniso- and by heterogametes.

Chrysophyta (Yellow-green, or Yellow-brown Algae)*

Plants unicellular or colonial, rarely filamentous; pigments contained in chloroplasts in which yellow, brown or golden-brown usually predominate; pigments are: 3 chlorophylls, 3 carotenes, as many as 7 (possibly 16) xanthophylls (not necessarily present in all forms); food storage in the form of oil or leucosin, the latter often giving the cell a metallic lustre; starch test with iodine negative; wall relatively thick, pectin, often silicified, or with siliceous scales, and in some Classes in 2 sections which adjoin or overlap in the midregion; motile cells and swimming reproductive elements furnished with 2 flagella of unequal length, or with 1 flagellum; rhizopodial (pseudopodial and amoeboid) extensions of the cell not uncommon in some families.

Euglenophyta (Euglenoids)

Cells solitary (gregarious in 1 family), swimming by 1, 2 or 3 flagella; a gullet present in the anterior end of the cell in many members, as is also a red eye-spot; chloroplasts few to many, variously shaped bodies (a number of genera colorless); pigments are: chlorophyll-a, -b, 4 carotenes, possibly 5 xanthophylls, and often red carotenoid (haematochrome); pyrenoids usually present, either on or in the chloroplasts, or free in the cell; food reserve in the form of an insoluble starch, paramylum, which is negative to the iodine test, and fatty substances; nucleus large and centrally located; cell membrane in the form of a pellicle, rigid or plastic with the cells metabolic, frequently striated; sexual reproduction (in the strict sense) lacking; vegetative reproduction by longitudinal cell division, and by encystment followed by multiplication of the cell.

Cryptophyta (Cryptophyceae, Cryptomonads)

Cells solitary (rarely colonial), mostly swimming, protozoanlike organisms with 2, often laterally inserted or subapical flagella unequal in length; chloroplasts few and large, brown, blue, or reddish, with pyrenoids commonly present; pigments are: 2 chlorophylls, 2 carotenes, 4 xanthophylls, phycocyanin and phycoerythrin (in some); food reserve in the form of solid starch or starchlike substances; iodine test for starch sometimes positive; cell membrane a firm periplast; a gullet commonly present in the anterior end; rarely with trichocysts; reproduction by longitudinal cell division; sexual reproduction unknown.

Pyrrhophyta (Dinoflagellates)

Cells solitary (rarely filamentous in a few marine genera); mostly swimming by 2 flagella of approximately equal length and mostly lateral in attachment (apical in 1 family), 1 flagellum wound about the cell in a transverse furrow, and 1 extended posteriorly from the point of attachment in a longitudinal furrow;

---

*According to some classification schemes this Phylum is composed of 3 Classes: Chrysophyceae (mostly protozoanlike yellow-green algae); Xanthophyceae (Heterokontae); Bacillariophyceae (Diatoms). The Bacillariophyceae constitutes such a unique group that some students recommend that it be assigned to a Phylum status.

cells mostly dorsiventrally flattened and differentiated, the longitudinal furrow extending along the ventral surface; cell wall (rarely absent) firm and simple, or formed of regularly arranged, polygonal plates (as in the so-called armored or thecate Dinoflagellata); pigments in chloroplasts include: chlorophyll-a, 2 carotenes, 4, possibly 6 xanthophylls (with peridinin and dinoxanthin being the most abundant), phycopyrrin often giving the cells (especially when in mass) a reddish color; food reserve starch, starchlike substances and oil; a pigment-spot (possibly an eye-spot) commonly present; reproduction by longitudinal division, by asexual zoospores, and sexual (known only in a few instances).

Rhodophyta (Red Algae)

Plants simple or branched filaments (unicellular in 1 questionable genus), and composed of multiaxial filaments to form thalli (especially in marine forms, pigments contained in chloroplasts including: chlorophyll-a and -d, 2 carotenes, 3 (possibly as many as 14) xanthophylls, phycoerythrin and phycocyanin (in freshwater genera phycoerythrin is reduced and the plants are commonly gray-green, violet-green or tancolored); food reserve in the form of floridean starch which is iodine-negative; walls relatively thick, often containing pores providing intercellular connections, highly mucilaginous; sexual reproduction by heterogametes, the egg contained in a carpogonium, the antherozoids as non-motile elements cut off from the tips of special branches; asexual reproduction by monospores, carpospores and tetraspores (in some genera); motility lacking in both vegetative and reproductive expressions; thalli often macroscopic in size.

Phaeophyta (Brown Algae)

A phylum almost entirely marine, including the brown sea weeds (kelps); essentially filamentous (some few microscopic) but mostly macroscopic, stout and leathery; pigments in chloroplasts, including: chlorophyll-a and -c, carotene-B and 7 xanthophylls (especially fucoxanthin); food reserve in the form of laminarin or soluble carbohydrates including alcohol (mannitol); pyrenoids sometimes present; reproduction vegetative by fragmentation, asexual by kidney-shaped, biflagellate zoospores with lateral attachment, or sexual by iso-aniso- and heterogametes.

Chloromonadophyta (Chloromonads)

An obscure and little-understood group composed of but a few genera and species; cells solitary, swimming by 1 or 2 flagella, apically attached; chloroplasts (when present) greenish with xanthophylls predominating over the chlorophyll, pigments including: chlorophyll-a, 2 carotenes and at least 3 xanthophylls; food reserve in the form of oils or fat; contractile vacuoles and a reservoir at the anterior end of the cell; cell contents with trichocysts radiately arranged immediately within the cell membrane or clustered near the anterior end; reproduction by cell division, sexual reproduction unknown.

Synopsis of the Algal Phyla

Most phyla of the algae are present in both marine and freshwater, although some occur more abundantly in one or the other of the two habitats. The Phaeophyta, for example are almost entirely marine, the Euglenophyta almost all freshwater in distribution. The following general key is presented to characterize the phyla and to facilitate a comparison.

1a Cells without chloroplasts, pigments blue-green, olive-green, or purplish, contained in thylakoids but seemingly distributed throughout the entire protoplast (although cells may be somewhat less colored in the central region); wall usually thin (often showing as a membrane only) and generally with a mucilaginous sheath (wide or narrow, watery or firm and definite); food reserve in the form of glycogen or a starchlike substance); iodine test for starch negative; no motile cells; no definite nucleus; no sexual reproduction. Constitute the Akaryonta (Prokaryota) . . . .
. . . . Blue-green Algae . . . . *Cyanophyta*

1b Cells with chloroplasts, the pigments not distributed throughout the protoplast; cell wall clearly evident (with rare exceptions, *Pyramimonas, e.g.,* Fig. 57); stored food in the form of starch, oils, or leucosin; starch test with iodine positive or negative; nucleus present. The Eukaryonta 2

2a Cells with grass-green chloroplasts (but see some species of *Euglena,* Fig. 13, or the filamentous alga *Trentepohlia,* Fig. 7 which, although possessing chlorophyll, have the green color masked by an abundance of a red carotenoid pigment [haematochrome]) 3

2b Cells with chloroplasts some other color, gray-green, brown, violet-green, yellow-green or sometimes purplish 5

3a Free-swimming, unicellular with numerous ovoid, star-shaped, or plate-like chloroplasts which are grass-green; food storage as clearly evident grains of insoluble paramylum (sticks or plates); iodine test for starch negative; 1, 2, or rarely 3 coarse flagella attached at the apex in a gullet; eye-spot usually evident .... *Euglenophyta*

3b   Organisms not as above............ 4

4a   Unicellular, without an eye-spot; chloroplasts numerous discs usually radially directed at the periphery of the cells; motile by means of 2 flagella inserted in an apical reservoir; trichocysts present immediately within the cell membrane (sometimes scattered or clustered); food reserve oil. Chloromonads .... *Chloromonadophyta*

4b   Unicellular, colonial, or filamentous; swimming or not (often free-floating); when swimming using from 2 to 4 fine flagella attached at the apex of the cell in a colorless region; chloroplasts 1 to several; usually with a conspicuous pyrenoid (starch-accumulating granule); iodine test for starch positive ..... Green Algae .... *Chlorophyta*

5a   Chloroplasts light olive-brown to dark brown; nearly all marine; essentially filamentous but occurring as thalli of macroscopic size (complexes of filaments), as well as microscopic, filamentous plants; stored food in the form of laminarin and alcohol; starch test with iodine negative; motile cells with 2 laterally attached flagella ..... Brown Algae .... *Phaeophyta*

5b   Plants marine or freshwater, but not occurring as brown thalli of macroscopic size 6

6a   Chloroplasts yellow-green to yellow- or golden-brown; food in the form of leucosin or oil; starch test with iodine negative; plants unicellular, colonial, or filamentous; sometimes swimming with apically attached flagella of unequal length, or with 1 flagellum; many forms, especially Diatoms, with siliceous walls; wall often in 2 adjoined or overlapping sections .... Yellow-green Algae .... *Chrysophyta**

6b   Chloroplasts not yellow-green or pale green but dark golden-brown, gray-green, violet-green (rarely blue or red); food in the form of oil or starchlike carbohydrates; iodine test for starch usually negative....................... 7

7a   Unicellular with dark, golden-brown chloroplasts; swimming by 2 laterally attached flagella; a conspicuous red (eye?) spot usually present; many forms with the wall composed of polygonal, cellulose plates; cell with a transverse and a longitudinal furrow; reserve food in the form of starch or oil ..... Dinoflagellates .... *Pyrrhophyta*

7b   Organisms unicellular or filamentous, not motile or if so, swimming by apical or subapical flagella (rarely lateral); chloroplasts red, brown, greenish-blue, violet-green or gray-green 8

8a   Plants non-motile; chloroplasts violet or gray-green, sometimes bluish-green in freshwater, red in marine forms; occurring as filamentous thalli of macroscopic size (microscopic in embryonic stages); food reserve starchlike carbohydrates, the iodine test for starch negative ..... Red Algae............ .... *Rhodophyta*

*See footnote on p. 10.

8b Organisms mostly motile and unicellular; chloroplasts 1 or 2 golden-brown or brownish but usually bluish or reddish bodies; organisms unicellular (rarely colonial); swimming by sub-apically attached flagella; food reserve starchlike carbohydrates; iodine test for starch positive in some . . . . Cryptomonads . . . . *Cryptophyta*

(This Phylum of the algae has several characteristics in common with the Dinoflagellata and in some systems of classification is included with the Pyrrhophyta.)

General References

Atkinson, G.F. 1890. Monograph of the Lemaneaceae of the United States. Ann. Bot., 4:177-229. Pls. 7-9.

Bourrelly, P. 1966. Les algues d'eau douce. Algues vertes. Boubée & Cie, Paris.

Bourrelly, P. 1968. Les algues d'eau douce. Algues jaunes at brunes. Boubée & Cie, Paris.

Bourrelly, P. 1970. Les algues d'eau douce. Les algues bleues et rouges. Les Eugléniens, Peridiniens et Cryptomonadines. Boubée & Cie, Paris.

Boyer, C.S. 1916. Diatomaceae of Philadelphia and Vicinity. Philadelphia.

Brunnthaler, J. 1915. Protococcales. In: A. Pascher. Die Süsswasserflora Deutschlands, Österreich und der Schweiz. Heft 5. Chlorophyceae 2:52-205. G. Fischer, Jena.

Burke, J.F. 1937. Collecting recent diatoms. Preparing recent diatoms. Mounting recent diatoms. New York Microsc. Soc. Bull., 1(3):9-12; 1(4):13-16; 1(5):17-30.

Chapman, V.J. 1952. The Algae. Macmillan Co., New York.

Collins, F.S. 1909 (1928). The Green Algae of North America. Tufts College Studies, Sci. Ser., 2(1909):79-480. Pls. 1-18. (Reprinted with Supplements 1 and 2 by G.E. Stechert Co., New York.)

Copeland, J.J. 1936. Yellowstone Thermal Myxophyceae. Ann. New York Acad. Sci., 36:1-232, 73 Figs.

Drouet, F. 1968. Revision of the Classification of the Oscillatoriaceae. Academy of Natural Sciences of Philadelphia.

Drouet, F. 1973. Revision of the Nostocaceae with Cylindrical Trichomes. Hafner Press, New York.

Drouet, F. and Daily, W.A. 1952. A Revision of the Coccoid Myxophyceae. Butler Univ. Bot. Stud., 12:1-218.

Eddy, S. 1930. The fresh-water armored or thecate Dinoflagellates. Trans. Amer. Microsc. Soc., 49:277-321.

Elmore, C.J. 1921. The Diatoms (Bacillarioideae) of Nebraska. Univ. Nebraska Stud., 21(1/4):1-214. 23 Pls.

Fleming, W.D. 1949. Cleaning and preparation of diatoms for mounting. Bull. Amer. Soc. Microscopists, 5(3):40-43.

Flint, L.H. 1947. Studies of fresh-water red algae. Amer. Jour. Bot., 34(3):125-131.

Flint, L.H. 1948. Studies of fresh-water red algae. Amer. Jour. Bot., 35(7):428-433.

Flint, L.H. 1949. Studies of fresh-water red algae. Amer. Jour. Bot., 36(8):549-552.

Flint, L.H. 1950. Studies of fresh-water red algae. Amer. Jour. Bot., 36:754-757.

Fritsch, F.E. 1935, 1945. The Structure and Reproduction of the Algae. Vols. I, II. Cambridge Univ. Press.

Geitler, L. 1930, 1931. Cyanophyceae. In: L. Rabenhorst's Kryptogamen-Flora von Deutschland, Österreich und der Schweiz. 14 Lf. 1(1930):1-228; lf. 2(1931):289-464. Figs. 1-131. Leipzig.

Gojdics, M. 1953. The Genus *Euglena*. Univ. Wisconsin Press, Madison.

Hansmann, E.W. 1973. Diatoms of the Streams of Eastern Connecticut. State Geol. & Nat. Hist. Surv., Bull. 106. 119 pp. 43 Figs.

Heering, W. 1914. Ulotrichales, Microsporales, Oedogoniales. In: A. Pascher. Die Süsswasserflora Deutschlands, Österreich und der Schweiz. Heft 6. Chlorophyceae 3:1-250. Figs. 1-384. G. Fischer, Jena.

Huber-Pestalozzi, G. 1955. Die Binnengewässer. Das Phytoplankton des Süsswassers. 4 Teil. Euglenophyceen. Stuttgart.

Huber-Pestalozzi, G. 1961. Die Binnengewässer. Das Phytoplankton des Süsswassers. 5 Teil. Chlorophyceae (Grünalgen). Ordnung: Volvocales. Stuttgart.

Hustedt, F. 1930. Die Kieselalgen. In: L. Rabenhorst's Kryptogamen-Flora von Deutschland, Österreich und der Schweiz. 7. Leipzig.

Irénée-Marie, Fr. 1939. Flore desmidiale de la région du Montréal. La Prairie, Canada.

Jacques, H.E. 1948. Plant Families—How to Know Them. Wm. C. Brown Company Publishers, Dubuque, Iowa.

Krieger, W. 1933-1939. Die Desmidiaceen. In: L. Rabenhorst's Kryptogamen-Flora von Deutschland, Österreich, und der Schweiz. 13 Abt. 1:1-712. Pls. 1-96; Abt. 1, Teil 2:1-117. Pls. 97-142. Leipzig.

Krieger, W. and Gerloff, J. 1962, 1965, 1969. Die Gattung *Cosmarium*. Lief. 1:1-112. Taf. 1-19; Lief 2:113-240. Taf. 23-42; Lief. 3-4:241-410. Taf. 43-71.

Lemmermann, E. 1913. Euglenineae. Flagellatae 2. In: A. Pascher. Die Süsswasserflora Deutschlands, Österreich, und der Schweiz. Heft 2:115-174. Figs. 181-377. G. Fischer, Jena.

Pascher, A. 1937-1939. Heterokonten. In: L. Rabenhorst's Kryptogamen-Flora von Deutschland, Österreich, und der Schweiz. XI:1-1097. Figs. 1-912. Leipzig.

Patrick, R. and Reimer, C.W. 1966, 1975. The Diatoms of the United States. Vols. I, II. Acad. of Natural Sciences of Philadelphia.

Prescott, G.W. Algae of the Western Great Lakes Area. IInd Ed. Wm. C. Brown Company Publishers, Dubuque, Iowa.

Prescott, G.W. 1968. The Algae: A Review. Houghton Mifflin Co., Boston.

Prescott, G.W., Croasdale, H.T. and Vinyard, W.C. 1972. Desmidiales. Part I. Saccodermae. Mesotaeniaceae. North Amer. Flora, Ser. II(6):1-84. New York Bot. Garden.

Prescott, G.W., Croasdale, H.T. and Vinyard, W.C. 1975. Synopsis of North American Desmids. Part II(1). Desmidiaceae. 1:1-275. Univ. Nebraska Press.

Printz, H. 1964. Die Chaetophoralen der Binnengewässer. W. Junk, Den Haag.

Round, F.E. 1965. The Biology of the Algae. St. Martin's Press, New York.

Schiller, J. 1933-1937. Dinoflagellatae. In: L. Rabenhorst's Kryptogamen-Flora von Deutschland, Österreich, und der Schweiz. X(1933). Teil 1:1-617; Teil 2, Lf. 1(1935):1-160; Lf. 2(1935):161-320; Lf. 3(1973):321-480; Lf. 4(1937):481-590. Leipzig.

Smith, G.M. 1920, 1924. Phytoplankton of the Inland Lakes of Wisconsin. I, II. Wisconsin Geol. & Nat. Hist. Surv. Bull 57. Madison.

Smith, G.M. 1950. Freshwater Algae of the United States. IInd Ed. McGraw-Hill Book Co., New York.

Tiffany, L.H. 1937. Oedogoniales. Oedogoniaceae. North American Flora, 11, Part 1. New York Botanical Garden.

Tiffany, L.H. and Britton, M.E. 1952. The Algae of Illinois. Univ. Chicago Press, Chicago.

Tilden, J.E. 1910. Minnesota Algae. I. Univ. Minnesota Press, Minneapolis.

Tilden, J.E. 1935. The Algae and Their Life Relations. Univ. Minnesota Press, Minneapolis.

Transeau, E.N. 1951. The Zygnemataceae. Ohio State Univ. Press, Columbus.

Ward, H.B. and Whipple, G.C. 1959. W.T. Edmondson, Ed. Fresh-Water Biology. John Wiley & Sons, New York.

West, W. and West, G.S. 1904-1912. A Monograph of the British Desmidiaeceae. Vols. I-IV. Ray Society, London.

West, W., West, G.S. and Carter, N. 1924. A Monograph of the British Desmidiaceae. Vol. V. Ray Society, London.

Whitford. L.A. and Schumacher, G.J. 1973. A Manual of the Fresh-water Algae in North Carolina. Sparks Press, Raleigh.

Wolle, F. 1887. Freshwater Algae of the United States. Vols. I, II. Bethlehem, Pa.

Wolle, F. 1892. Desmids of the United States and List of American Pediastrums. Bethlehem, Pa.

Wood, R.D. and Imahori, K. 1965, 1965. A Revision of the Characeae. Vols. I, II. J. Cramer, Stuttgart.

Pictured Key to the Common Genera of Freshwater Algae

1a Plants macroscopic, up to 40 or more cm high, growing erect, with stemlike axes bearing whorls of branches and forked, cylindrical 'leaves' clearly visible to the unaided eye. Figs. 1-3. Stoneworts. Characeae. 2

1b Plants microscopic, *or* if macroscopic with cellular structures and branches not clearly visible to the unaided eye; without whorls of branches clearly visible. 4

2a Branching unsymmetrical, with dense heads and whorls of short branches and scraggly longer ones; microscopically showing globular male organs (globules or 'antheridia') lateral beside the oval female organs (nucules or 'oogonia'); plants usually gray-green. Fig. 1 . . . . . .
. . . . *Tolypella*

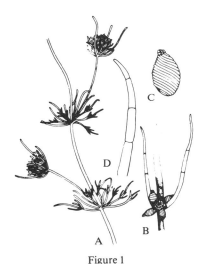

Figure 1

Figure 1 *Tolypella intricata* (Trentep.) v. Leonh. (A) portion of plant showing habit of branching and the heads of short branches in which the reproductive organs are located; (B) a node showing 4 "oogonia" and 1 "antheridium"; (C) an oogonium showing the 'crown' cells; (D) tip of branch.

There are about 10 species of the genus reported from the United States, but only 3 seem to be at all common. *Tolypella* species are widely distributed throughout the world but are more common in the southern hemisphere. In general these plants appear as scraggly *Chara*

(Fig. 2), but they grow singly rather than in dense beds. Like that genus also *Tolypella* species prefer hard-water lakes (or slowly flowing streams) and seem to be confined to rather shallow water. The Characeae is composed of 2 Tribes, Chareae and Nitellae. The former has an investment of 5 cells twisted around the female organ (oogonium or nucule), forming a 5-celled coronula at the apex. In the Nitellae the investing cells are delimited by a cross wall at the base of the coronula and these cells at the tip are again divided, hence there are 10 cells in the coronula, as in *Tolypella* and *Nitella*. In *Chara* the coronula is composed of 5 cells. The sex organs in *Tolypella* characteristically are borne in the dense heads of short branches; other branches are long and fingerlike. The plants are usually gray-green as are most species of *Chara* but they are not coated with lime as in that genus.

2b **Branching symmetrical with rather evenly spaced whorls of equal-length branches at stem nodes; globules ('antheridia') above or below the nucules ('oogonia')** 3

3a **Plants coarse and usually rough to the touch with lime; gray-green; ill-smelling (garlic or skunk odor); microscopically showing globules ('antheridia') lateral on the axis and below the oval nucules ('oogonia'), although some species are dioecious; coronula of oogonium composed of 5 cells; internodes of the main axis and branches in the whorls enclosed by columnar, corticating cells. Fig. 2 ...** .... *Chara*

Figure 2

Figure 2 *Chara* (A) *C. canescens* Lois.-Des., portion of a plant in which thornlike cells arising from the corticating elements give a spiny appearance; (B) *C. excelsa* Allen, a node of a 'leaf' on which there is an oval oogonium (nucule) above and a globular antheridium (globule) below. A monoecious species.

> (***Chara coronata*** **and** ***C. braunii*** **have no corticating cells and may be confused with** ***Nitella*.** **They can be identified by the habit of branching, by the location of the sex organs, and by the presence of 5 cells in the coronula).**

Most of the 42 species reported from the United States are world-wide in their distribution. Many species of *Chara,* however are endemic in Japan, Australia, South Africa, *etc.* They vary in length from 5 cm to a meter or more in length. They are found mostly in hard-water or alkaline lakes and slowly flowing streams in which calcium is abundant; sometimes in brackish water. Whereas they often occur in water a few inches to 4 feet in depth, they may be dredged from as deep as 20 meters (60 feet). Occasionally they are found almost terrestrial in spring seeps. In their physiology *Chara* plants cause lime to become deposited on stems and 'leaves', sometimes to the detriment of the plant which may sink under the weight. The encrustation is responsible for the common name "stonewort." Marl and other kinds of calcareous deposits may be formed largely by *Chara* over long periods of time.

3b **Plants delicate, usually not growing as erect as** *Chara,* **or if relatively stout never roughened with lime; dark green, not ill-smelling; microscopically with globules terminal on short pedicels within a cluster of branches and above the lateral nucules; axis and branches never corticated; coronula of 10 cells, the tips of the cells which are twisted around the oogonium having divided into 2 cells. Fig. 3**. . . . . *Nitella*

Figure 3

Figure 3 *Nitella.* (A) *N. tenuissima* (Desv.) Kütz., habit of plant; (B) *N. flexilis* (L.) C.A. Agardh, portion of plant showing habit of branching. Whorls of branches are often bi- or tri-furcate at the tips.

Species of *Nitella* are not seen as often as are charas because they usually grow more deeply, thriving in soft-water or acid lakes rather than in hard-water situations. Some species occur in bog lakes that are darkly stained with humic or tannic acids and are collected by dredging with a plant hook. Species like *N. tenuissima,* however, are to be found in shallow water almost buried in bottom sediments. The plants are greener than *Chara* because they are not encrusted with lime; are not ill-smelling; often have a glistening translucent appearance.

4a **(1) Cells** *containing* **chloroplasts with green predominating; or with other pigments predominating: yellow-green, golden yellow, brownish, reddish or bluish-green.** **5***

4b **Cells** *without* **chloroplasts, with pigments seemingly distributed and diffuse throughout the cell (or nearly so). Blue-green Algae. Cyanophyta** **565**

5a **Plants grass or leaf-green;** ** **or gray- to violet-green, or tawny-green. (Mostly Chlorophyta, Euglenophyta and the violet-green Rhodophyta). But see** *Botrydium* **(Fig. 326) and** *Vaucheria*

*See *Polytoma* (Fig. 32), one of several colorless members of the Chlamydomonadaceae; *Euglenomorpha* (Fig. 28) a colorless endozoophyte; *Dinobryon* (Fig. 357) which usually appears as empty cone-shaped loricas; *Peridinium, Ceratium* and other members of the dinoflagellates (Figs. 380, 385) which often occur as dead and empty cells which do not show the brown chloroplasts which characterize this flagellated Phylum, species and genera being differentiated by details of wall structures; *Trachelomonas* (Fig. 27) usually appearing as brown empty loricas of various shapes.

**Use the iodine test to identify and differentiate the Chlorophyta from all other groups [except that a few dinoflagellates (Pyrrhophyta) show a positive starch test].

(Fig. 301) which although densely green belong to the non-starch-producing Chrysophyta that are yellow-green or golden-brown 25

5b   Plants not grass-green or gray-green, but yellowish-green, apple-green or golden-yellow; *or* reddish either because of a chloroplast pigment, or haematochrome (carotenoid pigments); *or* rarely chloroplasts with a bluish tinge. (Look for genera of freshwater red algae [Rhodophyta] here.) 6

6a   Plants reddish with phycoerythrin, or orange-red with carotenoid pigments which partly or completely mask the green of the chloroplasts. (See *Ancylonema* (Fig. 226), a filamentous form sometimes greenish but more often violet or purplish) 7

6b   Plants not red, but yellow-green, golden brown, *or* with bluish chloroplasts or protoplasts; starch test negative 417

7a   Plants filamentous, erect or prostrate 8

7b   Plants solitary cells, or forming colonies invested by a common mucilage..... 11

8a   Thallus composed of closely adjoined, prostrate filaments forming an epiphytic disc on the leaves of land plants. Fig. 4 .... *Phycopeltis*

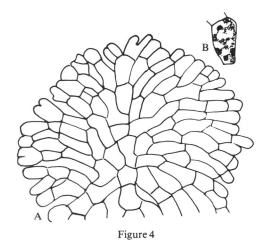

Figure 4

Figure 4   *Phycopeltis arundinacea* (Mont.) de Toni. (A) habit of dislike thallus; (B) single cell showing parietal, fenestrate chloroplast. (Chloroplasts sometimes appear as separate, irregularly shaped platelets.)

In this genus most species grow as a dislike layer of cells on the surface of leaves of higher plants. The thallus is rather regular in outline, produced by laterally appressed, branching filaments of rectangular cells (the branching habit often obscure). The cell contents are often orange or yellowish from carotenoid pigments (haematochrome) although actually there are green chloroplasts. Gametangial cells are sessile on the thallus whereas zoosporangia are stalked. Species are most common in humid, tropical and subtropical situations. Thus far the genus has not been reported from the United States but is to be expected in the Gulf states.

8b   Thallus with erect filaments, sometimes arising from a basal pseudoparenchymatous expanse or cushion of cells. (See *Kyliniella* [Fig. 312] sometimes reddish) 9

9a Filaments, erect, growing from a lobed, subepidermal cell in the stomatal chambers of leaves of higher plants, the erect stalks bearing a sporangium. Fig. 5 . . . .
. . . . *Stomatochroon*

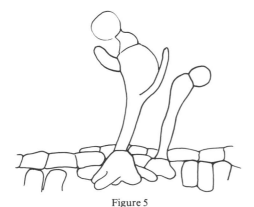

Figure 5

Figure 5 *Stomatochroon lagerheimii* Palm, showing upright branch from a subepidermal cell. Sporangium terminal. (Drawn from a specimen on a citrus leaf, Ecuador.)

The thallus in this genus is much reduced, consisting of an enlarged, lobed cell within the stomatal cavities of higher plant leaves, from which an erect stalk arises. The stalk or branch bears one or more sporangia. The stalk is brick-red with carotenoids. The genus is found in humid tropical regions, mostly on broad-leaved evergreens, but is known to occur on many kinds of plants which prefer sunny rather than densely shaded habitats. The basal cell is green, with the chloroplasts crowded in the lower part.

9b Thallus otherwise 10

10a Thallus composed of a subepidermal expanse of cells from which erect, hair-like branches arise, growing on fruits and leaves of land plants *(Magnolia, e.g.)* where they form circular, gray discs. Fig. 6 . . . . *Cephaleuros*

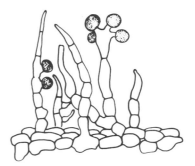

Figure 6

Figure 6 *Cephaleuros virescens* Kunze, diagram of a thallus as it grows under epidermis of host plant, with erect branches protruding externally and bearing sporangia.

Cephaleuros virescens Kunze occurs in tropical and subtropical parts of the world (including southeastern United States). It is parasitic or semiparasitic on the leaves and fruits of higher plants *(Magnolia, Thea,* citrus fruits, *Rhododendron).* Because of the discoloration and degeneration of host tissue in the vicinity of the parasite some damage is caused by this alga and a certain amount of economic loss results, especially to tea growers. Although the parasitized areas appear gray-green in color, individual filaments of the cushionlike thallus of the alga are reddish.

10b Filaments not growing from a subepidermal expanse, but occurring on wood, tree-trunks and leaves, the walls of the cells thick; apex of the filaments often capped with mucilage; unbranched or branched, with the branches arising at right angles to the main axis; sporangia terminal or lateral. (See also *Physolinum* [Fig. 279] which is usually reddish with carotenoids [haematochrome], sometimes included in *Trentepohlia*.) Fig. 7 . . . . *Trentepohlia*

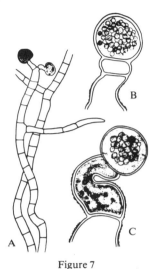

Figure 7

Figure 7   *Trentepohlia jolithus* (L.) Wallroth. (A) filament with 2 terminal sporangia; (B), (C) gametangium and a sporangium before being abscissed.

The species illustrated here and *T. aurea* are the 2 which are the most common of the 14 which have been reported from the United States. They grow on moist stones, dripping cliffs, and on the moist bark of trees, and on leaves. The characteristic orange color makes this plant conspicuous, especially when thalli form extensive patches, sometimes occurring as a felty-mat over large areas of rocky cliffs. In southern United States the moist sides of trees throughout extensive areas of the country side are colored orange by this alga. In humid situations of the tropics and subtropics the filaments become infested with a fungus to form the lichen *Coenogonium,* and in this condition the growth may be conspicuously fanlike on the sides of twigs and leaves. The carotenoid appears in the cell as a reaction to intense illumination.

Reproduction in *Trentepohlia* is by isogametes produced in a specialized cell, and by zoospores. Zoosporangia may be either lateral or terminal; are abscissed when mature and are carried by air currents to germinate in a moist place by releasing biflagellate zoospores.

11a (7) Thallus a globular, tubular or amorphous, gelatinous matrix enclosing many cells (young colonies few-celled) 12

11b Plants unicellular or incidentally clustered and gregarious, not enclosed in a gelatinous matrix 16

12a Chloroplast axial with a central pyrenoid 13

12b Chloroplast parietal, a plate or cup, with or without a pyrenoid 14

13a Cells oval to elliptic in mucilaginous tubes or strands, together forming a gelatinous mass of macroscopic proportions, color often more orange and yellowish then red. Fig. 8 . . . . . *Chroothece*

(Not *Chroothece* as assigned to the Cyanophyta.)

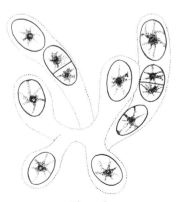

Figure 8

Figure 8 *Chroothece mobilis* Pasch. *et* Petrov., habit of portion of mucilaginous thallus showing cells with stellate chloroplasts and central pyrenoid.

This member of the Rhodophyta (Bangiales) grows mostly on damp soil and rocky cliffs. Here it forms gelatinous masses involving tubes of mucilage in which there are oval or subcylindrical cells. The stellate chloroplast is very striking. In mass the thallus is orange-colored or yellow-brown. This is not to be confused with a discarded blue-green alga name now referred to the genus *Coccochloris*.

13b **Cells spherical or nearly so, with a sheath which becomes confluent with the colonial mucilage; inhabiting damp soil, often more purple in mass than red. Fig. 9 . . . . *Porphyridium***

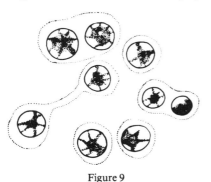

Figure 9

Figure 9 *Porphyridium cruentum* Näg., cells scattered or clumped in an amorphous mucilage.

If this is a member of the Rhodophyta it is the only one known that is one-celled. Plants occur in gelatinous masses on damp soil, especially in green houses, or on walls. The growth is usually purple or wine-colored. The cells have a stellate chloroplast. The species illustrated is fairly common and widely distributed. Another little-known species, *P. magnificum* Wood has been reported from Texas.

14a **(12) Cells in 2's and 4's at the periphery of a gelatinous matrix; chloroplast parietal but lobed, with a pyrenoid; pseudocilia and contractile vacuoles wanting. Fig. 10 . . . . . *Pseudotetraspora***

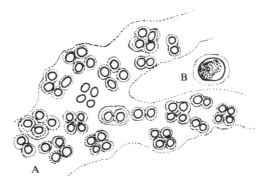

Figure 10

Figure 10 *Pseudotetraspora gainii* Wille. (A) habit of portion of thallus with cells in groups of 4; (B) a single cell showing parietal, lobed chloroplast with a single pyrenoid.

This genus occurs in both marine and freshwater situations, and in polar regions forms red or green snow. The colonial mass of spherical cells forms mucilaginous clumps and elongate strands; mostly microscopic.

14b **Cells not in 4's at the periphery of a colonial mucilage, but distributed throughout the matrix 15**

15a **Thallus mostly aquatic and free-floating, but sometimes attached; cells without individual sheaths apparent; with a lobed, parietal chloroplast and 2 contractile vacuoles; cells green but resting stages are red with carotenoids (haematochrome). Fig. 11 . . . . *Palmellopsis***

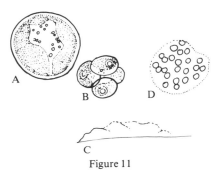

Figure 11

Figure 11 *Palmellopsis gelatinosa* Korsch. (A) single cell; (B) group of recently divided cells; (C) diagram of thallus on substrate; (D) colonial group of cells within mucilage. (A, redrawn from Korschikoff.)

The gelatinous colonies of this genus are amorphous, either planktonic or adherent to aquatic substrates. There is a cup-shaped chloroplast, a pyrenoid and 2 contractile vacuoles. The mucilage is not lamellate as in the similar genus *Palmella*. Thus far *Palmellopsis* is known only from Europe.

15b **Thallus a gelatinous, amorphous mass on soil; cells solitary or in pairs within the gelatinous sheath, the mucilage becoming confluent with the colonial matrix; chloroplast a parietal cup of the *Chlamydomonas* type, chlorophyll often masked by a red pigment; contractile vacuoles lacking. Fig. 12 ..... *Palmella***

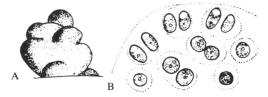

Figure 12

Figure 12 *Palmella miniata* Liebl. (A) habit of colony; (B) portion of colony showing arrangement of cells with individual sheaths within the colonial mucilage.

This plant forms lumpy, gelatinous masses, 2 to 8 mm in diameter on damp soil or on dripping rocks. The cells of *P. miniata* Leibl. are often red with (supposedly) carotenoids (haematochrome), whereas *P. mucosa* Kütz. (without individual sheaths) is always green. The oval cells, scattered throughout the mucilage may have indistinct, individual sheaths in some species. Like *Gloeocystis* (Fig. 86) which also has cup-shaped chloroplasts, *Palmella* may be confused with other gelatinous genera on soil.

16a **(11) Cells predominantly motile, with many individuals temporarily non-motile in the same habitat** **17**

16b **Cells predominantly non-motile, often encysted and stationary. Some times individuals emerge from the encysted condition and become motile in the same habitat.** **21**

17a **Chloroplasts many oval or irregular, platelike discs, or rarely diffused, the chlorophyll obscured by orange-red carotenoid (haematochrome); visible flagellum 1. Fig. 13** .... *Euglena**

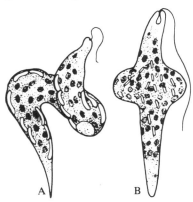

Figure 13

*One species of *Euglena* (*E. sanguinea* Ehr.) has a diffuse, blood-red color permanently; other species are brick-red when exposed to strong light, often forming powdery films on the water surface.

Figure 13 *Euglena.* (A) *E. convoluta* Korsch., showing lateral paramylum plates as seen on edge, one in flat view; (B) *E. elastica* Presc. Both of these are metabolic species, changing shape while swimming, whereas others are rigid and maintain a constant shape.

There is a gullet at the anterior end of *Euglena* and 1 or more contractile vacuoles. The eyespot is conspicuous (lacking in some species). Although usually green, these elongate, slowly moving organisms sometimes are colored red because of carotenoids. A pond or slough may have a bright red film produced by a *Euglena* 'bloom'. As with other red-colored green algae, the pigment is produced in response to intense light. *Euglena* species often are found in the psammon. About 60 species have been reported from the United States and many of these are world-wide in distribution. A few occur in marine waters.

17b Chloroplasts 1 or 2 parietal plates or a cup; flagella 2 . . . . **18**

18a Cells with a wide, sheathlike wall, the protoplast showing fine extensions to the wall margin; in the encysted stage the green chloroplast obscured by orange-red carotenoid (haematochrome); plants of rock pools and cemented basins. Fig. 14 . . . . *Haematococcus*

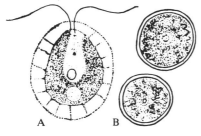

Figure 14

Figure 14 *Haematococcus lacustris* (Girod.) Rostaf. (A) swimming cell showing protoplast with processes extending to the outer wall; (B) cysts, brick-red with haematochrome.

Although motile, this organism is more frequently found as a conspicuous coating of red cysts in the bottom of shallow rock pools and cement basins. A common habitat is the garden bird-bath. In the swimming state it is readily identified by the fine fibrils which extend from the protoplast to the margin of the wide wall. The space between the wall margin and the protoplast is filled with colorless, watery mucilage. When not obscured by haematochrome the chloroplast can be seen to be parietal with many pyrenoids. At times, however, there are 2 chloroplasts, one posterior and one anterior, each plate with a pyrenoid. *Haematococcus* and *Chlamydomonas* have been found by culture studies to be mutually antibiotic.

18b Cells without such a wide, sheathlike wall **19**

19a Cells oval or pear-shaped, with a longitudinal furrow bordered by trichocysts; 1 reddish chloroplast. Fig. 15 . . . . *Rhodomonas*

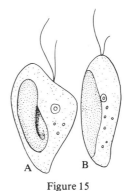

Figure 15

Figure 15 *Rhodomonas lacustris* Pascher. (A) cell in front view showing laminate chloroplast and flagella of different lengths; (B) lateral view of cell.

The cells of this flagellate are usually somewhat obversely pyriform or broadly oval. The two flagella are of unequal length. Whereas the single, broad, platelike chloroplast may be olive-green, it is usually red. A distinctive character is a longitudinal sulcus or furrow bordered by trichocysts. *Rhodomonas,* a member of the Cryptophyta is known from Ohio and Michigan streams.

19b Cells otherwise 20

20a Cells oval or broadly rounded, or asymmetrical at the anterior end, with 2 (or 1) parietal, olive-green chloroplasts which are often red; flagella 2, attached within an apical gullet. Fig. 16. ....
.... *Cryptomonas*

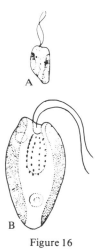

Figure 16

Figure 16 *Cryptomonas.* (A) *C. erosa* Ehr., a small species with 2 laminate chloroplasts; (B) *C. splendida* Czosn., showing flagella attached in a gullet.

There are 8 species of this flagellate reported from the United States, but there probably are many more. The organisms seem never to occur many together, are easily overlooked in dense mixtures of algae; are fast-moving like species of *Chroomonas.* The cells are somewhat pyriform, being broader at the anterior end. There are 2 yellowish-green chloroplasts which, especially at certain times of the year, are reddish, and they appear as 1 diffuse chloroplast. When the organisms are slowed in their movement and under proper optical conditions the characteristic gullet in the anterior end can be discerned, especially when the organism rotates on its axis. A somewhat similar-shaped flagellate is *Chilomonas* which has no chloroplasts but contains starch grains. See *Chroomonas* (Fig. 313) somewhat similar in shape which has flagella of equal length (the flagella of *Cryptomonas* are unequal), and a blue-green chloroplast.

20b Cells round, elliptic to oval, with 1 parietal, cup-shaped chloroplast and 1 or more pyrenoids (rarely the chloroplast parietal H-shaped); flagella not attached in an apical gullet but arising from blepharoplasts; eye-spot usually evident; plants of snow fields, red when encysted. Fig. 17. ..... *Chlamydomonas*

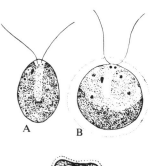

Figure 17

Figure 17 *Chlamydomonas.* (A) *C. polypyrenoideum* Presc.; (B) *C. sphagnicola*

Fritsch & Takeda, a species with bipapillate protrusions at the anterior end of the sheath; (C) *C. sanguinea* Lag., zygospore with ridged wall. (C, redrawn from Kol.)

Whereas this common genus is represented by approximately 507 described species, it is doubtful that they are all distinct taxa. Unless specimens are given careful study (sometimes involving culture) they may be confused with other biflagellated genera, or with the motile reproductive cells of other algae. The species of this genus are encountered more frequently than any other members of the Volvocales and are to be found in a great variety of habitats: eu- and tychoplankton, in small rock pools, rain barrels, laboratory aquaria. A favorable habitat is the barnyard pool or watering trough. *C. nivalis* (Bauer) Wille produces red snow at high altitudes.*

21a (16) Cells solitary, rarely 2 together, mostly terrestrial and on dripping rocks, enclosed in a much lamellated gelatinous sheath, the cells lying excentrically in the layers of mucilage. Fig. 18 . . *Urococcus*

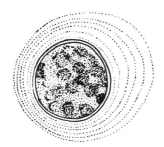

Figure 18

Figure 18 *Urococcus insignis* (Hass.) Kütz. (*Rufusiella insignis* [Hass.] Loeb.), solitary or 2 to 4 together in eccentric, lamellate sheaths.

Although placed originally in the Tetrasporales of the Chlorophyta this reddish-colored cell has been shown to be a stationary, non-flagellated member of the Gloeodiniaceae in the Pyrrhophyta. The cystlike cell apparently has many, small, crowded chloroplasts, brown-orange to reddish in color, and globules of oil. In reproduction motile gymnodinoid zoospores are produced. If species of *Urococcus* other than *insignis* are shown not to be *dinoflagellates* they will need to be transferred to another, renamed genus.

21b   Cells otherwise 22

22a   Cells epizoic in the hair scales of animals, especially the sloth; solitary or in clumps, gregarious. Fig. 19 . . . . *Cyanoderma*

Figure 19

Figure 19 *Cyanoderma bradypodis* Weber van Bosse. Clusters of cells in pseudo-filamentous arrangement within hair scales of Panama sloth; one cell enlarged to show disclike chloroplasts.

Cyanoderma bradypodis Weber van Bosse is a unique member of the Rhodophyta which is epizoic among the hair scales of the 3-toed sloth. The cells are scattered or in patches in which there is a tendency to form filaments. The hairs of the sloth may be pinkish because of the alga. In preserved specimens the chloroplasts seem to be several within the cell and disclike.

22b   Cells otherwise 23

Sphaerellocystis (Fig. 205) appears much like a non-motile *Chlamydomonas* in a wide, gelatinous sheath; has 2 contractile vacuoles but no eye-spot.

23a Cells inhabiting the tissues of higher plants, globular or flask-shaped with rhizoidal lobes and extensions. Fig. 20. . . . . *Rhodochytrium*

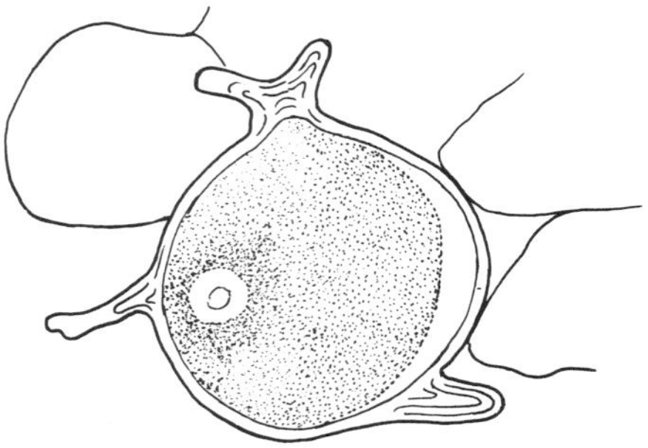

Figure 20

Figure 20 *Rhodochytrium spilanthidis* Lag., a large, red cell within the tissues of *Ambrosia* (Ragweed).

This curiously shaped, unicellular parasite, a member of the Chlorophyta, occurs on a greater variety of host plants than does the filamentous *Phyllosiphon* (Fig. 159), but seems to be most frequently found in ragweed, living within both stem and leaves. It is quickly identified by the red color and by the large number of starch grains. The chloroplast is massive and indefinite. Reproduction is by zoospores and isogametes.

23b Cells otherwise 24

24a Plants inhabiting snow fields and ice; cysts relatively small and smooth-walled; associated zygospores red with thick, knobby walls. Fig. 17 . . . . *Chlamydomonas*

24b Plants inhabiting rock pools and cemented basins; cells relatively large (up to 50 μm diam.), smooth-walled. Fig. 14 . . . . *Haematococcus*

25a (5) Organisms equipped with flagella, usually swimming in the vegetative state, solitary or colonial. (Preserved specimens should be examined for 2 or more minute protuberances at the anterior end of the cell which locate the position of flagella that may have been retracted or lost. Use 3-5% glycerine for mounting to slow down the movement of flagellated cells. [See Figs. 13, 17.]). 26

25b Organisms non-motile in the vegetative condition. (Check to be certain the organism is not a motile one at rest. See *Trachelomonas* [Fig. 27] which, although motile is commonly found as a non-motile brown shell [lorica] from which the flagellated protoplast has escaped.) Cells solitary or colonial, or filamentous forms 81

26a Cells solitary 27

26b Organisms colonial, 4 or more cells usually enclosed in a gelatinous sheath 68

27a Cells with a single flagellum, but housed in a vase-shaped lorica attached to filamentous algae or other substrates, the lorica open at the anterior end through which the flagellum protrudes; chloroplasts disclike. Fig. 21 . . . . . . *Ascoglena*

Figure 21

Figure 21 *Ascoglena vaginicola* Stein, cell in a sedentary lorica.

This genus and *Colacium* (Fig. 120) in the Euglenophyta are unique. Although provided with a flagellum they are attached and stationary. In *Ascoglena* the protoplast is enclosed by a reddish-brown lorica to which it is attached at the base by a protoplasmic strand. Like *Trachelomonas* (Fig. 27) the cell divides within the lorica, one portion swimming away to establish itself in a new sessile shell. Of the 2 known species, *Ascoglena vaginicola* Stein has been found in the United States, occurring in Ohio, Michigan and Montana.

27b Cells equipped with flagella, freely swimming 28

28a Cells shining green, motile by 2 flagella, 1 trailing and 1 directed forward; cells with *apical reservoir* and with trichocysts (in most forms) arranged at the periphery or in the apical region; iodine test for starch negative; food reserve fats. Chloromonadophyta . . . 29

28b Cells otherwise; grass-green or olive-green; flagella 1 to 4 but without 1 long,

trailing flagellum; trichocysts lacking, or if present (*Cryptochrysis,* Fig. 26) in rows on either side of a longitudinal furrow and not peripheral 31

29a Cells ovoid to pear-shaped, broadest at the posterior end; flagella apical; cells highly metabolic, flattened when seen in end or side view; trichocysts radially arranged immediately within the periphery. Fig. 22. . . . . *Vacuolaria*

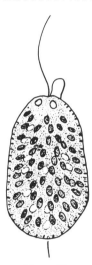

Figure 22

Figure 22 *Vacuolaria virescens* Cienk., showing peripheral trichocysts, oval chloroplasts and 2 anterior contractile vacuoles.

In this genus the pyriform cells are somewhat flattened as seen in side view. They are highly metabolic (changing shape readily when swimming). Like other members of the Chloromonadophyta, one of the 2 flagella trails, the other directed forward. There are numerous disclike chloroplasts, fusiform trichocysts within the periphery of the cell, but no eye-spot. Characteristically there are 2 apical contractile vacuoles (sometimes several small ones). *Vacuolaria* is usually found in ditches and swamps.

29b Cells shaped otherwise. **30**

30a Cells round in cross section, in front view oval, slightly narrowed posteriorly; trichocysts forming a cluster at the anterior end; apical reservoir circular; flagella somewhat lateral in attachment. Fig. 23 . . . . *Merotrichia*

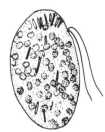

Figure 23

Figure 23 *Merotrichia capitata* Skuja, showing laterally attached flagella; trichocysts somewhat clustered in the anterior end. (Redrawn from Graffius.)

The cells in this genus are broadly ovate, usually somewhat larger anteriorly, with 2 long, laterally attached flagella. As is typical, 1 of the flagella trails. The trichocysts are slender and needlelike, scattered through the cell but definitely crowded and clustered near the anterior end. Like most members of the Chloromonadophyta this genus occurs in bogs and swamps.

30b Cells oval to nearly circular, in some species narrowed posteriorly and forming a short caudus, flattened when seen in side view; flagella apical; apical reservoir triangular; trichocysts radiately arranged at the periphery of the protoplast. Fig. 24 . . . . *Gonyostomum*

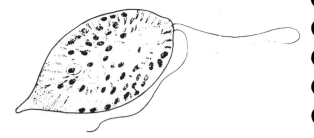

Figure 24

Figure 24 *Gonyostomum semen* (Ehr.) Stein, showing numerous ovoid chloroplasts and the peripheral trichocysts. It is normal for 1 flagellum to be directed forward, the other trailing.

Although rare, this genus is widely distributed, and occurs more frequently than other chloromonads, in acid lakes and bogs. Under the microscope *Gonyostomum semen* (Ehr.) Stein reminds one of a flat, green bottle, usually is quiet in the microscope field but moves with sudden, jerky motions for short distances. Among the peripheral chloroplasts are slender trichocysts which throw out threads upon stimulation. The forward directed flagellum is at least twice the length of the cell. The apical gullet is pyramidal. In addition to *G. semen* another species *G. latum* Iwan., circular in outline, has been found in southeastern United States.

31a (28) Chloroplasts in the form of parietal, longitudinal plates (or sometimes diffuse); flagella 1 or 2. (See *Scherffelia*, (Fig. 48) however, in which some species show longitudinal chloroplasts but 4 flagella.) **32**

31b Cells with chloroplasts otherwise, 1 cup-shaped chloroplast, with 2 parietal chloroplasts, *or* with numerous disclike (rarely ribbonlike) chloroplasts; red eyespot usually present. Fig. 13. . . . . **33**

32a Cells cylindrical, with broadly rounded, truncate poles; 1 flagellum directed forward; with 2 laminate, longitudinal, grass-green chloroplasts. Fig. 25 . . . . . .
. . . . *Monomastix*

Figure 25

Figure 25 *Monomastix opisthostigma* Scherf. with 2 laminate chloroplasts and a basal pigment-spot.

In this genus the cells are definitely short- cylindric and broadly rounded at both poles. The chloroplasts are 2, yellowish, broad, parietal plates with a single large pyrenoid each. In the anterior end there is a contractile vacuole above which 1, relatively long and forward-directed flagellum arises. In the posterior region is a cluster of elongate trichocysts, and laterally a red eye-spot. The genus was originally assigned to the Cryptophyta but is now placed in the Prasinophyceae of the Chlorophyta.

32b Cells oval in front view, dorsiventrally flattened, often lobed and truncate at the anterior end, with a longitudinal furrow on either side of which is a row of circular trichocysts; flagella 2, of unequal length; chloroplasts 2 lateral plates, olive-green in color but frequently brownish or yellow-brown. Fig. 26 . .
. . . . *Cryptochrysis*

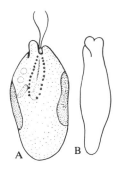

Figure 26

Figure 26 *Cryptochrysis commutata* Pascher. (A) front view showing gullet with marginal trichocysts, and parietal chloroplasts; (B) side view of the same cell showing compression.

This biflagellate has 2 laminate chloroplasts which may be olive-colored or brownish. (The chloroplasts are obscure in preserved material.) The flagella are short and equal in length. The species known from Massachusetts has a bilobed anterior end with the flagella attached in a median depression.

33a (31) Cells with numerous, disclike (rarely ribbonlike) chloroplasts*; food reserve in the form of variously shaped, colorless (white) paramylum bodies (See Fig. 29.) which do not stain blue-black with iodine; slow-moving by 1 (usually) or more stout flagella (See Fig. 31.); a red eye-spot usually evident; cells mostly longer than wide, round in cross section in most species, sometimes flattened or strap-shaped. Division Euglenophyta 34

33b Cells with 1 cup-shaped or star-shaped chloroplast, (but see *Polytoma* [Fig. 32],

*The questionable euglenoid *Cryptoglena* has 2 parietal chloroplasts. See Smith, 1950, p. 355 for a description.

colorless), usually containing 1 or more conspicuous pyrenoids, *or* rarely with 2 parietal chloroplasts; food reserve starch, iodine test positive; lens-shaped eye-spot usually evident; actively swimming with fine, often obscure flagella, 1 to 8 in number. (Examine carefully for stublike flagellar remains. Add 5% glycerine to microscope mount to slow down organisms for observations.) Motile Chlorophyta (in part) 40

34a Cells enclosed in a lorica or shell with an apical opening through which the flagellum projects 35

34b Cells not enclosed in a lorica 36

35a Lorica vaselike, tubular, sessile on filamentous algae or other substrates, brownish because of iron deposits; enclosed cell fusiform or elongate-elliptic, metabolic; flagellum 1. Fig. 21 . . . . *Ascoglena*

35b Lorica of various shapes, enclosing a flagellated cell, not sessile. (The loricas are more often than not found empty, brown or tan.) Fig. 27 . . *Trachelomonas* (See also 44a, p. 35).

(The genus *Strombomonas* includes the *Trachelomonas* species which have a pale tan or nearly colorless lorica.)

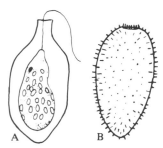
Figure 27

Figure 27 *Trachelomonas*. (A) *T. ampulla* Playf. showing euglenoid cell within lorica; (B) *T. conica* Playf. with a spiny-walled lorica.

There are several hundred species of *Trachelomonas,* each showing a differently shaped and decorated shell or lorica. The amount of iron present in the lorica determines the intensity of color. The species with colorless or yellowish loricas have been transferred by some students to the genus *Strombomonas.* *Trachelomonas* species are found intermingled with other algae in shallow water of ditches and bogs, or among aquatic weeds near lake shores.

36a (34) Cells with 3 flagella; endozoic in the digestive tracts of aquatic invertebrates. Fig. 28 . . . . *Euglenamorpha*

Figure 28

Figure 28 *Euglenamorpha hegneri* Wenrich, two shapes.

This unusual euglenoid may be either colorless or pigmented (in light); occurs in the intestinal tract of frogs and tadpoles. This is the only known tri-flagellated member of the phylum. There are many oval plastids and an eye-spot.

36b Cells with 1 or 2 flagella; not endozoic 37

37a Cells nearly round or broadly fusiform in outline in front view and appearing disclike or flattened when seen from the side, but with a longitudinal ridge so that the cells are somewhat triangular in end view; cells usually twisted at the posterior end where in some species there is a caudus (tail-piece); paramylum in the form of 1, few or several 'doughnut' rings or discs; periplast usually showing longitudinal, spiral striations; flagellum 1. Fig. 29 . . . . *Phacus*

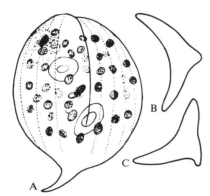

Figure 29

Figure 29 *Phacus*. (A) *P. curvicauda* Swir., in front or ventral view showing eye-spot, chloroplasts and 2 ring-shaped paramylum bodies; (B), (C) *P. triqueter* (Ehr.) Duj. as seen in end view, the triangular shape being produced by a longitudinal flange on the dorsal surface of the cell.

Although some species are spirally twisted and 'top'-shaped, some are flat or at most only slightly saucer-shaped or pancakelike, with a long or short tail-piece. The cells appear to be flat as seen in front view, but from the side or top they are somewhat triangular because of a longitudinal flange. The rings of paramylum are conspicuous, sometimes so large as to fill nearly the entire diameter of the cell. There are numerous oval chloroplasts and a red eye-spot.

Phacus commonly appears in the same habitat with *Euglena*.

37b Cells shaped differently, fusiform, subcylindric to pear-shaped; paramylum bodies not shaped as above; flagella 1 or 2 . . . . **38**

38a Cells highly metabolic (changeable in shape while swimming), but assuming an oval or pear shape, with a truncate apex; flagella 2; anterior gullet conspicuous; numerous disclike chloroplasts. Fig. 30. . . . . *Eutreptia*

Figure 30

Figure 30 *Eutreptia viridis* Perty. (Redrawn from Lemmermann.)

The cells in this genus are fusiform when normally extended, and bluntly truncate at the anterior end. Posteriorly they are abruptly narrowed into a caudal extension. There is an eye-spot and a gullet with small, adjacent vacuoles in the anterior end. Under favorable optical conditions a granular swelling can be discerned at the base of each flagellum. The periplast (cell membrane), as in many euglenoids, is spirally striated.

38b Flagellum 1; cells shaped otherwise . . 39

39a Cells round, oval or pear-shaped in front view, round in end view; fixed in shape when swimming; paramylum in the form of 2 (or 4) lateral rings folded along the periplast; tail-piece, if present, in the form of a short, sharp projection from the broadly rounded posterior end. Fig. 31. . . . . *Lepocinclis*

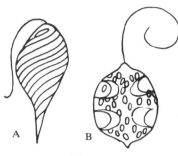

Figure 31

Figure 31 *Lepocinclis.* (A) *L. acuta* Presc., Showing spirally striated periplast; (B) *L. glabra* fa. *minor* Presc., showing 4 lateral, bandlike paramylum bodies.

There are many species of this genus, world-wide in distribution, never forming pure growths, often intermingled with species of *Euglena* and *Phacus.* In optical section the cells are round and in many the periplast is spirally striated. On either side of the cell is a folded ring of paramylum, sometimes 2 such rings on each side. Most species have a long or short caudus. The cells swim more actively than *Euglena.*

39b Cells elongate, fusiform or nearly cylindrical; round, or (in a few species) somewhat flattened when seen in end view; fixed in shape when swimming, or metabolic; flagellum 1, forked at the base, each portion arising from a bleph-

aroplast; chloroplasts oval plates, or (rarely) diffuse or ribbonlike; paramylum in the form of 1 or 2 large, or several small, rods or sticks; tail-piece sometimes present, formed by the gradual narrowing of the cell. Fig. 13 . . . . *Euglena*

40a (33) Cells colorless (similar to *Chlamydomonas* in shape and flagellation). Fig. 32 . . . . *Polytoma**

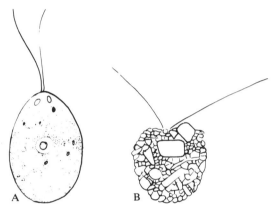

Figure 32

Figure 32 *Polytoma.* (A) *P. obtusum* Pascher; (B) *P. granuliferum* Lackey, the wall impregnated with minute grains of sand. (B, redrawn from Lackey.)

This colorless chlamydomonad often has an eye-spot. There are 2 contractile vacuoles in the anterior end and there is the same neuromotor apparatus that is found in *Chlamydomonas.* There are at least 36 species recognized, differentiated by a combination of cell-shape and internal characteristics.

40b Cells with chloroplasts, pigmented; starch test positive 41

*One dubious species, *P. granuliferum* Lackey is dark, almost black; has an investment to which sand grains are attached (Fig. 32a).

41a Protoplasts at a considerable distance within the cell wall and connected to it by fine, radiating processes; before encysting cells are green, but cysts have a mass of red pigment (haematochrome) that partly or entirely obscures the green color; forming rust-colored growths in rock pools and cemented basins. Fig. 14 . . . . *Haematococcus*

41b Cells not as above; free-swimming, not colored red by carotenoids. 42

42a Cells with a definite, cellulose wall (sometimes thin); in some with a gelatinous sheath, or with a lorica (a shell exterior to the cell or protoplast) 43

42b Cells without a definite cellulose wall (focus carefully) with a cell membrane only; lorica lacking 60

43a Cells with 4 flagella 57

43b Cells with 2 flagella 44

44a Cells enclosed in a shell-like lorica which in some species is externally roughened 45

44b Cells not enclosed in a lorica, but some forms enclosed by a gelatinous sheath 52

45a Lorica smooth or punctate. 46

45b Lorica externally granulate 48

46a Lorica colorless, rectangular in front view, with the corners produced into horns. Fig. 33 . . . . *Pteromonas*

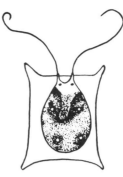

Figure 33

Figure 33 *Pteromonas aculeata* Lemm. Other species are truncately oval as seen in front view.

This genus takes its name from the winged appearance of the envelope. Although there are 7 species known, the most common perhaps is *P. aculeata* Lemm., recognized by the rectangular appearance in front view. Like *Phacotus* (Fig. 38) *Pteromonas* commonly occurs in the plankton of rivers. See *Scotiella* (Fig. 204) a genus which has been regarded by Pascher as belonging to *Pteromonas,* even though not flagellated in the vegetative condition. (See note on p. 112 under *Scotiella*.

46b Lorica globular or obversely pyriform 47

47a Lorica globular, not granular but punctate, appearing rough or granular (focus sharply); the enclosed cell a different shape. Fig. 34 . . . . *Dysmorphococcus*

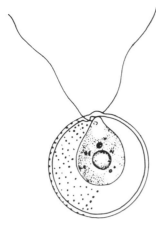

Figure 34

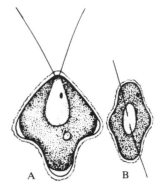

Figure 35

Figure 34 *Dysmorphococcus variabilis* Takeda, showing the lorica with a sample of the small pores.

This genus has a broadly oval lorica which is different in shape from the enclosed protoplast that lies some distance within the shell. The lorica has numerous, small pores, giving a dotted appearance and there are 2 separate pores through which the flagella extend. An eye-spot is present, and 1 or more pyrenoids in the parietal, cup-shaped chloroplast.

47b   Lorica obversely pyriform and 3-lobed when seen in front view, compressed when seen from the side, in end view oval but with 2 lateral protuberances, one on either side; the cell closely fitting the shape of the lorica. Fig. 35 . . . . *Cephalomonas*

Figure 35 *Cephalomonas granulata* Higin. (A) front view showing eye-spot; (B) apical view showing flagella attachment. (Redrawn from Higinbotham.)

These cells have a colorless lorica which is close-fitting. The anterior end of the lorica is narrowed and conical, abruptly narrowed posteriorly to form a broadly rounded lobe. After emerging, the flagella are widely divergent.

48a   (45) Lorica cordate in front view, irregularly granular, not dark-colored, somewhat quadrate when seen from the side, with a lobe at each angle and a fifth, median posterior lobe; the cell within the lorica elliptic. Fig. 36 . . . . . .
. . . . *Wislouchiella*

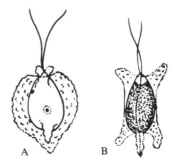

Figure 36

Figure 36 *Wislouchiella planctonica* Skvor. (A) front view showing shape of cell within lorica; (B) side view showing armlike extensions of lorica.

Named for the biologist Wislouch, this biflagellated organism is identified readily by the oddly-shaped lobes or processes of the wall which extend in several planes. The lorica is conspicuously granular. The species illustrated is rare but widely distributed in the United States.

48b Lorica shaped and decorated otherwise........................... 49

49a Lorica irregularly lobed, lumpy, with low, irregularly arranged granules, the lorica usually 4-lobed in end view; enclosed protoplast broadly oval to elliptic and about the same shape as the lorica. Fig. 37.............. *Thoracomonas*

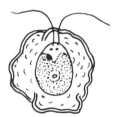

Figure 37

Figure 37 *Thoracomonas feldmannii* Bourr. (Redrawn from Bourrelly.)

These cells have a transparent, gelatinous lorica that has brown lumps because of knotted iron deposits. The cytology of the cell within is much like that of *Chlamydomonas* but there may be many pyrenoids; the eye-spot is median in the cell.

49b Lorica shaped and decorated otherwise........................ 50

50a Lorica and its cell approximately the same shape as seen in front view, the lorica often brown, bivalved, the seam showing as a suture in side view; irregularly granular. Fig. 38 .... *Phacotus*

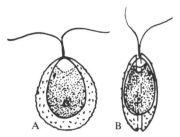

Figure 38

Figure 38 *Phacotus lenticularis* (Ehr.) Stein. (A) lorica and contained cell in front view; (B) lorica in side view showing the suture of the two sections.

This genus is relatively rare, but is often abundant in collections from habitats where it occurs. It is both eu- and tychoplanktonic and may occur in rivers as well as lakes. The lorica and the protoplast are both egg-shaped in front view, much narrowed when seen from the side when the bivalve nature of the lorica is evident. There is a distinct space between the cell and the lorica. Three species have been reported from the United States.

50b Lorica shaped and decorated otherwise........................ 51

51a Lorica thin, without an obvious apical pore, with spirally arranged granules, the protoplast the same shape as the lorica. Fig. 39 .... *Granulochloris*

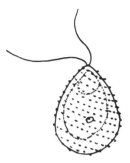

Figure 39

Figure 39 *Granulochloris seriata* Pash. *et* Jaho., lorica with spiral granulations. (Redrawn from Pascher & Jahoda.)

This genus is well-named because of the highly granular nature of the lorica. The organisms are distinctive with the spiral arrangement of the granules. The protoplast is similar to *Chlamydomonas.* There are about 13 species known from the United States.

51b Lorica wall relatively thick, with a broad apical pore, ornamented with prominent calcium granules; the lorica globular, the protoplast obversely pear-shaped. Fig. 40. . . . . *Coccomonas*

Figure 40

Figure 40 *Coccomonas orbicularis* Stein, showing a granular lorica containing an elliptic cell containing an eye-spot, a cup-shaped chloroplast and pyrenoid. (Redrawn from Conrad.)

The lorica is thick-walled and has calcareous and iron impregnations and granules which are

irregularly arranged. The protoplast is decidedly narrowed at the anterior end in most species. Upon cell division the lorica splits and the 2 daughter cells are liberated. There are at least 11 species known.

52a (44) Cells fusiform, sometimes elongate and narrowly pointed at the apices, without an external sheath. Fig. 41 . . . . *Chlorogonium*

Figure 41

Figure 41 *Chlorogonium elongatum* Dang. Other species are much more elongate-fusiform.

The cells in this genus are all more elongate than any other members of the Volvocales. Usually found in swamps and shallow ponds, they sometimes appear abundantly in laboratory aquaria. Of the 30 known species, 4 have been reported from the United States.

52b Cells shaped otherwise, oval, elliptic, with a smooth, even sheath, or with a lobed sheath; without a shell or lorica 53

53a Cells with a lobed sheath, or with a sheath bearing extensions 54

53b Cells with a smooth, even sheath; protoplast smooth-walled 55

54a Cells with 4 reflexed, pointed lobes (3 in view); quadrate in end view. Fig. 42. . . . . *Brachiomonas*

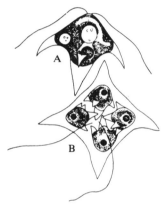

Figure 42

Figure 42 *Brachiomonas westiana* Pascher. (A) vegetative cell; (B) zoospore-formation. (Redrawn from West.)

This genus is marine but often occurs in brackish water and tide pools. The 2 flagella arise from the narrow anterior end of the cell which is 4-lobed and quadrate when seen in end view; cells with an extended, conelike posterior. About 6 species are known.

54b Cells with an irregularly lobed and lumpy sheath, the protoplast often withdrawn from the wall. Fig. 43 . . . . *Lobomonas*

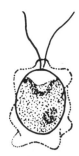

Figure 43

Figure 43 *Lobomonas rostrata* Hazen, the outer wall lobed and lumpy.

The irregular, lumpy appearance of the sheath possessed by cells in this genus is one of its chief characteristics. The organisms appear in the same habitat with *Haematococcus* (Fig. 14) *i.e.,* temporary rain water pools and cemented basins. There are 10 species but apparently only 1 has been reported from the United States.

55a (53) Protoplast with an envelope decidedly narrowed anteriorly, not the same shape as the envelope. Fig. 44. . . . . *Sphaerellopsis*

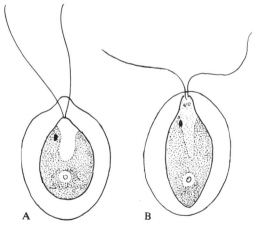

A   B

Figure 44

Figure 44 *Sphaerellopsis.* (A) *S. gloeocysti-formis* (Dill) Gerloff; (B) *S. fluviatilis* Pascher.

This genus probably should be classified under *Chlamydomonas* (Fig. 17) although specialists separate its 20 species on the basis of the very wide, gelatinous sheath which is different in shape from that of the protoplast. It occurs in the tychoplankton of lakes and ponds. See also *Smithsonimonas* (Fig. 45). *Sphaerellopsis* occurs mostly as red cysts in snow.

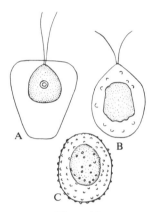

Figure 45

Figure 45 *Smithsonimonas abbotii* Kol. (A) young cell with smooth wall, a cup-shaped chloroplast and large, central pyrenoid; (B) an older cell with warts beginning to appear on the wall; (C) resting stage with outer wall densely warty. (Redrawn from Kol.)

The cells are oval, elliptic or nearly round, with or without a thin gelatinous sheath; chloroplast variable but mostly a parietal cup, with 1 or more pyrenoids.

55b **Protoplasts not decidedly narrowed anteriorly; protoplast the same shape as the gelatinous sheath (when present) 56**

56a **With 1 or more pyrenoids. Fig. 17 . . . . . .**
. . . . *Chlamydomonas*

56b **Cells similar in shape to the sheath; chloroplast similar to that of *Chlamydomonas* but without a pyrenoid. Fig. 46 . . . . *Chloromonas***

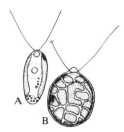

Figure 46

Figure 46 *Chloromonas.* (A) *Ch. gerloffii* Ettl with a parietal plate type of chloroplast; (B) *Ch. serbinowi* var. *minor* Ettl with parietal, irregularly shaped discs. (Redrawn from Ettl.)

This genus long has been considered as synonymous with *Chlamydomonas* by students of the algae. But one investigator (Gerloff in 1962) assigned the chlamynomonad species without a pyrenoid to *Chloromonas.* Thus far one species has been reported from the United States, occurring in permanent snow fields. See note under *Scotiella* (p. 112).

57a **(43) Cell enclosed in a yellowish lorica which is ornamented with prominent granules or other markings; lorica strongly flattened and sometimes pear-shaped in lateral view. Fig. 47 . . . .**
. . . . *Pedinopera*

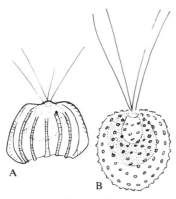

Figure 47

Figure 47 *Pedinopera*. (A) *P. rugulosa* (Playf.) Pascher; (B) *P. granulosa* (Playf.) Pascher. (Redrawn from Pascher.)

The pores and granules which ornament the wide, yellowish sheath help to identify this genus. Sometimes there are longitudinal ridges. There are 5 species, differentiated on the basis of lorica markings, only 1 of which has been reported from the United States.

57b Cells without a lorica **58**

58a Cells oval in front view, strongly flattened when seen from the side; sheath with 2 flaring, winglike lateral processes apically; chloroplasts 2, lateral and parietal. Fig. 48 . . . . *Scherffelia*

Figure 48

Figure 48 *Scherffelia cornuta* Conrad with apical, winglike extensions. (Redrawn from Conrad.)

This genus is distinct because of the lateral wings which arise near the apex of the cell. There are some 12 species, differentiated primarily by the form of the wings. It is to be expected in the United States; thus far known only from Europe.

58b Cells shaped otherwise, with a single chloroplast **59**

59a Cells broadly oval to somewhat cordate, with a slightly depressed apex, decidedly compressed as seen in end view; the 4 flagella arising in 2 separate pairs apically. Fig. 49 . . . *(Platymonas) Tetraselmis*

Figure 49

Figure 49 *Platymonas elliptica* G.M. Smith, 4 flagella in 2 pairs.

Although ordinarily found in brackish water, this genus contains at least 1 species that appears in fresh water. To make identification the cells should be seen from the top or side to determine whether they are flattened. The eyespot may be anterior or median. It is thought by some that this name is synonymous with *Platymonas,* a genus which occurs in both fresh and salt water.

59b Cells ovate or cordate, round in end view, the flagella all arising from one point. Fig. 50 . . . . *Carteria*

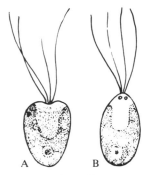

Figure 50

Figure 50 *Carteria,* two shapes. (A) *C. cordiformis* (Carter) Dies.; (B) *C. klebsii* Dill.

Like *Polytoma* (Fig. 32) this genus is characterized by having 4 flagella, but the cells are round when seen in end view. The chloroplast is variable in shape and may not appear as shown in the illustration. It may be a thin plate along the wall, cup-shaped and covering most of the wall except at the anterior end, or H-shaped. This is a large genus of some 70 species, separated into 5 subgenera on the basis of the chloroplast shape and the presence or absence of pyrenoids. One study has indicated that *Carteria ovata* Jacobs. is a diploid phase of a *Chlamydomonas* species.

60a (42) Cells with a single flagellum, directed posteriorly. Fig. 51. . . . .
. . . . *Pedinomonas*

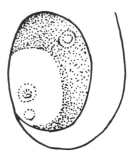

Figure 51

Figure 51 *Pedinomonas rotunda* Korsch. with posteriorly directed flagellum. (Redrawn from Korschikoff.)

Cells in this genus have no wall. They are circular or oval and possess a single, long flagellum that is directed posteriorly so that the organism swims backward. The chloroplast lies basally and along one side. There is a pyrenoid but no eye-spot. There may be 4 species of this genus, but only 1 has been reported from the United States, Ohio and Montana.

60b Cells with 2 to 8 flagella. 61

61a Cells with 2 flagella 62

61b Cells with 4 or 8 flagella. 65

62a Cells obversely pear-shaped or oval, broadly rounded anteriorly, but with longitudinal folds, 4- to 6-lobed or stellate when seen in end view. Fig. 52. . . . .
. *Stephanoptera*

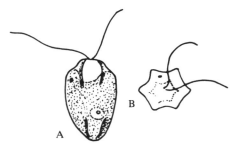

Figure 52

Figure 52 *Stephanoptera gracilis* (Artari) G.M. Smith. (A) front view showing cup-shaped chloroplast and pyrenoid; (B) vertical view showing lobes of cell and point of flagella. attachment.

These pyriform, motile cells without a wall

have longitudinal folds (wings) and when seen from the end they are 6-radiate and starlike. The chloroplast also has longitudinal flanges which extend into the folds of the membrane. The eye-spot is rodlike. One species has been found in the United States, occurring in saline water.

62b Cells not obversely pear-shaped, without longitudinal folds; not stellate in end view. **63**

63a Cells narrowed anteriorly, egg-shaped or pear-shaped in front view, round in end view; cells often reddish with haematochrome; species of brackish or saline waters. Fig. 53. . . . . *Dunaliella*

Figure 53

Figure 53 *Dunaliella salina* Teodor., two cell shapes.

The species illustrated and 2 others are apparently the only ones described for the genus thus far, but they are widely distributed in saline and brackish inland waters. The chloroplast of these pyriform cells lies in the posterior, broader part. *Dunaliella* belongs to a family which is characterized by the lack of a true cell wall, having only a cell membrane. Areas in Great Salt Lake are colored red by this organism.

63b Cells shaped otherwise; not inhabiting brackish and saline waters **64**

64a Cells transversely reniform (bean-shaped) with flagella of decidedly different length, attached apically but both trailing; eye-spot black, anterior; chloroplasts sometimes brownish or bluish-green. Fig. 54 . . . . *(Nephroselmis) Heteromastix*

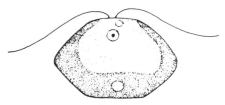

Figure 54

Figure 54 *Heteromastix angulata* Korsch. with nearly equal flagella.

This name now includes in synonymy the genus *Nephroselmis*. The cells are transversely angularly oval with the chloroplast lying along the posterior wall. The 2 flagella are unequal in length, a feature difficult of determination. There is a large, conspicuous pyrenoid and a less conspicuous eye-spot (sometimes lacking). Although classified in the Volvocales of the Chlorophyta, some students of the algae believe that the genus *Monomastix* of the Cryptophyta is synonymous.

64b Cells irregularly spheroid, decidedly flattened when seen from the side, with an apical depression; flagella attached below the apex and directed at a deflexed angle; chloroplast a complete, parietal band with 2 pyrenoids; eye-spot red, median rather than anterior. Fig. 55 . . . . *Mesostigma*

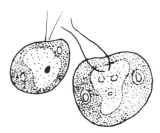

Figure 55

Figure 55 *Mesostigma viridis* Lauterborn. (Redrawn from Bourrelly.)

Cells in this genus are broadly oval to nearly circular in outline but are much flattened when seen from the side. The chloroplast completely encircles the cell. There are 2 conspicuous pyrenoids. The flagella are attached in a subapical depression. Characteristic of the Polyblepharidaceae there is no cellulose wall. Apparently only 2 species are known.

65a (61) Cells oval in front view, round in end view, with 8 apical flagella and 2 or 4 contractile vacuoles; chloroplast cup-shaped with a large basal pyrenoid. Fig. 56 . . . . *Polyblepharides*

Figure 56

Figure 56 *Polyblepharides fragariiformis* Hazen, showing 3 of the contractile vacuoles at the anterior end. (Redrawn from Hazen.)

This genus is identified by having a tuft of flagella (usually 8) which are relatively short.

There are several contractile vacuoles at the anterior end. If there is an eye-spot it lies lateral about midway in the cell. There may or may not be a pyrenoid. Thus far only 2 species have been reported from the United States, occurring along the Atlantic seaboard.

65b Cells with 4 flagella **66**

66a Cells pear-shaped or elliptic, with an apical depression and with longitudinal folds; 4-lobed when seen in end view, the flagella attached one in each depression between the lobes. Fig. 57 . . . . *Pyramimonas*

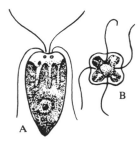

Figure 57

Figure 57 *Pyramimonas tetrarhynchus* Schmar. (A) front view showing chloroplast, pyrenoid and contractile vacuoles; (B) end view showing points of flagella attachment.

This is *Pyramidomonas* of some authors, including about 25 species, all of which are 4-lobed when seen in end view. There is a flagellum attached in each of the 4 depressions between the lobes. The eye-spot is variable in position. The 4 species reported from the United States have a wide distribution and are to be expected in ponds, pools and roadside ditches.

66b Cells shaped otherwise, without longitudinal folds **67**

67a **Cells curved, bean-shaped or wormlike, with 4 long flagella directed posteriorly or anteriorly; chloroplast an elongated plate. Fig. 58. . . . . *Spermatozoopsis***

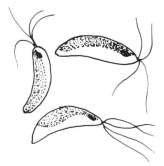

Figure 58

Figure 58 *Spermatozoopsis exultans* Korsch., three cell shapes.

These cells are curved or crescent-shaped with 4 flagella, but sometimes only 2. The chloroplast is a parietal plate without a pyrenoid; the eyespot is relatively large and conspicuous. Of the 2 known species 1 has been reported from the United States.

67b **Cells broadly oval and broadly rounded anteriorly, round in end view, with 4 flagella arising from one point; chloroplast cup-shaped with a basal pyrenoid. Fig. 59 . . . . *(Tetrachloris) Quadrichloris***

Figure 59

Figure 59 *Quadrichloris carterioides* (Pascher *et* Jahoda) Fott. (Redrawn from Fott.)

This name is synonymous with *Tetrachloris.* The cells are very similar to *Pyramimonas* in one view but are round in cross section. The membrane is not rigid and the organisms change shape while swimming. There are 5 species known, but thus far none has been reported from the United States.

68a **(26) Cells arranged in a plane, forming a flat or twisted plate 69**

68b **Cells arranged within a spherical, spheroidal, oval or somewhat cubical colony 70**

69a **Colony horseshoe-shaped, flat or twisted, the colonial sheath with 2 or 3 posterior lobes. Fig. 60 . . . . *Platydorina***

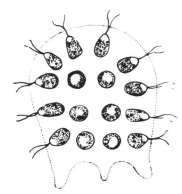

Figure 60

Figure 60 *Platydorina caudatum* Kofoid. The flagella of the cells in the center of the colony are directed vertically to the colony plane.

Although rare and seldom seen, this genus can be identified readily by the flattened, slightly twisted, horseshoe-shaped colony. It is found in the same habitats as favor other members of the Volvocales. Originally found in Illinois, *P. caudata* Kofoid has appeared in 8 other states but never from outside the United States.

69b Colony a circular or subquadrangular plate. Fig. 61 . . . . *Gonium*

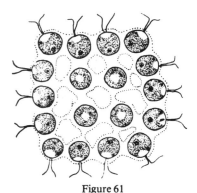

Figure 61

Figure 61 *Gonium pectorale* Muell., an 18-celled colony showing openings in the colonial mucilage.

The number of cells in this disclike colony may vary according to species, from 4 to 32 or 64. The rectangular plates tumble over and over in motion. The flagella of the inner cells are directed at right angles to the colony plane; those of the lateral cells are directed outward in the colony plane. Eight species are known from the United States. In sexual reproduction each cell becomes metamorphosed to act as an isogamete. Apparently some species are homothallic and others heterothallic (plus and minus gametes formed in different strains or clones).

70a **(68) Colony globular or pyriform, or quadrangular, without an investing colonial mucilage** 71

70b **Colony globular or ovoid; cells enclosed by a mucilaginous sheath** 75

71a **Colony pyriform, narrowly tapering posteriorly; cells radiately arranged. Fig. 62** . . . . *Raciborskiella*

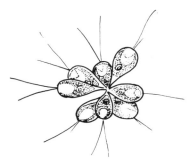

Figure 62

Figure 62 *Raciborskiella uroglenoides* Swir. (Redrawn from Swirenko.)

This genus with 2 species is distinct because of the colonial arrangement of the radiating cells that taper to a fine point posteriorly. Only 1 species is known, from freshwater.

71b **Colony otherwise, composed of oblong, elliptic or pyriform cells, not radially arranged** 72

72a **Colony small, composed of 2 or 4 cells arranged with long axes parallel and all directed the same. Fig. 63** . . . . *(Pascheriella) Pascherina*

Figure 63

Figure 63 *Pascherina tetras* (Korsch.) Silva, cells showing vacuoles near the base of the flagella.

This name is synonymous with *Pascheriella*. The organism is rare, occurs in rain water pools and catch basins of temporary duration. Unlike most members of this group of the algae the chloroplast is a laminate, parietal plate rather than being cup-shaped. *Pascherina* apparently has been found only in one station in the United States, California.

72b Colonies of more numerous cells, 8 to 16, with long axes not parallel 73

73a Cells with 4 flagella arising from a protuberance at the broad anterior end; cells compactly clustered in tiers of 4, with the anterior ends directed the same way. Fig. 64 .... *Spondylomorum*

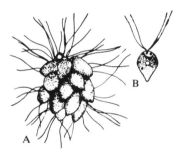

Figure 64

Figure 64 *Spondylomorum quaternarium* Ehr. (A) colony; (B) single cell showing posterior eye-spot and subflagellar vacuoles.

In this genus the cells have a 'huddled' appearance with the broad anterior ends all directed the same way. The cells have a tendency to arrange themselves in tiers. The eye-spot is posterior rather than anterior. Whereas the flagella are conspicuous it is difficult to determine that there are 4 on each cell. It has been suggested that this genus should be united with *Uva* (Fig. 66).

73b Cells with 2 flagella; cells not in tiers of 4 74

74a Cells pyriform to rhomboid, with narrow anterior ends and long flagella all directed the same; 8 quadrately arranged in 2 series, 4 widely-spaced cells above 4 others, the groups of 4 interconnected by gelatinous strands. Fig. 65 ... . *Corone*

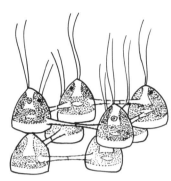

Figure 65

Figure 65 *Corone bohemica* Fott, colony with cells in 2 tiers.

In this quadrangular, 8-celled colony the cells are all directed the same way; bear 2 very long flagella. Thus far only 1 species is known, from Czechoslovakia.

74b Cells huddled in a cluster with anterior ends and flagella directed the same way. Fig. 66 .... *(Pyrobotrys) Uva*

Figure 66

Figure 66 *Uva gracilis* (Korsch.) Bourr., a colony of individuals showing the posterior eye-spot.

Pyrobotrys is a synonymous name for this genus as well as for *Chlamydobotrys*. Like *Spondylomorum* (Fig. 64) the cells of this colonial alga are closely grouped; have their anterior ends all directed the same way. There are 2 long flagella and a conspicuous eye-spot, anterior or posterior. The chloroplast is somewhat massive and is without a pyrenoid. Three species have been reported from the United States.

75a **(70) Colony spheroidal or oval; cells crowded, somewhat pyriform, with the broad ends all directed outwardly. Fig. 67** . . . . *Pandorina*

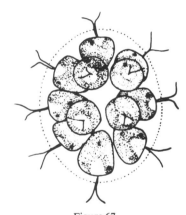

Figure 67

Figure 67 *Pandorina morum* Bory. Cells are pyriform and often are more compactly arranged than shown in the illustration.

This is a tumbling and rolling colony which usually is more oval than spherical, with the cells rather compactly arranged. The flagella extend from the broad anterior end of the pyriform cells in a more or less parallel fashion at first and then flare widely as they emerge from the colonial sheath. Often colonies are found in which each cell has undergone cleav-

age to form daughter colonies. *Pandorina morum* Bory is world-wide in distribution, whereas *P. charkowiensis* Korsch., also found in the United States, is relatively rare. The cells of the latter are compressed-spheroid; have a chloroplast with longitudinal ridges.

75b **Colony globular or broadly ovoid; cells not crowded but evenly spaced.** 76

76a **Colonies large, up to 1 mm in diameter and containing as many as 50,000 cells. Fig. 68** . . . . *Volvox*

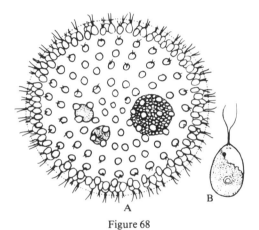

Figure 68

Figure 68 *Volvox tertius* Meyen. (A) colony showing 1 daughter colony and 2 eggs; (B) cell with parietal chloroplast and eye-spot.

This globular colony contains thousands of individuals and usually can be seen with the unaided eye. It occurs in water that is rich in nitrogenous matter and sometimes produces "blooms" of short duration during summer months. It is often accompanied by *Pleodorina* (Fig. 72) and *Eudorina* (Fig. 71). Reproduction is oogamous and when colonies are mature they may contain several, much enlarged egg cells and packets (plakea) of motile sperm. Some species, however, are heterothallic and produce either one or the other of the gametes. Frequently the colonies will be found to have 1 to several daughter colonies produced by the cleavage of selected vegetative cells.

76b Colonies composed of from 8 to 128 cells. 77

77a **Cells fusiform, with sharply pointed lateral processes or extensions of the protoplast, arranged in a circle within an ovoid, gelatinous sheath. Fig. 69** . . . . *Stephanosphaera*

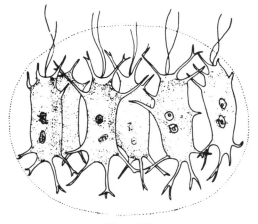

Figure 69

Figure 69 *Stephanosphaera pluvialis* Cohn. Oval colony with cells forming a median band. The chloroplasts commonly contain 2 pyrenoids.

Like *Pascherina* (Fig. 63) this organism occurs in small, temporary pools in rocks, sometimes at high altitudes. The cells are all directed one way, lying in an equatorial belt within the oval, gelatinous sheath, with flagella all directed forward.

77b **Cells round or ovoid, without lateral processes** 78

78a **Cells arranged at the equator of a spheroidal or oval gelatinous sheath; cells in 2 tiers, 8 or 16 in number. Fig. 70** . . . . *Stephanoon*

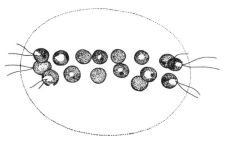

Figure 70

Figure 70 *Stephanoon askenasyii* Schew., oval colony with cells forming a median belt.

This genus is a clear, oblate, gelatinous sheath in which 1 or 2 tiers of spherical cells encircle the equator. The cells are very similar to other volvocoids but possess 2 flagella of unusual length. Each cell may form a daughter colony as in *Pandorina* (Fig. 67).

78b **Cells not arranged at the equator of a gelatinous sheath, but distributed throughout** 79

79a **Cells all the same size within the colony, arranged within the periphery of the colonial mucilage, but showing a tendency to occur in tiers. Fig. 71** . . . . . *Eudorina*

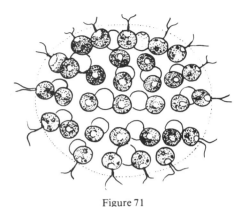

Figure 71

Figure 71 *Eudorina elegans* Ehr. In this species the cells have a tendency to be arranged in transverse bands or tiers.

Unlike *Pandorina* (Fig. 67) the cells are round or slightly oval and are rather evenly spaced within a globular, colonial mucilage. The cells usually show a tiered arrangement. *Eudorina elegans* Ehr., the most common species with world-wide distribution, occurs with other colonial Volvocales, in both eu- and tychoplankton. *E. unicocca* G.M. Smith is known also from the United States and from Panama. It is differentiated by the colony showing a more definite polarity, with the sheath forming lobes at the posterior end.

79b Cells of two sizes within the same colony, the smaller located in one sector or at one pole. 80

80a Colony globular, composed of from 32 to 164 cells; larger reproductive cells at the anterior sector, and with from 4 to 12 smaller, vegetative cells at the opposite pole; cells not enclosed by individual sheaths. Fig. 72 . . . . . *Pleodorina*

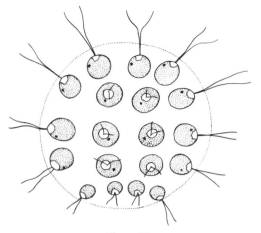

Figure 72

Figure 72 *Pleodorina illinoisensis* Kofoid, with 4 smaller vegetative cells. *P. californica* Shaw colonies have about half of the cells smaller and strictly vegetative.

The colony of *Pleodorina* usually consists of 132 or 164 cells; is globular and the cells are spherical. The two sizes of cells, the smaller confined to the posterior pole, identify this genus. In reproduction *Pleodorina* is oogamous like *Volvox* (Fig. 68). It is commonly found in the same habitats with *Volvox* and *Eudorina* but never seems to form 'blooms'. The common species illustrated is widely distributed in the United States.

80b Colony spheroidal, of from 16 to 124 cells, somewhat flattened where 2 or 4 smaller, vegetative cells are located, each cell enclosed by an individual sheath. Fig. 73 . . . . *Astrephomene*

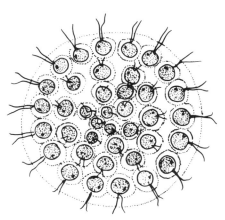

Figure 73

Figure 73 *Astrephomene gubernaculifera* Pocock. Posterior view of colony showing the smaller "rudder" cells.

This is a globular colony of 128 or 164 spherical cells, each individual being enclosed by its own sheath. There is a cup-shaped chloroplast but no pyrenoid; an eye-spot; two apical flagella which arise separately. The colony is distinctive by being somewhat flattened at the posterior pole where there is a group of 4 (or several) smaller cells with their flagella so directed that they produce what is called a rudder. There is but one species, originally described from South Africa, occurring in the United States.

81a (25) **Plant a** *filament* **(cells in continuous or interrupted linear series, with or without branches;** *or* **a flat, horizontal expanse;** *or* **an attached cushion of filaments with branches closely appressed laterally so that the filamentous plan is not clearly evident (See** *Apatococcus* **[Fig. 123].);** *or* **a branched, tubular, threadlike thallus without cross walls (coenocytic). 291**

81b **Plant** *not* **filamentous, not as above, but solitary cells;** *or* **a colony of 2 or more cells, often enclosed by mucilage or by old mother-cell walls. (See** *Oocystis* **[Fig. 146].);** *or* **cells variously adjoined to one another. (See** *Scenedesmus* **[Fig. 76],** *Pediastrum* **[Fig. 128].). . . . 82**

82a **Cells solitary or gregarious but not forming colonies of adjoined cells 183**

82b **Cells adjoined to form a colony or incidentally clustered;** *or* **arranged within a colonial sheath or copious mucilage 83**

83a **Colony composed of 2 cells adjoined along the entire length of one cell wall 84**

83b **Colony composed otherwise 86**

84a **Colony composed of 2 trapezoidal cells, adjoined along their bases. Fig. 74. . . . .** *Euastropsis*

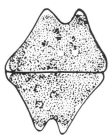

Figure 74

Figure 74 *Euastropsis richteri* (Schmid.) Lag.

There is but one species of this unique genus in

which the colony is always composed of but 2 trapezoidal cells that are similar in shape to those of *Pediastrum* (Fig. 128). The genus is rather rare but widely distributed; common in arctic Alaska.

84b   Colony composed otherwise 85

85a   Cells oval, the wall beset with needlelike spines with thickened bases. Fig. 75. . . .
. . . . . *Dicellula*

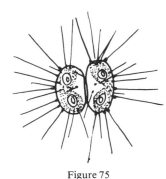

Figure 75

Figure 75 *Dicellula planctonica* Swir. (Redrawn from Korschikoff.)

Cells in this genus occur in pairs, side by side. The walls in 1 species are furnished with slender needles which have a buttonlike base. Only 2 species are known but as yet not reported from the United States.

85b   Cells oval or fusiform; walls smooth or with 1 or 2 curved spines, not thickened at the base, or with teeth; colony composed of from 4 to 12 cells, adjoined along their longitudinal walls in a linear series. Fig. 76 . . . . *Scenedesmus*

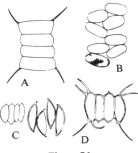

Figure 76

Figure 76 *Scenedesmus*. (A) *Sc. quadricauda* (Turp.) Bréb.; (B) *Sc. bijuga* var. *alternans* (Reinsch) Hansg.; (C) *Sc. incrassatulus* var. *mononae* G.M. Smith; (D) *Sc. opoliensis* P. Richter. (C, redrawn from Tiffany.)

There are numerous species and varieties of this genus, occurring in both tycho- and euplankton. The cells are oval, fusiform, or crescent-shaped according to species. In some there are 2 forms of cells in the same colony (coenobe), the individuals at the outside of the row often being lunate, rather than oval, and different in the form of spines. Although usually in a row of 4, some species have cells in a double, alternating series; or there may be a single row of 12 cells. Occasionally *Scenedesmus* cells occur solitary, especially when cultured. Besides being widely distributed in nature, *Scenedesmus* cells appear frequently in laboratory aquaria, coloring the water green, and they are commonly found in the psammon. Under unnatural conditions of culture *Scenedesmus* species show all manner of variability in respect to cell shape and presence or absence of spines. Also under culture conditions *Scenedesmus* may adopt a zoospore method of reproduction rather than by the usual autospores. Perhaps the most common species is *S. quadricauda* (Turp.) Bréb.

86a (83) Colony composed of cells invested by a common mucilage (but see *Oocystis* [Fig. 146] with the cells enclosed by old mother-cell walls, not by mucilage) 87

86b Colony composed of an aggregate of cells often adjoined, not enclosed by mucilage but may be enclosed by or attached by old mother-cell walls or fragments of them. (In some forms the cells may be solitary, e.g., *Chaetosphaeridium* [Fig. 117]). 133

87a Colony attached or adherent, sometimes on soil, mosses or in snow 88

87b Colony free-floating, sometimes entangled among other algae but not growing attached. (Frequently algae which are normally attached become separated from their substrates. Look for attaching stalks, discs or other evidence of the organisms having been attached.) 100

88a Colony a number of branched, gelatinous stalks arising from an attaching disc; cells fusiform, in 2's or 4's at the end of the strands. Fig. 77 ....
.... *Tetracladus*

Figure 77

Figure 77 *Tetracladus mirabilis* Swir., cells at the ends of branching gelatinous strands. (Diagrammed from Swirenko.)

In this genus cells occur in 4's (sometimes in 2's) at the end of thick, branched, gelatinous stalks which arise from an attaching base. There is 1 parietal chloroplast with a pyrenoid. Autospore-formation is the only known method of reproduction. *T. mirabilis* Swir. is the only species described thus far.

88b Colony otherwise. 89

89a Colony in the form of compact packets of angular cells enclosed in a close, gelatinous sheath; commonly found among epidermal cells of aquatic plants, but sometimes found free-living. Fig. 78. .... *Chlorosarcina*

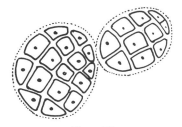

Figure 78

Figure 78 *Chlorosarcina consociata* (Klebs) G.M. Smith. The chloroplast is parietal and cup-shaped.

The cells are usually angular from mutual compression and ordinarily are enclosed by a gelatinous sheath. There are 5 species but only 1, *C. consociatus* (Klebs) G.M. Smith seems to be common in the United States; found in duckweed thalli *(Lemna)*. Old, colorless plants of *Lemna trisulca* often show the green flecks of this and other endophytic algae such as *Chlorochytrium lemnae* (Fig. 160).

89b Colony otherwise, not endophytic in the tissues of aquatic plants **90**

90a Colonial mucilage saclike or balloon-shaped, sometimes intestiniform; macroscopic **91**

90b Colony microscopic; pyriform, or composed of gelatinous strands **92**

91a Colonies amorphous, on moist soil, dripping rocks (rarely under water); colonial mucilage forming lumpy (usually macroscopic) masses; enclosed cells round or oval, each enclosed by one or more individual sheaths. Fig. 12 . . . . *Palmella*

91b Thallus gelatinous, balloonlike, saccate or intestiniform (macroscopic); cells in 2's and 4's at the periphery and often showing pseudocilia (long hairlike, false flagella). Fig. 79 . . . . *Tetraspora*

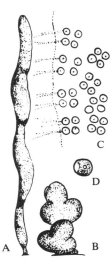

Figure 79

Figure 79 *Tetraspora.* (A) *T. cylindrica* (Wahl.) C.A. Agardh, habit of colony;

(B) *T. gelatinosa* (Vauch.) Desvaux, habit of colony; (C) portion of colony margin to show arrangement of cells and pseudocilia; (D) single cell showing cup-shaped chloroplast.

Early in the spring, or throughout the summer in cold, running water, gelatinous, balloonlike or intestiniform strands of *Tetraspora* may be found attached to submersed substrates. At times thalli 2 or 3 feet long develop. Most of the dozen or so recognized species are macroscopic when fully developed, but a few appear as microscopic, floating colonial clumps. Especially when plants are young, favorable optical conditions will disclose long, fine, shadowy pseudocilia which extend through the colonial mucilage and far beyond. Compare *Tetraspora* with *Gloeocystis* (Fig. 86) which has larger and fewer cells in a gelatinous matrix; no pseudocilia.

92a (90) Colony pyriform, microscopic, attached at the narrow end to filamentous algae or aquatic plants; cells in pairs or in 4's, with clearly evident pseudocilia. Fig. 80 . . . . *Apiocystis*

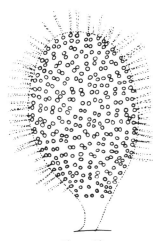

Figure 80

Figure 80 *Apiocystis brauniana* Näg. Diagram of a colony to show cell arrangement and pseudocilia.

This plant is always in the form of a microscopic, balloonlike sac growing attached to filamentous algae or to aquatic plant stems. The cells are arranged somewhat irregularly in 2's and 4's. The pseudocilia are usually conspicuous throughout the life of the colony. *Apiocystis brauniana* Näg. is relatively common and widely distributed, but in the Alaskan Arctic there is a rare species in which the gelatinous sac has a very thick and lamellated sheath and the thallus grows to almost macroscopic size.

92b   **Colony formed otherwise 93**

93a   **Thallus consisting of branched, gelatinous strands which are freely anastomosing, enclosing spherical cells, sparse in number, in linear pairs; (the thallus faintly colored); macroscopic. Fig. 81........................**
       .... *(Schizodictyon) Gloeodendron*

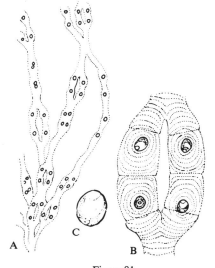

Figure 81

Figure 81   *Gloeodendron catenatum* (Thomp.) Bourr. (A) habit of thallus; (B) portion of thallus enlarged to show lamellate sheaths; (C) single cell with parietal chloroplast. (B, diagrammed from Thompson.)

The anastomosing, gelatinous tubes which comprise the thallus of the genus are conspicuously lamellate transversely. The oval or globular cells are in linear pairs, sparsely distributed throughout the length of the tube. *Schizodictyon* is synonymous with this genus. There are 2 species.

93b   **Colony otherwise............... 94**

94a   **Mucilaginous matrix indefinite in shape, without a bounding sheath, intermingled in mosses or on dripping rocks; cells cylindrical or subcylindric, relatively large with showy chloroplasts and conspicuous pyrenoids. (See *Cylindrocystis* [Fig. 82].) 95**

94b   **Mucilaginous masses with a definite boundary; cells elongate and subcylindric, relatively small and numerous within the colony. (See *Gloeocystis* [Fig. 86].) 96**

95a   **Cells cylindrical, some species slightly constricted in the midregion; chloroplasts 2, axial, stellate. Fig. 82 .... *Cylindrocystis***

(Some species have been transferred to the desmid genus *Actinotaenium*.)

Figure 82

Figure 82 *Cylindrocystis brebissonii* Menegh. with 2 stellate chloroplasts and large pyrenoids; central nucleus.

Although some of the half-dozen species of this Saccoderm desmid have very slightly constricted cells, they are mostly unconstricted and have a wall composed of 1 piece. There are no wall pores characteristic of the true or Placoderm desmids. (See Fig. 174.) There is a star-shaped chloroplast with a large pyrenoid in each half (or end) of the cell. The genus is not confined to acid habitats as is true for most of the desmids, but may occur in alkaline bogs, among wet mosses or on stones especially in alpine regions. Some species have been transferred to *Actinotaenium* which is not separated herein from the desmid genus *Cosmarium* (Fig. 187).

95b Cells oval, watermelon-shaped, with a single parietal chloroplast, not constricted in the midregion (some species with narrowly cylindrical cells); contents purplish with phycobilins. Fig. 83 . . . . *Mesotaenium*

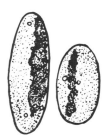

Figure 83

Figure 83 *Mesotaenium de greyii* Turner Fa., showing variation in cell shape and the single lamellate chloroplast.

These are oval or oblong cells; have an axial, platelike chloroplast and a wall that is all in 1 piece (Saccoderm desmids). Although some of the 10 species occurring in this country are free-floating, others such as *M. macrococcum* (Kütz.) Roy et Biss. usually occur among mosses and in gelatinous masses on rocky seeps, often at high altitudes.

96a (94) Cells cylindrical or elongate-oval, numerous, scattered throughout an amorphous but evident mucilage (rarely free-floating), the cells usually with their long axes parallel. Fig. 84 . . . . *Coccomyxa*

Figure 84

Figure 84 *Coccomyxa dispar* Schmidle. The end of a gelatinous strand showing oval or bacilliform cells with a parietal chloroplast.

There are 24 recorded species of this genus; only 1 seems to have been reported from the United States, forming gelatinous masses of varying extent on damp soil, on wet wood, on old fungus sporophores; rarely is free-floating. It has been found also attached to wooden flumes of swiftly flowing water. In subaerial habitats they are intermingled and closely associated with either unicellular or colonial algae and may show a great variation in size within the same colony.

96b Cells shaped or arranged otherwise . . 97

97a Cells elliptical or nearly spherical, arranged in 2's and 4's within an amorphous mucilage; mother-cell wall enclosing a family of cells, all within a common mucilaginous sheath; colonies may be microscopic but also appear as gelatinous masses on subaerial substrates, moist soil, *etc.* Fig. 85 . . . .
. . . . *Palmella* stage of *Chlamydomonas*

Figure 85

Figure 85 *Palmella* stage of *Chlamydomonas* sp. Cells have a cup-shaped chloroplast, with pyrenoid, and a red eye-spot which sometimes is not evident.

Chlamydomonas (Fig. 17) often becomes quiescent under unfavorable conditions, or as a normal stage in the life history. The cells adhere to a substrate, lose the flagella, but continue to divide, enclosing themselves in mucilage, and thus form extensive gelatinous masses. The cells are arranged in 2's and 4's, the groups surrounded by a mucilaginous sheath, thus somewhat resembling *Palmella* (Fig. 12). In this form *Chlamydomonas* is often mistaken for *Gloeocystis* (Fig. 86). Often cells in the *Palmella* stage retain an eye-spot. Laboratory aquaria often contain *Chlamydomonas* in this palmelloid stage. The cells become actively motile again by developing new flagella.

97b Cells spherical (oval in some species), scattered throughout the colonial mucilage or in groups of 4; orange-colored oil globules sometimes present; eye-spot lacking 98

98a Cells enclosed by distinct concentric sheaths of mucilage, scattered, single or in 2's and 4's within amorphous, colonial mucilage (often free-floating); chloroplast a parietal cup. Fig. 86 . . . .
. . . . *Gloeocystis*

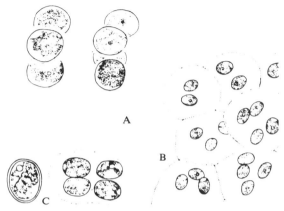

Figure 86

Figure 86 *Gloeocystis.* (A) *G. gigas* (Kütz.) Lag.; (B) *G. ampla* (Kütz.) Lag.; (C) *G. major* Gerneck, showing dense, cup-shaped chloroplast.

There are several species of *Gloeocystis* and all are very common. They are not very distinctive plants and therefore many small, round or oval green cells, especially when enclosed in mucilage, belonging to other genera may be mistaken for this genus. The concentric layers of mucilage about the cells provide a helpful character for identification. *Gloeocystis ampla* (Kütz.) Lag. does not always have mucilage in layers, however, but this free-floating species is identified by having oblong or oval cells.

98b Cells with or without evident individual sheaths; mucilage not in concentric layers. 99

99a Chloroplast cup-shaped, not covering the entire wall; cells all the same size within the colonial mucilage which is lumpy and usually macroscopic; cells with individual sheaths; occurring on soil. Fig. 12 . . . . *Palmella*

99b Colonial mucilage thin and sometimes not evident; cells of variable sizes; gregarious, in or on soil; chloroplast covering almost the entire wall. Fig. 87 . . . . *Chlorococcum**

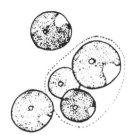

Figure 87

Figure 87 *Chlorococcum humicola* (Näg.) Rab. Cells are sometimes enclosed in a mucilaginous sheath.

Although this plant occurs on soil it reproduces by swimming reproductive cells (zoospores). One species, *C. infusionum* (Schrank) Menegh. is aquatic and is differentiated from *C. humicola* (Näg.) Rab. by the fact that its cells are all uniform in size and shape. Old, wet bones, and rocks under dripping water are favorable places for both species. Unless it is *Desmococcus* (*Pleurococcus* [Fig. 122]) *C. humicola* is probably the most widely distributed alga species in the world.

100a (87) Colonial sheath fusiform, definite in shape (margin lacey in some species); cells fusiform, rarely solitary but usually in 4's. Fig. 88 . . . . *Elakatothrix*

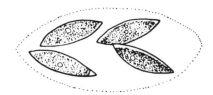

Figure 88

Figure 88 *Elakatothrix viridis* (Snow) Printz in a mucilaginous sheath showing parietal, laminate chloroplast and pyrenoid.

In this genus the cells are somewhat 'cigar'-shaped and occur end to end in pairs or side by side within a fusiform, gelatinous sheath. *E. gelatinosa* Wille has both ends of the cells pointed, whereas other species have the adjoined poles truncate. In 1 of the 5 species found in the United States, *E. americana* Wille the gelatinous, colonial sheath is irregularly lacy or fringed.

100b Colony not fusiform but globular, globular-quadrate, cubical, or elongate and threadlike 101

101a Colony regularly spherical or oval, *or* a rectangular plate 117

101b Colony shaped otherwise; irregular strand-like, or cubical (some young *Tetraspora* [Fig. 79] may be somewhat sperical in colony shape). 102

102a Colony of 4 cells or multiples of 4 in one plane, interconnected by gelatinous strands, the cells bearing a scalelike fragment of the old mother-cell wall. Fig. 89 . . . . *Coronastrum*

*See Appendix for other unicellular green algae, p. 268.

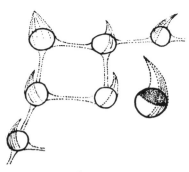

Figure 89

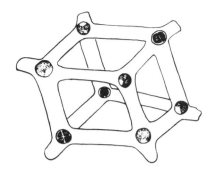

Figure 90

Figure 89 *Coronastrum aestivale* Thompson. (Redrawn from Thompson.)

This is a very rare alga and the species illustrated is the only one reported from the United States of the 3 that are known. The arrangement of the cells and their distinctive, winglike scale make identification certain.

Figure 90 *Pectodictyon cubicum* Taft. (Redrawn from Taft.)

There are 2 species known for this genus. Since its original discovery in Ohio by Taft, *P. cubicum* has appeared in several midwest and western states and in Florida. It occurs in the euplankton of lakes as well as in the tychoplankton of small ponds.

102b   Colony formed otherwise. 103

104b   Colony formed otherwise. 105

103a   **Colony of 8 egg-shaped cells, the narrow anterior ends all directed the same, arranged in 4's in 2 tiers and interconnected by narrow gelatinous strands (cells equipped with flagella which may be lost or retracted and not evident). Fig. 65. . . . .*Corone***

103b   Colony formed otherwise. 104

104a   **Colony a hollow cube of 8 round cells arranged one at each corner and interconnected by stout, gelatinous strands. Fig. 90 . . . .*Pectodictyon***

105a   **Cells spherical, arranged in a linear series within long, gelatinous, pseudo-filamentous strands which often branch and anastomose. Fig. 91 . . . .*Palmodictyon***

Figure 91

Figure 91 *Palmodictyon viride* Kütz., cells in more than 1 series within an anastomosing, gelatinous sheath.

Palmodictyon viride Kütz. and *P. varium* (Näg.) Lemm.* are fairly common in mixtures of algae from shallow ponds or the tycho-plankton of lakes, but never appear in any abundance. The former species has cells en-closed in individual sheaths, whereas the latter is without cellular sheaths. Some strands of co-lonial mucilage are simple, others irregularly branched and sometimes anastomose.

105b   Cells arranged otherwise 106

106a   Cells cylindrical, elongate-ellipsoid or elliptical..................... 107

106b   Cells other shapes 110

107a   Cells elliptical, grouped in 2's and 4's and partly enclosed in remains of the mother-cell wall; multiples of 4 all en-closed in a gelatinous matrix; 1 parietal chloroplast. Fig. 92 .... *Lobocystis*

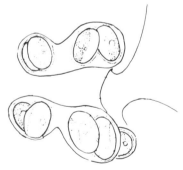

Figure 92

Figure 92   *Lobocystis dichotoma* Thompson. (Redrawn from Thompson.)

Cells in this colonial genus are arranged in pairs, enclosed by a sheath and occurring at the ends of dichotomous or V-shaped strands. The entire colony is usually found enclosed by a gelatinous sheath. There is 1 parietal chloro-plast (or 2), each with a pyrenoid. *L. dichotoma* Thompson has been found in Kansas.

107b   Cells shaped or arranged otherwise 108

108a   Cells oval or watermelon-shaped with an axial, platelike chloroplast; cells few within a watery, amorphous mucilage. Fig. 83 .... *Mesotaenium*

108b   Cells shaped otherwise, or with different chloroplasts.................... 109

109a   Cells cylindrical, rarely slightly constricted in the midregion, with 2 stel-late, axial chloroplasts; enclosed in a watery, inconspicuous mucilage; (cells more often than not solitary). Fig. 82 .. .. *Cylindrocystis*

109b   Cells bacilliform or elongate-elliptic, many within a copious mucilage; chlo-roplast a parietal plate. Fig. 84 .... *Coccomyxa*

110a   (106) Colony forming stringy, intes-tiniform strands, sometimes branched and anastomosing; cells in 2's and 4's. Fig. 79 .... *Tetraspora*

110b   Colony shaped otherwise........ 111

111a   Colony composed of a few (2 to 4) oval cells, with 1 parietal chloroplast, en-closed by an irregularly shaped, lamel-late, gelatinous sheath. Fig. 93 .... *Dactylothece*

*Some students of the algae believe that these are 2 names for the same species.

Figure 93

Figure 93 *Dactylothece confluens* (Kütz.) Lag.

Cells in this genus are shaped as in *Mesotaenium* (Fig. 83) but are much smaller (not more than 3-4 μm in diameter) and have a laminate, parietal chloroplast. The cells are enclosed in mucilage somewhat formless) and occur as thin, expanded masses on moist rocks. *Dactylothece confluens* (Kütz.) Hansg. is the only species reported from North America. The species can be identified by the few number of cells involved in a colony.

111b   **Colony composed of round, pyriforn or oval cells; gelatinous sheath not stratified** 112

112a   **Semicircular fragments of old mother-cell wall partly enclosing daughter cells, or lying about in the mucilage; cells often showing pseudocilia. Fig. 94** . . . . *Schizochlamys*

   (*Schizochlamys gelatinosa* A. Braun is held to be synonymous with the genus *Placosphaera*.)

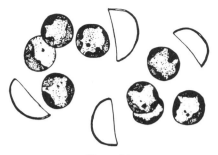

Figure 94

Figure 94 *Schizochlamys gelatinosa* A. Braun. enclosed in an amorphous mucilage.

At times the free-floating, gelatinous masses of this plant may be of macroscopic proportions, and so abundant that they may be scooped from the surface of the water by hand. More frequently the growths are less extensive and small aggregates of cells occur intermingled with other algae in shallow water situations. The fragments of old mother-cell walls scattered about the cells help in its identification. Under favorable optical conditions a tuft of long, fine pseudocilia are discernible, 4 to 8 from each cell. The cells (often oblate-spheroid) have 1 or 2 platelike chloroplasts, parietal but may appear massive and indistinct. This is an anomalous plant, placed in the Tetrasporales because of the gelatinous colony, with indefinitely arranged cells, and because of the pseudocilia. The presence of oil rather than starch as a reserve food, and the method of cell division raise a question as to its true affinity.

112b   **Semicircular fragments of old mother-cell walls not present** 113

113a   **Colonies saclike or irregularly globose; cells arranged in 4's. Fig. 79** . . . . *Tetraspora*

113b   **Colonies shaped otherwise; cells not arranged in 4's** 114

114a Cells ovoid or cuneate, compactly arranged in semiopaque mucilage which (in one species especially) is brown or yellow and so obscures the cells; colonies frequently compounded by interconnecting strands of tough mucilage between clusters of cells. Fig. 95 . . . .
. . . . *Botryococcus*

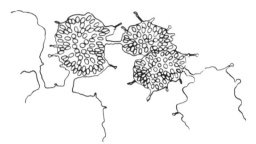

Figure 95

Figure 95 *Botryococcus braunii* Kütz., showing an expression in which a colonial complex is formed by interconnecting strands of tough mucilage. Colonies may appear solitary and as a dark, yellow-brown lump in which individual cells scarcely can be seen, if at all. The color occurs because of the brown oil within the mucilage, the cells containing green chloroplasts.

Specimens of this genus, especially *Botryococcus braunii* Kütz. can be identified by the golden-brown, amorphouse lumps of mucilage containing oil, with cells in the colony scarcely visible. In *B. protuberans* West & West and *B. sudeticus* Lemm. the mucilage is less opaque and in the former the cells are in small clumps at the ends of tough, mucilaginous strands. Individual cells have a yellowish sheath and are somewhat radiately arranged. Occasionally *B. braunii* develops growths of 'bloom' proportions. This genus is considered to be responsible for a curious type of coal in Scotland.

114b Cells round, oval or pyriform, not arranged in semiopaque mucilage; (mucilage of juvenile *Botryococcus* (Fig. 95) is colorless, however). 115

115a Cells round or oval, in 2's and 4's within gelatinous, lamellate sheaths; cells with individual sheaths in some species; colonies solitary or aggregated. Fig. 86 . . . .
. . . . *Gloeocystis*

115b Cells shaped and arranged differently 116

116a Gelatinous colony globular; cells wedged-shaped, in radiating clusters; 1 parietal chloroplast without a pyrenoid. Fig. 96 . . . . *Gloeoactinium*

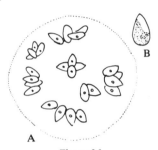

Figure 96

Figure 96 *Gloeoactinium limneticum* G.M. Smith. (A) colony; (B) single cell with parietal chloroplast.

The colonial sheath is wide and ample, enclosing clusters of elliptic or wedge-shaped (cuneate) cells that are mostly grouped in 4's. There is but one species.

116b Colony irregular; cells pyriform, sharply pointed posteriorly, radiately arranged. Fig. 97 . . . . *Askenasyella*

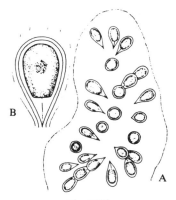

Figure 97

Figure 97 *Askenasyella chlamydopus* Schmidle. (A) colony with cells in radiate arrangement; (B) single cell. (Redrawn from Schmidle.)

The gelatinous matrix which encloses the pyriform cells in this genus may be nearly globular, but more often is irregular in outline. The cells are arranged in a radiate fashion, with the narrow end directed inwardly. The parietal chloroplast is in the broad, anterior end. A fine protoplasmic thread extends from the posterior end of the cell to the chloroplast. Plants are either tychoplanktonic or adherent.

117a (101) **Cells incised or constricted in the midregion to form semicells (cell halves), the cells often interconnected by fine, gelatinous strands. Fig. 98.** . . . . *Cosmocladium*

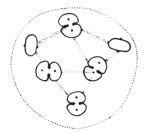

Figure 98

Figure 98 *Cosmocladium tuberculatum* Presc., a colonial desmid in which cells are interconnected by gelatinous strands.

This is a *Cosmarium*-like genus of desmids in which the cells are colonial, 6 or more enclosed in a colorless, gelatinous sheath. It belongs to the Placoderm group of the Desmidiaceae, those which have the wall in 2 sections that adjoin in the midregion where there is usually an incision (sinus) and an isthmus between the 2 semicells. There are but few species in the genus. Like the majority of desmids it occurs in soft or acid-water bogs. Under properly reduced illumination the fine, often double, gelatinous strands interconnecting the cells can be seen.

117b **Cells not constricted to form semicells** **118**

118a **Chloroplast star-shaped, the radiating processes with their outer ends flattened against the wall. Fig. 99.** . . . . *Asterococcus*

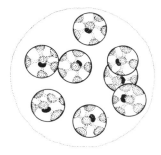

Figure 99

Figure 99 *Asterococcus superbus* (Cienk.) Scherf., a colony of cells showing stellate chloroplast and central pyrenoid.

Two species of this genus are known from North America. *Asterococcus superbus* (Cienk.) Scherf. may occur as a single cell or in small colonies of from 2 to 8 relatively large cells. This species shows the star-shaped chloroplast more clearly than does *A. limneticus* G.M. Smith. The latter has smaller cells, larger in number within the colony. As the name indicates, this species occurs in the euplankton.

118b Chloroplast not axial and star-shaped **119**

119a Cells arranged in groups of 4 at the ends of branching and somewhat radiate, mucilaginous strands (focus carefully into the colony). (See Fig. 100.). **120**

119b Cells not arranged at the ends of branching strands **121**

120a Cells appearing both reniform (bean-shaped) and ovoid in the same colony. Fig. 100. . . . . *Dimorphococcus*

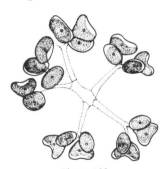

Figure 100

Figure 100 *Dimorphococcus lunatus* A. Braun, depending on position of cells they are either oval or cordate.

In this plant the cells are in clusters of 4, 2 of which when seen in top view appear oval, whereas 2, as seen from the side are reniform or somewhat crescent-shaped, hence the name. *Dimorphococcus lunatus* A. Braun is often abundant in soft-water lakes, whereas the other species known from this country, *D. cordatus*

Wolle is less frequently found. Both species may occur in open-water plankton or in the tychoplankton near shore. A character that is helpful in identification is a negative one, the absence of a conspicuous gelatinous sheath enclosing the colony.

120b Cells spherical or broadly oval, all the same shape within the colony. Fig. 101 . . . . *Dictyosphaerium*

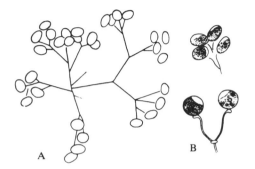

Figure 101

Figure 101 *Dictyosphaerium pulchellum* Wood. (A) habit of colony; (B) individual cells at ends of radiating, branched fibrils.

Cells are in clusters of 4 as in *Dimorphococcus* (Fig. 100), but all the same shape, round or oval. The interconnecting fibers, the remains of old, mother-cell walls, are finer than in that genus. Sometimes an indefinite gelatinous envelope can be discerned. The chloroplast is a cup-shaped, parietal layer covering most of the wall.

121a (119) Cells globular **122**

121b Cells other shapes, ovoid, fusiform, crescent-shaped, or bean-shaped . . . **127**

122a Cells spherical, in 2's and 4's at the periphery of amorphous mucilage; cells with 1 parietal chloroplast and a pyrenoid, often forming red or green snow. Fig. 10 . . . . *Pseudotetraspora*

122b Cells arranged otherwise 123

123a Cells with a distinct sheath; colonial mucilage lamellate. Fig. 86 . . . .
. . . . *Gloeocystis*

123b Cells without distinct sheaths; colonial mucilage not lamellate 124

124a Colony spherical, the mucilage including fine, radiating fibrils; cells in packets of 4 distributed throughout the mucilage. Fig. 102 . . . . *Radiococcus*

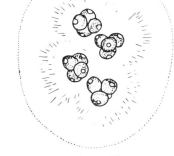

Figure 102

Figure 102 *Radiococcus nimbatus* (de Wild.) Schmidle. (Redrawn from Thompson.)

The cells are cruciately arranged in clusters of 4. The colony sheath is ample and nearly always shows very fine, radiating fibrils from the cell clumps toward the periphery of the envelope. There is 1 of the 2 known species in the United States. The genus *Eutetramorus* (Fig. 103) may be a synonym of this genus.

124b Colony without radiating fibrils. . . . 125

125a Colony a clear globe of mucilage, with cells in 4's arranged at or near the periphery. Fig. 103. . . . . . *Eutetramorus*

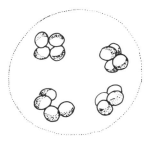

Figure 103

Figure 103 *Eutetramorus globosus* Walton, quartets of cells widely dispersed in a gelatinous sheath.

In this genus the cells are in quartets but not cruciately arranged as in *Radiococcus* (Fig. 102) and the sheath is homogeneous, without radiating fibrils. There is but 1 species.

125b Cells arranged otherwise 126

126a Colony of from 32 to 64 cells scattered throughout the colonial mucilage, ordinarily with several daughter colonies of smaller cells intermingled; chloroplast cup-shaped. Fig. 104 . . . .
. . . . *(Palmellocystis) Sphaerocystis*

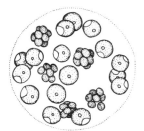

Figure 104

Figure 104 *Sphaerocystis schroeteri* Chodat, a colony with mature cells and daughter colonies within a gelatinous sheath.

This genus, of which there is only 1 species in North America, should be compared with *Planktosphaeria* (Fig. 105). There is 1 cup-shaped chloroplast, and the colony, in addition to undivided cells, almost invariably shows a cluster of small cells formed by the cleavage of a parent cell into 4 or 8 daughters. There is a wide, gelatinous sheath but the individual cells do not show the lamellate envelope of *Gloeocystis* (Fig. 86), a genus which may be confused with *Sphaerocystis*. Young cells of *Planktosphaeria* may have 1 cup-shaped chloroplast but as the cells mature, several, angular chloroplasts characterize the genus.

126b Colony of from 8 to 16 cells, with polygonal chloroplasts, each with a pyrenoid; colony not containing clusters of daughter cells. (Young cells have only 1 or 2 chloroplasts.) Fig. 105 . . . . *Planktosphaeria*

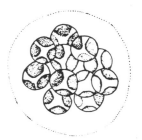

Figure 105

Figure 105 *Planktosphaeria gelatinosa* G.M. Smith, a colony of young cells.

This plant, only 1 species known, occurs mostly in the tychoplankton among other algae, but occasionally is collected in open water. Unless care is used, this plant may be confused with *Gloeocystis* (Fig. 86), *Sphaerocystis* (Fig. 104) or other spherical algal cells within a mucilaginous sheath. When mature the cells are recognizable by their angular, 5-sided chloroplasts, each of which contains a pyrenoid. The colonial sheath is often very thin and difficult of determinations.

127a (121) Cells lunate or sickle-shaped, tapered and pointed at the apices, if blunt definitely U-shaped. 128

127b Cells bean-shaped, oval, or spindle-shaped, the apices sometimes tapered but rounded or blunt at the poles 129

128a Cells lunate, sharply curved, the apices sometimes nearly touching, irregularly arranged within a gelatinous sheath. Fig. 106. . . . . *Kirchneriella*

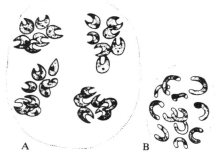

A   B

Figure 106

Figure 106 *Kirchneriella.* (A) *K. lunaris* (Kirch.) Moebius, acicular cells; (B) *K. obesa* var. *major* (Ber.) G.M. Smith, the colonial sheath often indistinct.

The crescent-shaped or arcuate cells of this genus are enclosed in a mucilage which sometimes is indistinct. *Kirchneriella,* of which there are 6 species in the United States, usually occurs in open water plankton. The species are differentiated by degrees of curvature and variations in stoutness. The cells are mostly so sharply curved that their apices nearly touch, whereas in *Selenastrum* (Fig. 124), with which *Kirch-*

Follicularia is 1-celled or colonial, the cells eccentrically arranged in a gelatinous, lamellate sheath; has chloroplasts similar to *Planktosphaeria.*

neriella should be compared, the cells are symmetrically crescent-shaped, are definitely arranged, and are not enclosed by a gelatinous sheath.

128b **Cells symmetrically crescent-shaped, in groups of 4, 2 of which face one another (concave margins apposed); colonial sheath usually evident. Fig. 107 . . . .**
. . . . *Tetrallantos*

Figure 107

Figure 107 *Tetrallantos lagerheimii* Teiling.

Tetrallantos lagerheinii Teil. is a rare species, the only one known for the genus, but is widely distributed. The characteristic arrangement of the cells is determined at the time that they are formed in groups of 4 within the mother-cell. After the mother-cell wall breaks down to release the daughter cells, fragments of the wall persist as interconnecting or radiating threads within the colonial mucilage which often is very thin and difficult of determination.

129a **(127) Cells curved, sausage-shaped with rounded ends, adjoined to one another by fine threads; gelatinous sheath wide. Fig. 108 . . . . *Tomaculum***

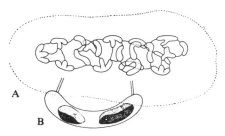

Figure 108

Figure 108 *Tomaculum catenatum* Whitford. (A) colony; (B) single cell with parietal chloroplasts.

These sausage-shaped cells, sometimes with lateral, lobelike extensions, are interconnected by threadlike strands. Upward of 20 such joined cells are enclosed in a hyaline, gelatinous sheath to form free-floating colonies. There are 1 or 2 parietal chloroplasts, each with a pyrenoid. The single species has been found only in North Carolina.

129b **Cells shaped or arranged otherwise 130**

130a **Cells fusiform or spindle-shaped . . . 132**

130b **Cells ovate, bean-shaped or oblong 131**

131a **Cells oval, somewhat irregularly arranged in 4's, forming a flat plate. Fig. 109 . . . . *Dispora***

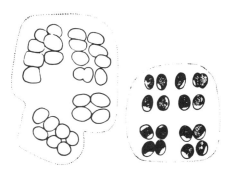

Figure 109

Figure 109   *Dispora crucigenioides* Printz.

There is only 1 species of *Dispora* in the United States and apparently is rare; 2 others have been reported from Europe. It is sometimes difficult to differentiate these flat, platelike colonies from some species of *Crucigenia* (Fig. 130), especially *C. irregularis* Wille in which the cells are irregular in arrangement, although tending to occur in 4's.

131b   Cells bean-shaped or oblong, reproducing by autospores which are retained within the enlarged mother-cell wall; (the wall may gelatinize and appear as a mucilage sheath in some instances). Fig. 110 . . . . *Nephrocytium*

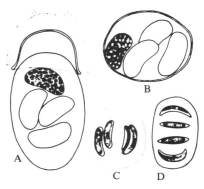

Figure 110

Figure 110   *Nephrocytium*. (A) *N. ecdysis-cephanum* West; (B) *N. obesum* W. & W.

(*Oonephris obesa* [W. & W.] Fott); (C) *N. limneticum* (G.M. Smith) G.M. Smith; (D) *N. lunatum* West.

Most of the 8 species of this genus which occur in the United States are reniform or bean-shaped, often with 1 convex lateral wall and 1 flattened or slightly concave. They occur in mixtures of algae in the tychoplankton, although 2 species, *N. agardhianum* Naeg. and *N. limneticum* G.M. Smith are usually found in the euplankton, or in relatively soft-water lakes. The former has elongate, almost vermiform cells which are curved or spirally twisted. *N. obesum* West & West has been transferred to *Oonephris obesa* (W. et W.) Fott. It has a massive, spongy chloroplast.

132a   (130) Cells in linear pairs, 1 or several such pairs within a common mucilaginous investment (cells with long axes approximately parallel); sometimes are found solitary in collections. Fig. 88 . . . . *Elakatothrix*

132b   Cells arranged in parallel bundles, reproducing by autospores (daughter colonies forming within the mother-cell). Fig. 111 . . . . *Quadrigula*

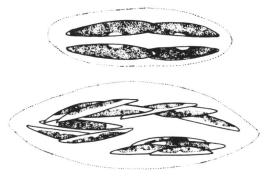

Figure 111

Figure 111   *Quadrigula chodatii* (Tanner-Fullman) G.M. Smith, fusiform cells with a parietal chloroplast, within a fusiform, gelatinous sheath.

There are 5 species found in the euplankton of lakes, all of which have elongate-fusiform or 'cigar'-shaped cells. The species illustrated has cells with narrowed, bluntly rounded poles, whereas the others have pointed apices. The cells occur in rather compact bundles of 4, all lying parallel in the colony.

133a   (86) **Cells in compact packets of 4 or in multiples of 4, not enclosed in mucilage. Fig. 112.** . . . . *Chlorosarcinopsis*

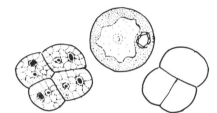

Figure 112

Figure 112 *Chlorosarcinopsis minor* (Gerneck) Herndon. (Redrawn from Herndon.)

The 8 or 10 species in this genus may be either submersed or on moist subaerial habitats. Whereas cells are often found to be solitary they normally form cubical packets of spheroidal individuals with some walls flattened by mutual compression. The chloroplast is parietal, covering most of the wall. Species must be differentiated by culturing.

133b   **Cells not in compact packets of 4** . . . 134

134a   **Cells (or some of them in the colony) bearing long, gelatinous bristles or scales, or hairs. (See Fig. 117.)** 135

134b   **Cells without gelatinous bristles, with or without spines which may be longer or shorter than the cell diameter. (See Fig. 131.)** 140

135a   **Cells arranged in quadrate colonies of 4, interconnected by stands, each cell bearing a scalelike fragment of mother-cell wall. Fig. 89** . . . . *Coronastrum*

135b   **Colony formed otherwise.** 136

136a   **Cells in clusters of from 2 to 6 (rarely solitary), bearing more than 1 hair** . . 137

136b   **Cells bearing but 1 hairlike bristle** . . 139

137a   **Hairs with a gelatinous sheath at the base. Fig. 113** . . . . *Conochaete*

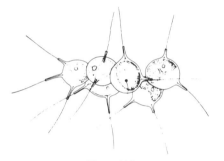

Figure 113

Figure 113   *Conochaete comosa* Klebahn.

The globose cells of this genus are epiphytic and are enclosed by a gelatinous envelope from which 2 or 3 setae arise, the bases sheathed.

137b   **Hairs without a gelatinous sheath at the base** 138

138a   **Cells several together (sometimes solitary) in a lamellate sheath, each cell with several long bristles. Fig. 114.** . . . . *Polychaetophora*

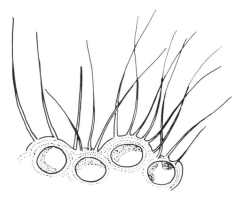

Figure 114

Figure 114 *Polychaetophora lamellosa* West & West.

In this genus the cells are usually clumped, or occur in a short, linear series, enclosed in a lamellate, mucilaginous matrix which bears several gelatinous bristles. The chloroplast is a circular plate, apparently without a pyrenoid.

138b   Cells in a non-lamellate sheath, with 2 or 3 bristles (rarely 1). Fig. 115 . . . .
. . . . *Oligochaetophora*

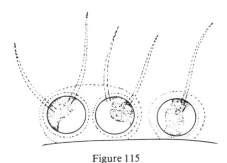

Figure 115

Figure 115 *Oligochaetophora simplex* G.S. West.

The epiphytic cells bear 2 or 3 simple, un-branched, spinelike bristles. The cells occur solitary or in clumps of from 2 to 4, enclosed by a common gelatinous sheath. Compare with *Conochaete* (Fig. 113) and *Chaetosphaeridium* (Fig. 117).

139a   (136)   Cells forming an attached, compact cluster within the mother-cell wall which bears a branched hair that has no sheath. Fig. 116 . . *Dicranochaete*

Figure 116

Figure 116 *Dicranochaete reniformis* Heiron.

Although this curious plant usually occurs as single cells, the individuals may be clustered as a result of recent cell division. It grows on fila-mentous algae and submersed aquatic plants; apparently is very rare. The unique branched setae on the sheath which arise from the lower side of the cell make identification certain.

139b   Cells loosely arranged side by side in a cluster, each bearing an unbranched hair which has a basal sheath. Fig. 117 . . . . .
. . . . *Chaetosphaeridium*

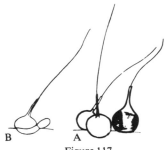

B   A

Figure 117

Figure 117 *Chaetosphaeridium globosum* (Nordst.) Klebahn. (A) group of cells; (B) one cell showing tubelike utricle which may extend from one cell to another and so form a loose colony.

The globular, hair-bearing cells of this genus occur either singly or in aggregates of from 2 to 8, living epiphytically on larger algae or aquatic plants. Frequently the cells are loosened from their substrate and are found floating free. This genus and *Dicranochaete* (Fig. 116) are anomalous members of the Coleochaetaceae, in some classification systems, but are probably best placed in their own respective families.

140a (134) **Cells attached at the ends of branching, gelatinous stalks or in elongate cups which telescope one another, the cups formed from old, mother-cell walls** **141**

140b **Cells not at the ends of branching gelatinous strands nor in cuplike sheaths** **143**

141a **Cells oval, single or in pairs, within cuplike remains of old, mother-cell walls, the cups telescoping one another, forming branching chains. Fig. 118** ....
.... *Ecballocystis*

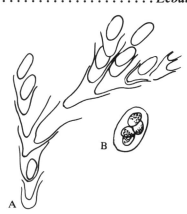

Figure 118

Figure 118 *Ecballocystis fritschii* Iyengar. (A) colony with cells enclosed in remains of old sheaths; (B) cell with autospores. (Redrawn in part from Iyengar.)

In this genus (monospecific) the oval cells are partly enclosed by cup-shaped remains of old, mother-cell walls. The cups are telescoping and several generations are arranged to form branching chains. There are from 2 to 4 parietal chloroplasts, each with a pyrenoid.

141b **Cells shaped or arranged otherwise; growing epiphytically on microcrustacea** **142**

142a **Cells ellipsoid or somewhat fusiform; chloroplasts 1 or 2 longitudinal bands; attaching stipes slender. Fig. 119** ......
.... *(Chlorangium) Chlorangiella*

Figure 119

Figure 119 *Chlorangiella pygmaea* (Ehr.) Silva (*Chlorangium stentorinum* [Ehr.] Stein).

According to some students, this genus is synonymous with the name *Chlorangium*. It is a swimming member of the Tetrasporales which loses its flagella and becomes sedentary. Sometimes the Chlorangiellaceae is included in the Volvocales. The organism becomes attached, anterior end down, to small crustacea and insect larvae by means of a gelatinous stalk. After attaching the cell continues to divide, each new cell forming a stalk so that a branched colony is produced. Some small animals such as *Cyclops* may go swimming about with a veritable plume of green cells on the antennae. Such hosts may appear green to the unaided eye because of the large numbers of attached algae. Compare *Chlorangiella* with *Colacium* (Fig. 120), a member of the Euglenophyta.

142b   Cells ovate to oblong, or ovoid; chloroplasts numerous, ovoid discs; stalks relatively stout, branched, with a cell at the apex of each branch. Fig. 120 . . . .
. . . . . *Colacium*

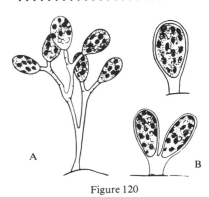

Figure 120

Figure 120 *Colacium.* (A) *C. arbuscula* Stein; (B) *C. vesiculosum* Ehr.

Microscopic animals, especially crustaceans may appear green because of the large numbers of *Colacium* individuals attached to them, either singly or in plumelike clusters. Like *Chlorangiella* (Fig. 119) this organism is a motile, green cell belonging to the Colaciaceae in the Euglenophyta. There is a large number of disclike chloroplasts and a conspicuous, red eye-spot. The cells are at the ends of branched stalks, anterior end downward. The rather specific association of the algal cells with the animal host invites speculation as to how this relationship was established, and maintained.

143a   (140) Cells ellipsoid or spindle-shaped, attached end to end, forming loose, branching chains. Fig. 121. . . . .
. . . . *Dactylococcus*

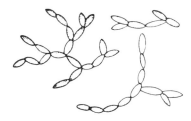

Figure 121

Figure 121 *Dactylococcus infusionum* Näg. (Redrawn from G.M. Smith.)

This anomalous genus is known from soil collections of algae. The characteristic chainlike arrangement of cells develops as the cells are cultivated, whereas they are probably solitary in nature.

143b   Cells not arranged in branching chains 144

144a   Cells globose or flattened along some sides from mutual compression; in packets or pseudofilaments; forming green films on moist substrates; chloroplast a parietal plate with pyrenoid indistinct. Fig. 122. . . *(Protococcus; Pleurococcus)* Desmococcus

(See Fig. 123, *Apatococcus,* similar to *Desmococcus* but without a pyrenoid.)

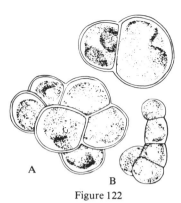

Figure 122

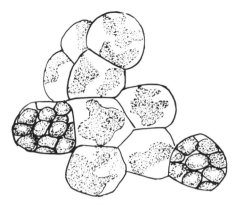

Figure 123

Figure 122 *Desmococcus viridis* (Ag.) Brand. (*Protococcus viridis* Ag.). (A) clump of cells; (B) filamentous tendency in cell arrangement.

This genus like *Apatococcus* (Fig. 123) has been and will continue to be confused with *Protococcus* and *Pleurococcus*. Honest attempts have been made to dissociate and clarify the taxonomic names. *Desmococcus* occurs on moist substrates (trees, boards, rocks) and is associated with *Apatococcus*. The cells occur in irregular packets but show a distinct tendency toward the formation of short filaments. The cells have a parietal chloroplast with a pyrenoid. Reproduction occurs readily by cell division and by aplanospores; cells are easily distributed by air currents, birds and insects. Trees with soft, spongy bark invariably have *Desmococcus* as well as *Apatococcus,* especially on the north, more moist side. Conifers seldom have these algae coating the bark.

In *Apatococcus* (Fig. 123) the cells form packets and have the same shape and arrangement as in *Desmococcus*. There is a broad, somewhat lobed to massive chloroplast without a pyrenoid. Cell division is carried on actively whereas aplanospore-formation is not uncommon. There is not a tendency toward filament-formation as in *Desmococcus* (Fig. 122). Plants occur in the same habitats as the latter; synonymous with some *Pleurococcus* and *Protococcus* as reported in the literature.

Figure 123 *Apatococcus lobatus* (Chod.) Boye-Petersen, clump of cells showing pseudo-filamentous tendency; some cells forming aplanospores.

144b   **Cells differently shaped, not producing films on subaerial substrates** 145

145a   **Cells crescent-shaped, or sharply acicular (needle-shaped)**. 146

145b   **Cells some other shape**. 147

146a   **Cells strongly crescent-shaped, closely clustered, but not entangled. Fig. 124** .. .. *Selenastrum*

Figure 124

Figure 124 *Selenastrum gracile* Reinsch, the chloroplast is laminate and parietal, covering most of the wall.

These gracefully curved cells occur in clusters of from 4 to 32, with a tendency to have the convex or 'outer' walls approximated. The curvature of the 'outer' and 'inner' walls of the crescent are more nearly equal than in the somewhat similarly shaped cells of *Kirchneriella* (Fig. 106), a genus which has cells irregularly arranged within a gelatinous envelope. Five species are commonly found in the United States, mostly differentiated by size of cell and degree of curvature. Mixtures of algae from shallow water often include *Selenaatrum*, but they may occur in the euplankton.

146b Cells straight, or acicular, or only slightly crescent-shaped, in some species loosely entangled about one another (frequently solitary rather than colonial). Fig. 125. . . . . *Ankistrodesmus*

(See also *Closteriopsis* (Fig. 126) a long, slender, needlelike cell with a chloroplast that has a row of pyrenoids.)

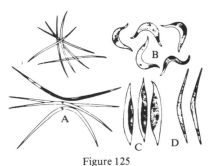

Figure 125

Figure 126

Figure 125 *Ankistrodesmus.* (A) *A. falcatus* (Corda) Ralfs; (B) *A. convolutus* Corda; (C) *A. braunii* (Näg.) Brunn.; (D) *A. fractus* (W. & W.) Brunn. The solitary species of this genus have been transferred to *Raphidium* by one author.

Figure 126 *Closteriopsis longissima* Lemm.

Although there are about 12 species of this genus common in the United States, *Ankistrodesmus falcatus* (Corda) Ralfs is probably the one most frequently found. It occurs as solitary or loosely clustered needles or slightly fusiform cells intermingled with other algae. Sometimes this species forms almost pure growths in artificial pools or laboratory aquaria. One species, *A. spiralis* (Turn.) Lemm. has needle-shaped cells that are spirally twisted about one another to form bundles. Some species, those which occur as solitary cells and with a characteristic method of liberating autospores, have been transferred to the genus *Raphidium*.

147a (145) Cells attached either along their side or end walls to form definite patterns, nets, plates, triangular clusters or short rows 148

147b Cells attached otherwise, if adjoined by lateral walls then not forming definite patterns 167

148a Cells cylindrical, 1 cell attached to 2 others at end walls repeatedly to form a network. Fig. 127 ...... *Hydrodictyon*

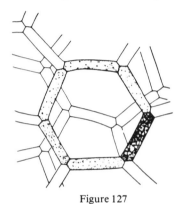

Figure 127

Figure 127 *Hydrodictyon reticulatum* (L.) Lag.

This is the familiar "water net" which often grows in such dense mats in lakes and small ponds or irrigation ditches as to become a troublesome weed. This unique alga is able to reproduce very rapidly because each cell of the net in turn produces a cylindrical net or a sheetlike expanse of cells (depending on the species) within it, which upon escape enlarges enormously, each cell again forming a daughter colony. The nets are of macroscopic size and there is a report of one found more than 2 feet in length. It is thought that the first written record referring to a specific alga is of *Hydrodictyon* in ancient Chinese literature.

148b Cells some other shape, not attached to form a network 149

149a Cells arranged to form flat, circular or rectangular plates 150

149b Cells not arranged to form flat plates ..
.... 153

150a Cells forming circular plates (sometimes irregularly subcircular), the marginal cells usually different in shape from those within. Fig. 128 ..... *Pediastrum*

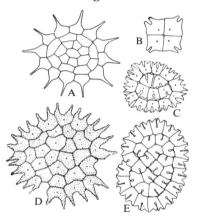

Figure 128

Figure 128 *Pediastrum*. (A) *P. simplex* (Meyen) Lemm.; (B) *P. tetras* (Ehr.) Ralfs; (C) *P. biradiatum* var. *emarginatum* fa. *convexum* Presc.; (D) *P. boryanum* (Turp.) Menegh.; (E) *P. obtusum* Lucks.

Although there are many species of this genus the cells all have a great similarity, varying only slightly in shape and wall markings. They can be identified as *Pediastrum* by their platelike arrangement. In some species the plate is continuous whereas in others there are interstices and fenestrations. Commonly the cells at the margin differ in shape from those within. One species, *P. tetras* (Ehr.) Ralfs contains only 4 cells in the colony, but in other species there may be as many as 32 or 64 (always some multiple of 2). Rarely a 2-celled colony may be seen in which instance it might be mistaken for *Euastropsis* (Fig. 74). *Pediastrum* cell walls are highly resistent to decay and many species are found in fossil or semifossil condition. The author knows of no instance in which *Pediastrum* cells have been found parasitized by aquatic fungi or bacteria. Colonies frequently occur in the psammon. Arctic lakes may exhibit remarkable developments of *Pediastrum* species, practically a 'bloom', with colonies attaining huge proportions.

150b **Cells not arranged to form circular plates**. **151**

151a **Cells triangular, or ovoid, forming quadrangular plates and bearing 1 or more spines. Fig. 129** . . . . . . *Tetrastrum*

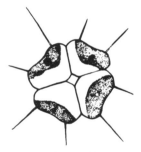

Figure 129

Figure 129 *Tetrastrum heterocanthum* (Nordst.) Chodat.

There are 9 species of this genus reported from the United States, varying in cell shape and spine length. The thallus consists of a flat plate of 4 cells that bear from 1 to 4 slender spines each on the outer free walls. The cells are oval, heart-shaped or angular from mutual compression.

151b **Cells rectangular or trapezoidal, or if oval never with spines** **152**

152a **Cells rectangular, oval or trapezoidal, the outer walls entire (not incised); arranged in 4's to form quadrate plates, or in multiples of 4. Fig. 130** . . . *Crucigenia*

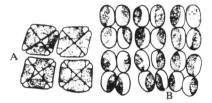

Figure 130

Figure 130 *Crucigenia.* (A) *C. tetrapedia* (Kirch.) W. & W.; (B) *C. rectangularis* (A. Braun) Gay.

These cells (like *Tetrastrum* [Fig. 129]) occur in 4's but usually form multiple colonies or groups of rectangular plates. There are about 15 species in the United States, differentiated by the shape of the cells.

152b **Cells trapezoidal, the outer free walls deeply incised, forming oval to somewhat angular plates (only 4 cells present in the plates of one species). Fig. 128** . . . . *Pediastrum*

153a **(149) Cell wall with spines** **154**

153b **Cell wall without spines** **158**

154a **Cells ellipsoid or oval; spines numerous, needlelike** **155**

154b **Cells ovoid or spherical; spines few (1 to 4); cells definitely arranged and definite in numbers (usually more than 2)** . . . **156**

155a Cells oval, side by side in pairs; the wall beset with numerous, slender spines with a thickened base. Fig. 75 . . . . . *Dicellula*

155b Cells ellipsoid, spines numerous and needlelike, the cells arranged side by side because of interlocking of spines (usually cells solitary). Fig. 131 . . . . . *Franceia*

(See also *Bohlinia,* Fig. 131A.)

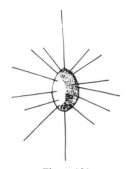

Figure 131

Figure 131 *Franceia droescheri* (Lemm.) G.M. Smith.

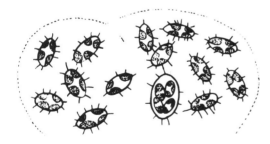

Figure 131A

Figure 131A *Bohlinia echidna* (Bohlin) Lemmer.

The species illustrated here and 2 others occur in the euplankton of lakes but rather rarely. The cells are solitary but are often clustered because of the interlocking of the needlelike spines which arise from all sides of the cell

walls. *Chodatella* (Fig. 194) has a similarly shaped cell but with the needlelike spines confined to the poles, or to the midregion of the cell. In *Franceia* the spines are uniformly distributed.

Although sometimes found solitary, cells of *Bohlinia* are mostly embedded in an amorphous mucilage. Like *Franceia* the wall bears some spines but are much shorter. Some authors consider *Bohlinia* to be synonymous with *Franceia. Bohlinia echidna* (Bohlin) Lemm. has been reported from the United States.

156a (154) Cells ovoid or fusiform, arranged side by side in 1 row or in 2 alternating rows; spines on the wall mostly short, arising from the poles of the cells, but sometimes from the face of the cells also. Fig. 76 . . . . *Scenedesmus*

156b Cells spherical, in groups of 4 or in multiples of 4 to form compound colonies; outer walls bearing long, slender spines 157

157a Colony triangular; spines 1 to 7. Fig. 132 . . . . *Micractinium*

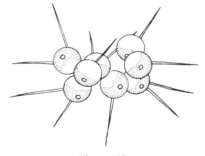

Figure 132

Figure 132 *Micractinium pusillum* Fres., pyramidal clusters of cells, each bearing 1 or 2 spines.

This rare alga occurs in the euplankton, having clusters of 4 round cells arranged in the form of a small pyramid. Each cell bears 1 to several long, tapering spines. Another species which is seldom seen is *M. quadrisetum* (Lemm.) G.M. Smith, having oval or elliptic cells.

157b   Entire colony pyramidal, with the outer free walls of the cells bearing a single, stout spine. Fig. 133. . . . . *Errerella*

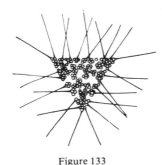

Figure 133

Figure 133   *Errerella bornhemiensis* Conrad.

Cells of this plant, of which there seems to be but a single species, are arranged to form a 3-dimensional pyramid. It is known only from the euplankton and apparently is very rare, although it has been reported from several parts of the United States.

158a   (153) Cells spherical or polygonal, arranged to form hollow, spherical or many-sided colonies; cells adjoined by interconnecting protuberances of the mucilaginous cell sheaths. Fig. 134 . . . .
. . . . . *Coelastrum*

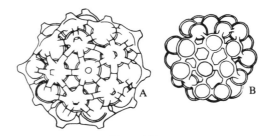

Figure 134

Figure 134   *Coelastrum.* (A) *C. cambricum* Archer; (B) *C. microporum* Näg. (A, redrawn from G.M. Smith.)

As the genus name suggests, cells of this plant are arranged to form a hollow colony. In some species the cells are closely associated and the hollow feature is discerned with difficulty, whereas in others the cells are clearly separated in a peripheral layer and interconnected by prominent 'arms' of the mucilaginous sheath that encloses each cell. There are about 14 species known from the United States. These differ in respect to cell shape and the length of the processes of the sheath which in some species produces marginal protuberances. *Coelastrum microporum* Näg. is especially common in both eu- and tychoplankton.

158b   Cells not forming hollow colonies, not so adjoined 159

159a   Cells fusiform, radiating from a common center. Fig. 135 . . *Actinastrum*

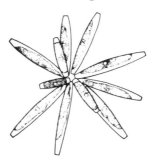

Figure 135

Figure 135   *Actinastrum hantzschii* Lag.

These 'cigar'-shaped cells radiate from a common center to form colonies not enclosed in a gelatinous sheath. The species illustrated is more common in the plankton than *A. gracillimum* G.M. Smith in which the cells are not so definitely fusiform.

159b   Cells shaped otherwise, not forming a colony of radiating cells **160**

160a   Cells ellipsoid to fusiform, adjoined end to end, forming chainlike series. Fig. 121 . . . . *Dactylococcus*

160b   Cells not forming chains **161**

161a   Cells ovoid, ellipsoid, or fusiform, adjoined by their lateral walls to form a row of 4 (or 8) in a single series, or a double series in which the cells of the 2 rows alternate. Fig. 76. . . . *Scenedesmus*

161b   Cells globular or variously shaped, not attached side by side in one plane . . . **162**

162a   Cells oval or fusiform, arranged side by side in a curved plane; cells with or without spines. Fig. 76 . . . *Scenedesmus*

162b   Cells arranged otherwise or shaped differently **163**

163a   Cells fusiform or trapezoidal, adjoined with their long axes parallel about a common center. Fig. 136 . . *Tetradesmus*

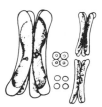

Figure 136

Figure 136   *Tetradesmus smithii* Presc.

This plant resembles some *Scenedesmus* species (Fig. 76) but differs in having cells with their long axes parallel and quadrately arranged in a packet rather than in 1 plane. There are 3 species reported from the United States. *T. wisconsinense* G.M. Smith has trapezoidal cells.

163b   Cells shaped or attached otherwise . . **164**

164a   Cells crescent-shaped to somewhat triangular, the poles extended into horns, cruciately arranged with the convex margins apposed. Fig. 137. . . . . *Lauterborniella*

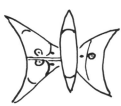

Figure 137

Figure 137   *Lauterborniella elegantissimum* Schmidle. (Redrawn from Schmidle.)

In this genus flattened, triangular-crescent cells are arranged about a common center with their concave walls outward. The outer angles are extended to form sharp poles. There is a parietal chloroplast and 1 pyrenoid.

164b Cells shaped or attached otherwise. . 165

165a Cells oval, showing some polarity, the 'apical' more broadly rounded than the 'posterior', arranged in clusters of 4 with long axes parallel, but the cells somewhat staggered; walls often ribbed. Fig. 138 . . . . *Enallax*

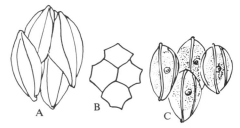

Figure 138

Figure 138. *Enallax*. (A) *E. alpina* Pascher, habit of cell arrangement; (B) *E.* sp., end view of colony; (C) *E.* sp., showing chloroplasts and pyrenoids. (A, B redrawn from Pascher.)

This genus is very similar to *Scenedesmus* (Fig. 76) but is differentiated by the bunched or clustered arrangement of the cells (not side by side in 1 plane). The cells are somewhat staggered or tiered in their arrangement.

165b Cells shaped or arranged otherwise 166

166a Cells crescent-shaped, in groups of 4, 2 with concave sides toward one another, the other 2 cells in another plane with the poles at one end only in juxtaposition. Fig. 107 . . . . *Tetrallantos*

166b Cells sickle-shaped, fusiform or crescent-shaped, twisted about one an-

other, or in bundles with long axes parallel. Fig. 125 . . . . *Ankistrodesmus*

167a (147) Cells adjoined by gelatinous strands or threads formed from remains of old, mother-cell walls 168

167b Cells not adjoined by strandlike remains of mother-cell walls 173

168a Colony cubical or square in outline 169

168b Colony some other shape 170

169a Cells spherical, at the corners of a hollow cube, formed by gelatinous tubes. Fig. 90 . . . . *Pectodictyon*

169b Cells globular or pyriform, in 4's, forming a square, interconnected by gelatinous strands, the cells bearing scale-like fragments of the old, mother-cell wall. Fig. 89 . . . . *Coronastrum*

170a Cells spindle-shaped, in clusters of 4-8-16 at the ends of radiating, gelatinous strands. Fig. 139 . . . . *Actidesmium*

Figure 139

Figure 139 *Actidesmium hookeri* Reinsch.

This rare plant occurs in the tychoplankton of

shallow pools; also found commonly in the Arctic. The star-shaped clusters of cells at the end of radiating (sometimes dichotomously branched) gelatinous strands render it certain of identification. The clusters of cells result from the fact that zoospores formed in the cells remain clustered at the tip of the mother-cell. The chloroplast is parietal or diffuse and does not contain a pyrenoid, as far as known.

170b **Cells shaped or arranged otherwise . . . .**
. 171

171a **Cells globose, in clumps of 4-8, the groups held together by looplike fragments of the old, mother-cell wall. Fig. 140 . . . .** *Westella*

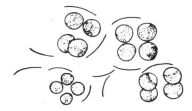

Figure 140

Figure 140 *Westella botryoides* (West) de Wild., showing fragments of mother-cell walls.

This plant should be compared with *Dictyosphaerium* (Fig. 101), which it may resemble superficially. *Westella* has no gelatinous envelope and although there are strands left by the old, mother-cell wall, they do not produce the regular, radiate fibrils characteristic of *Dictyosphaerium*. Two species only are known from the United States.

171b **Cells in 4's at the ends of fine, radiating strands which are the remains of mother-cell walls 172**

172a **Cells globular or somewhat oval, in 4's**

at the ends of fine, dichotomously divided strands. Fig. 101 . . . .
. . . . *Dictyosphaerium*

172b **Cells in 4's, both oval and reniform in the same cluster, at the ends of (often indistinct) radiating (sometimes double), gelatinous strands. Fig. 100 . . . .**
. . . . *Dimorphococcus*

173a **(167) Cells pyriform, bean-shaped, the outer angles of the cells bearing horns; cells at the ends of radiating, stout gelatinous strands, forming a globular colony. Fig. 141 . . . .** *Sorastrum*

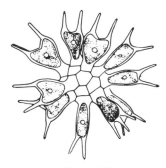

Figure 141

Figure 141 *Sorastrum americanum* (Bohlin) Schmidle.

There are only 4 species of this genus reported from the United States, of which *Sorastrum spínulosum* Näg. is probably the most common, occurring in the euplankton. This species has relatively stout, short spines at the angles of the cells, and the basal pedicel is scarcely developed so that the colony appears as a compact cluster.

173b **Cells other shapes, not so arranged, without horns, or with spines different from above 174**

174a Cells clustered, spherical or oval, in 2's and 4's, separated from one another by semiopaque masses of dark mucilage which form X-shaped bands over the colony. Fig. 142 . . . . *Gloeotaenium*

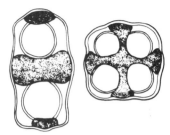

Figure 142

Figure 142 *Gloeotaenium loitelsbergerianum* Hansgirg, 2- and 4-celled colonies.

This monotypic genus is unique in appearance because of the dark bands of mucilage that occur between the cells of the colony, sometimes forming a cross-shaped pattern. There are 2 or 4 cells tightly enclosed within the mother-cell wall. The chloroplast is dense, massive and usually obscured by numerous starch grains. *G. loitelsbergerianum* Hansg. is rare but widely distributed over the world; occurs in the tychoplankton.

174b Cells not separated from one another by masses of dark mucilage 175

175a Cells spherical, clustered (rarely solitary), with thick walls which show refractive or refringent alveolae. Fig. 143 . . . . *Keriochlamys*

Figure 143

Figure 143 *Keriochlamys styriaca* Pascher. (Redrawn from Pascher.)

The globular or semiglobular cells in this genus are readily identified by the thick wall which has alveolae that refract the light. Each cell has a parietal, platelike chloroplast and 1 pyrenoid. The genus is to be expected throughout the United States but thus far has been reported only from Kansas.

175b Cells arranged otherwise, without alveolar walls. 176

176a Cells with long, needlelike spines . . . 177

176b Cells without long, needlelike spines 178

177a Cells spherical, with slender spines, needlelike throughout their length; colonial only because of entangling of spines; chloroplast a parietal plate with a large, distinctly bean-shaped pyrenoid. Fig. 144 . . . . *Golenkinia*

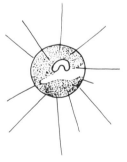

Figure 144

Figure 144 *Golenkinia radiata* Chodat, showing reniform pyrenoid.

There are only 2 or 3 species of this genus reported, all of them common in tow samples from open water of lakes, *Golenkinia radiata* (Chod.) Wille has long, needlelike spines, 2 or 3 times the diameter of the cell in length, whereas *G. paucispina* West & West has more numerous spines that are about equal to the cell diameter. There is a parietal chloroplast with a distinctly curved or U-shaped pyrenoid.

177b   Cells spherical, with spines that are thickened at the base or which taper from base to apex; chloroplast with a circular pyrenoid. Fig. 145 . . . . *Golenkiniopsis*

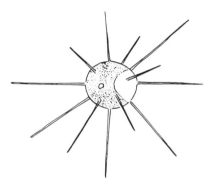

Figure 145

Figure 145   *Golenkiniopsis solitaria* Korsch., showing cup-shaped, parietal chloroplast and spherical pyrenoid.

This genus, very similar to *Golenkinia* (Fig. 144), has been established on the basis of its circular type of pyrenoid, and by the fact that the spines are broader at the base, tapering to a fine point. (Compare with Fig. 144.)

178a   (176) Cells enclosed by old mother-cell wall 179

178b   Cells not enclosed by mother-cell wall 180

179a   Cells somewhat bean-shaped or kidney-shaped, *or* broadly elliptic (old mother-cell wall often appearing as a muci-laginous sheath. (See note at 131b and Fig. 110.) Fig. 110 . . . . . . *Nephrocytium*

179b   Cells elliptic, lemon-shaped, or sometimes subcylindric, 1 or more generations of mother-cell walls enclosing daughter cells (autospores). Fig. 146 . . . . *Oocystis*

(Compare with *Rayssiella* [Fig. 147] which has a unique way of liberating autospores; is similar in many respects to *Oocystis*.)

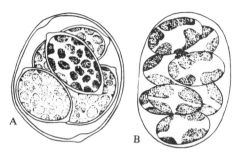

Figure 146

Figure 146   *Oocystis*. (A) *O. eremosphaeria* G.M. Smith; (B) *O. borgei* Snow.

There are several species of this genus common in both euplankton and tychoplankton. They are differentiated by the presence or absence of nodules at the poles, and by the number of chloroplasts. Two or 3 generations of cell walls may be enclosed within an original mother-cell wall which enlarges so that it often appears as a gelatinous sheath and therefore may be misleading as a taxonomic character.

The autospores in the genus *Rayssiella* are arranged at 2 poles in the mother-cell. Upon release the mother-cell wall forms loops which interconnect the clusters of daughter cells. Only 1 species is known, *R. hemispherica* Edel. & Presc. Fig. 147.

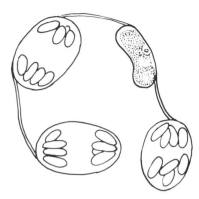

Figure 147

Figure 147 *Rayssiella hemisphaerica* Edel. & Presc., one vegetative cell and autospore-formation in others.

180a (178) Cells spherical, occurring evenly distributed, often with interspersed colonies of small cells within the gelatinous sheath (sheath sometimes difficult to discern), chloroplasts several angular plates. Fig. 105 . . . . . . *Planktosphaeria*

180b Cells variously shaped but not with distributed clumps of daughter cells in the colony, often densely aggregated; chloroplast 1, parietal 181

181a Cells spherical or angular from mutual compression when occurring in clumps subaerial 182

181b Cells fusiform or needle-shaped; aquatic. Fig. 125 . . . . . *Ankistrodesmus*

182a Cells spherical, gregarious but not adjoined, sometimes solitary. Fig. 87 . . . . *Chlorococcum**

182b Cells in dense clumps, forming a film or layer on moist substrates; cells spherical or angular from mutual compression, sometimes forming pseudofilaments. Fig. 122. . . *(Protococcus; Pleurococcus) Desmococcus***

183a (82) Cells attached, consisting of a euglenoid, flagellated organism enclosed in a brown, vase-shaped lorica with a wide, anterior opening. Fig. 21 . . . . *Ascoglena*

183b Cells free-floating, attached, or endophytic; without a sessile lorica 184

184a Cells fusiform, crescent-shaped or sickle-shaped, with sharply pointed or narrowly rounded and tapering apices 185

184b Cells some other shape. 194

185a Cells with 2 axial chloroplasts, bearing longitudinal ridges, a chloroplast in each horn of the cell; pyrenoids conspicuous, usually in an axial row 186

185b Cells with 1 chloroplast, or with parietal chloroplasts not arranged as above 187

186a Cells furnished with a stout spine at either pole. Fig. 148 . . . *Spinoclosterium*

*See Appendix for other spherical cells, p. 268.
**See Bourrelly 1966.

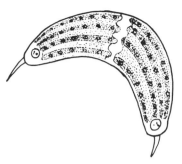

Figure 148

Figure 148 *Spinoclosterium curvatum* Bernard.

Spinoclosterium, apparently with but a single species (and a variety) is disjunct in its distribution throughout the world, but is very rare. *S. curvatum* Ber. has been found in southeast United States, in the glaciated region of central North America, and from the mountains of the far west. It was originally reported as *Closterium cuspidatum* by Bailey in southeastern United States, later reported from the East Indies. Some authors retain this genus under *Closterium,* but the distinctive polar spines seem to justify its separation, otherwise the diagnosis of the genus *Closterium* would need emendation. The chloroplasts are prominently ridged and there are terminal vacuoles as in *Closterium.* Zygospores never have been reported and it is essential that reproduction in this genus be learned to complete its diagnosis. Plants occur in sparse numbers in *Sphagnum* bogs and soft-water habitats. The variety *spinosum* Presc. with cells inflated near the poles was described from Michigan.

186b Cells crescent-shaped, variously bowed, but in some species nearly straight, without apical spines. Fig. 149. . . . .
. . . . . *Closterium*

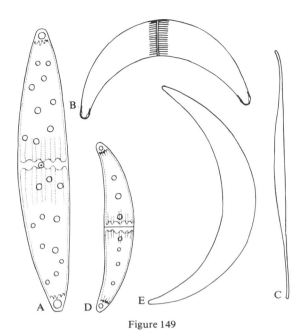

Figure 149

Figure 149 *Closterium.* (A) *Cl. lunula* var. *biconvexum* Schm., showing ridged chloroplasts and terminal vacuoles; (B) *Cl. loefgrenii* Borge, with girdle band and striated wall; (C) *Cl. kuetzingii* Bréb.; (D) *Cl. leibleinii* Kütz.; (E) *Cl. parvulum* var. *majus* West.

This is one of the few Placoderm desmids in which there is no incision or sinus in the midregion. The wall is in 2 sections, however, and the cell contents are arranged as 2 semicells, with a chloroplast in each. Whereas some species may be nearly straight, *Closterium* is characteristically bowed or crescent-shaped, sometimes inflated in the midregion, and sometimes with the concave and convex margins showing different degrees of curvature. The poles may be narrowed to fine points or may be broadly rounded (sometimes recurved, or slightly inflated). The two chloroplasts are often ridged with laciniate margins. Each contains from 1 to several pyrenoids. In the poles of the cell is a conspicuous vacuole in which there are from 1 to many vibrating granules of gypsum (showing Brownian movement). Cell division in some species leaves transverse bands or girdles. The wall may be smooth or striate; is often brownish because of iron deposits.

187a   (185) Cells only slightly crescent-shaped, usually straight or nearly so and sometimes S-shaped or sigmoid, with the poles drawn out into fine points 188

187b   Cells definitely crescent-shaped, the poles abruptly sharp-pointed or drawn out into fine points; (but see *Ankistrodesmus* [Fig. 125]) 191

188a   Cells attached by a slender stipe to other algae or to microfauna. Fig. 150 . . . . . .
. . . . *Characium*

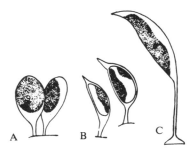

Figure 150

Figure 150   *Characium.* (A) *Ch. debaryanum* (Reinsch) de Toni; (B) *Ch. ornithocephalum* A. Braun; (C) *Ch. rostratum* Reinsch.

There are numerous species in this genus, differentiated by shape of cell and by length or stoutness of a stalk. Some are very minute and are easily overlooked, whereas others are fre-

quently seen, growing singly or in associations so as to form patches on filamentous algae or on small animals. The genus *Characiopsis* (Fig. 330) contains species shaped like some of those in *Characium,* and care should be used in determining the color of the chloroplast, the presence or absence of a pyrenoid, and the presence of starch as a reserve food. *Characiopsis* has a yellow pigment predominating, the starch test is negative and there is no pyrenoid; is a member of the Chrysophyta.

188b   Cells not attached by a stipe 189

189a   Cells fusiform, the poles extended to form setalike spines 190

189b   Cells crescent-shaped or sigmoid, tips of cells narrowed to fine points, not setalike but narrowly pointed or very narrowly rounded; chloroplast not extending the full length of the cell. Fig. 151 . . . . *Ourococcus*

Figure 151

Figure 151   *Ourococcus bicaudatus* Grob. (Redrawn from G.M. Smith.)

This rare plant might be confused easily with *Ankistrodesmus* spp. (Fig. 125) but the cells are usually stouter; are not nearly so needlelike. It is closely related to *Elakatothrix* (Fig. 88) in the Order Tetrasporales because the cells retain the ability to divide vegetatively to form new individuals, whereas in the Order Chlorococcales, which includes forms similar to *Ourococcus,* the cells cannot undergo division but must form new individuals within the wall of the parent cell.

190a Cells fusiform, straight or slightly curved with each pole extended into a long, stout spine. Fig. 152 . . *Schroederia*

Figure 152

Figure 152 *Schroederia setigera* Lemm.

The long, stout spine at the poles of straight, fusiform cells help to differentiate this genus. When the cells are young there is a single laminate chloroplast and 1 pyrenoid but in age there is a multiplication within the cell to form several platelike chloroplasts each with a pyrenoid. In reproduction the contents of the cell undergo cleavage to form zoospores. There is 1 species known for the United States. Another species *S. judayi* G.M. Smith has been transferred to the genus *Ankyra* (Fig. 153).

190b Cells fusiform or sub-cylindirc, bent or curved, the apical pole extended into a long spine, the posterior with a stout, narrow extension which is forked or bi-fid at the tip; the cell wall separating medianly to release zoospores. Fig. 153 . . . . *(Schroederia judayi) Ankyra*

Figure 153

Figure 153 *Ankyra judayi* (G.M. Smith) Fott.

In this genus somewhat curved, fusiform cells have a long, straight spine at one pole and a narrowed extension 'posteriorly' which is forked at the tip. There is 1 parietal chloroplast and a pyrenoid.

191a (187) Cells slender, needlelike or sometimes fusiform (usually in clusters, but may be solitary), often only slightly crescent-shaped; chloroplast parietal (the outline often discerned with difficulty). Fig. 125 . . . . (*Raphidium* p.p.) *Ankistrodesmus*

191b Cells stouter, not needlelike; definitely crescent-shaped; the chloroplast parietal 192

192a Cells bearing a stout spine at either end. Fig. 154. . . . . *Closteridium*

Figure 154

Figure 154 *Closteridium lunula* Reinsch.

In this genus the cells are lunate or nearly straight; are somewhat similar to some species of *Closterium* (Fig. 149) but can be differentiated readily by the fact that there is but a single, parietal chloroplast that nearly covers the cell wall, and by the absence of terminal vacuoles characteristic of *Closterium*. There are 3 species reported from the United States.

192b Cells without a stout spine at the poles 193

193a Cells enclosed in a mucilage (usually in clusters but sometimes solitary); in some species the curvature is so great that the tips of the cells nearly touch. Fig. 106 . . . . *Kirchneriella*

193b Cells not or rarely enclosed in mucilage (See Fig. 107.), usually in clusters but sometimes solitary; curvature of the inner margin nearly that of the outer; tips of the cell not almost touching. Fig. 124 . . . . *Selenastrum*

194a (184) Living in the tissues of higher plants, or in animals or their eggs. . . 195

194b Not living in the tissues of plants nor in animals (but may be in the mucilage of other algae) 201

195a Globular cells inhabiting the envelope of salamander or frog eggs. (Occasionally desmids and many other unicells incidentally occur also.) Fig. 155 . . *Oophila*

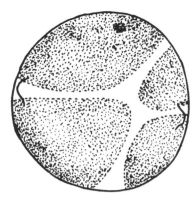

Figure 155

Figure 155 *Oophila amblystomatis* Lambert, a cell showing fragmented chloroplast and 2 wall pits.

Oophila amblystomatis Lambert occupies the mucilage of frog and salamander eggs, often so abundantly as to give the egg masses a distinct green color. It is a unicellular member of the Chlorococcales. Pits in the wall are not understood.

195b Endozoic or endophytic in other organisms 196

196a Cells globose, numerous and minute, within the bodies of Protozoa, sponges, *Hydra, et al.* Fig. 156 . . . . *(Zoochlorella) Chlorella**

*See Appendix for other unicellular green algae, p. 268.

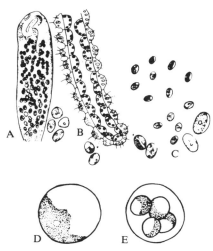

Figure 156

Figure 156 (A) *Ch. (Zoochlorella) parasitica* Brandt within *Ophrydium,* a colonial ciliate; (B) *Ch. (Zoochlorella) conductrix* Brandt in *Hydra;* (C) *Ch. ellipsoidea* Gerneck, two cells enlarged to show parietal chloroplasts; (D) *Ch. vulgaris* Beij., cell with chloroplast and (E) a cell with autospores.

Cells in this genus are small, oval or spherical, occurring singly or gregarious, especially on soil and on moist subaerial substrates; whereas some species occur as inhabitors of Protozoa, *Hydra,* sponges and other microfauna. Freshwater sponges are invariably green because of *Chlorella,* often referred to as *Zoochlorella.* A few species are fairly well-defined but mostly they must be cultured to make identifications. There is a thin, cup-shaped or platelike chloroplast. Like other members of the Chlorococcales, reproduction takes place by internal cell division, in this instance forming non-motile autospores. *Chlorella* is a genus which lends itself well to culturing for studies in photosynthesis, production of proteins and in the production of an antibiotic, chlorellin. It has been

(and is) used in mass culture in an investigation of its possible use as a source of human food. *Palmellococcus* has been referred to *Chlorella* as a subgenus.

197a   Cells globular or angular, with thick cyst walls, occurring in (or *on,* in the plasmodial stage) the cells of *Sphagnum;* chloroplast massive and green, although often colored yellow or golden, a member of the Xanthophyceae in the encysted stage. Fig. 157 . . *Chlamydomyxa*

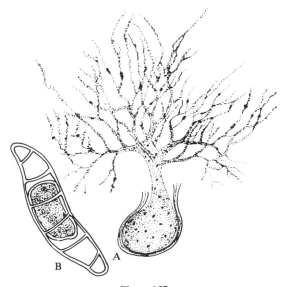

Figure 157

Figure 157 *Chlamydomyxa labrynthuloides* Archer. (A) vegetative thallus with pseudopodia; (B) cyst within a *Sphagnum* cell. (Diagrammed from Archer.)

This curious organism (often greenish) occurs as a plasmodium over and in *Sphagnum.* The protoplast forms cysts within the empty, water-storage cells of *Sphagnum* leaves, which are thick-walled and usually yellow or golden-tan in color. The somewhat globular cysts have been shown to be multinucleate (as is the plasmodium). The cysts eventually form smaller, 2-nucleated cysts in which zoospores are formed. The organism is of uncertain position but is considered as a member of the Xanthophyceae.

197b Plants otherwise **198**

198a Plant a branched tube with a globose swelling at the tips, growing on and among the cells of *Sphagnum*. Fig. 158. . . . . *Phyllobium*

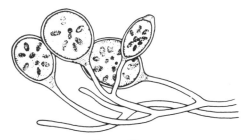

Figure 158

Figure 158 *Phyllobium sphagnicola* G.S. West, a few colorless strands with terminal akinetes among cells of *Sphagnum*.

The thallus of this genus consists of almost colorless, branched threads growing within and among the leaves of *Sphagnum*. The contents of the tube accumulate at the tips of branches to form thick-walled akinetes. There are numerous ellipsoid chloroplasts, radially arranged in the apical swellings.

198b Plant formed otherwise; not living in *Sphagnum* **199**

199a Plant a much-branched, coenocytic tube (multinucleate and without cross walls), growing in leaves of Araceae such as the Indian Turnip. Fig. 159 . . *Phyllosiphon*

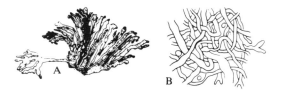

Figure 159

Figure 159 *Phyllosiphon arisari* Kühn. (A) habit of thallus within host tissue; (B) portion of thallus showing tangled filaments. (A, redrawn from Just.)

This branched, tubular plant is non-cellular; forms green patches in the leaves of higher plants which become discolored. It may be more widely distributed than appears at present, but so far it is known only from northern and eastern sections of the United States, but occurs in Europe and tropical regions of high humidity. The coenocytic character and the lack of cross walls places *Phyllosiphon* in the Siphonales.

199b Plant not a branched, coenocytic tube. **200**

200a An irregularly shaped, flasklike cell in the tissues of *Ambrosia* (ragweed), and in other plants; cells usually red. Fig. 20 . . . . *Rhodochytrium*

200b An irregularly oval, thick-walled cell in the tissues of *Lemna* (duckweed). Fig. 160 . . . . *Chlorochytrium*

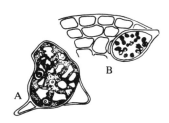

Figure 160

Figure 160 *Chlorochytrium lemnae* Cohn. (A) cell showing netlike chloroplast; (B) cell within host tissue. (A, redrawn from Bristol Roach.)

This genus occurs in both marine and freshwater hosts. In addition to the species illustrated there are 4 others known from inland waters of the United States, the most common being *Chlorochytrium lemnae* Cohn. Plants of *Lemna trisulca* when allowed to become colorless in the laboratory, frequently show green spots locating the position of the endophytic alga. The cell is diploid, the only haploid stage being the gametes which are produced within the encysted cells. The chloroplast is usually a parietal reticulum when the cells are mature, but is cup-shaped when young.

201a (194) **Cells in the form of swollen vesicles on moist soil, with colorless, subterranean, rhizoidlike extensions. Fig. 161** . . . . *Protosiphon*

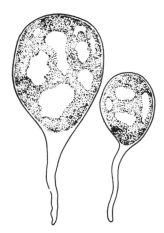

Figure 161

Figure 161 *Protosiphon botryoides* (Kütz.) Klebs, showing clathrate, parietal chloroplast.

Protosiphon is a 1-celled, coenocytic and bulbous plant, 0.5 mm in diameter, inhabiting moist soil. Confusingly, it appears with *Botrydium* (Fig. 326) a member of the Xanthophyceae which is also balloonlike, but larger (1 to 3 mm in diameter), and does not contain starch. There is a reticulate, sheetlike chloroplast (unlike the numerous disclike chloroplasts of *Botrydium*) which covers most of the wall. The sac has a colorless, rhizoidal extension that penetrates the soil. This genus is placed in its own family within the Siphonales.

201b **Cells not as above** 202

202a **Cells attached, either sessile or on a stalk** 203

202b **Cells not attached, free-floating; *or* on moist soil, on snow, sometimes forming a green film** 212

203a Cells bearing a seta or hair 204

203b Cells without setae 205

204a Setae simple (unbranched). Fig. 177 . . .
. . . . *Chaetosphaeridium*

204b Setae branched. Fig. 116 . . . .
. . . . *Dicranochaete*

205a (203) Cells on a slender stalk, or with the basal portion of the cell narrowed to form a stalk 210

205b Cells on a stout, thick stalk or a broad basal disc; cells globular, or oval, or elongate-elliptic 206

206a Cells globular, or irregularly globose 207

206b Cells elongate, long-oval or somewhat fusiform 208

207a Cells globular, sedentary, with anterior end downward (usually on *Cladophora* (Fig. 294) on a broad, brown, short-pedicel and enclosed in a brown sheath; protoplast filling the cell; chloroplast cup-shaped and located in the upper (posterior) portion of the cell; possessing a red eye-spot when young. Fig. 162 . . .
. . . . *Malleochloris*

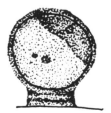

Figure 162

Figure 162 *Malleochloris sessilis* Pascher. (Redrawn from Pascher.)

This rare plant is to be sought on filamentous algae such as members of the Cladophoraceae. The sheath that encloses the cell is often reddish. Reproduction (similar to other Tetrasporales) is by swimming spores and by isogametes. There is a cup-shaped or urnlike chloroplast, a pyrenoid and 2 contractile vacuoles.

207b Cells irregularly globular, sedentary on a broad attaching base, stalk practically wanting; with a thick, brown sheath enclosing a protoplast that does not fill it, the protoplast containing a cup-shaped chloroplast and an eye-spot as well as a contractile vacuole (giving the general appearance of a shelled zoospore). Fig. 163 . . . . *Chlorophysema*

Figure 163

Figure 163 *Chlorophysema contractum* Pascher. (Redrawn from Pascher.)

The lorica of cells in this genus is thick and brown. It is attached by a broad disc at the base of a very short (scarcely evident) pedicel. There are some 7 species, both epiphytic and epizoic.

208a (206) Cells oval, elongate, with from 2 to several chloroplasts and enclosed in and attached by cup-shaped remains of old, mother-cell walls, successive generations of cell walls forming branching chains, the cells 1 or 2 together in the extremities. Fig. 118 . . . . *Ecballocystis*

208b Cells shaped and growing differently. . .
. . . . 209

209a Cells oval to elongate-elliptic, attached by relatively broad, branching stalks to microcrustacea; chloroplasts many oval discs. Fig. 120 . . . . *Colacium*

209b Cells elongate-oval to fusiform, pointed anteriorly, narrowed posteriorly to a broad, attaching disc which is formed external to the cell wall; chloroplasts usually several to many narrow, parietal plates. Fig. 164 . . . . *Characiochloris*

Figure 164

Figure 164 *Characiochloris characioides* Pascher, showing scattered contractile vacuoles.

In this genus the cells are similar to *Chlorangiella* (Fig. 119) but usually has many parietal chloroplasts and 2 or more contractile vacuoles. There is no stalk but the cell (oval to fusiform) arises directly from a disc which is exterior to the cell wall. The plants are either epiphytic or epizoic; a member of the Tetrasporales.

210a (205) Cells oblong or fusiform, gregarious (sometimes solitary), on a slender attaching stipe which is often branched so that "plumes" are formed; epizoic on microcrustacea. Fig. 119 *(Chlorangium) Chlorangiella*

210b Cells shaped or growing differently . . . .
. . . . 211

211a Cells globular, attached by a slender, tapering stipe to the mucilage of other colonial algae *(Coelosphaerium, Anabaena)*; chloroplast parietal, lying along the upper (outward) wall. Fig. 165 . . . . .
. . . . *Stylosphaeridium*

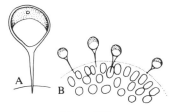

Figure 165

Figure 165 *Stylosphaeridium stipitatum* (Bach.) Geit. & Gimesi. (A) cell with apical chloroplast; (B) habit of growth on a blue-green colony.

This curious epiphyte is found in abundance when it occurs at all, as minute 'hat pins' in the mucilage of colonial or filamentous blue-green algae. Planktonic species of *Anabaena,* and *Coelosphaerium* (Fig. 477) at certain times of the year, and usually when they are in 'bloom' condition may be densely beset with the epiphytes. There is a parietal chloroplast occupying the anterior wall.

211b Cells elongate-ovoid or fusiform (often curved), *or* if globular with the chloroplast parietal along the lateral walls; stalk as long as or shorter than the cell. Fig. 150 . . . . *Characium*

212a (202) Cells elongate-fusiform, cylindrical or rod-shaped, crescent-shaped, slightly curved or straight; several to many times longer than their diameter 213

212b Cells oval, circular (or nearly so), pyramidal, trapezoidal, or star-shaped, isodiametrically angular, not more than 3 times the diameter in length. 235

213a Cells with narrowed apices, sometimes sharply pointed 214

213b Cells with broadly rounded or truncate apices 219

214a Chloroplasts 2, axial, one in either horn of a crescent-shaped cell which may be only slightly curved. Fig. 149. . . . .
. . . . *Closterium*

(See also *Spinoclosterium* Fig. 148.)

214b Chloroplasts otherwise; cell not so shaped. 215

215a Cells decidedly fusiform, one or both poles extended into setae or sharp points 216

215b Cells not broadly fusiform. 218

216a Cells actually globular but enclosed in a fusiform sheath with longitudinal ridges. Fig. 166. . . . . *Desmatractum*

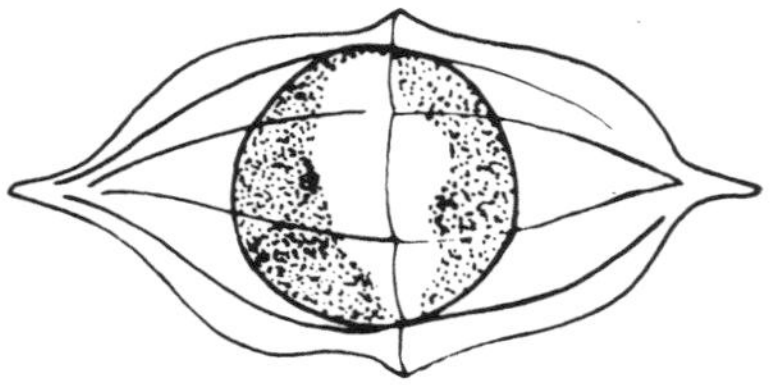

Figure 166

Figure 166 *Desmatractum bipyramidatum* (Chod.) Pascher.

This genus, with 2 species occurring in the United States, is rather rare, but widely distributed. It occurs in the euplankton of lakes and streams, often appearing as a minute seed pod with a single, globular seed. The wall is very wide and transparent and so forms a sheathlike envelope. There is a broad, parietal chloroplast and 1 pyrenoid. Both autospores and zoospores are used in reproduction.

216b Cells themselves fusiform; without such a sheath. 217

217a Setae formed by a narrowing of the cell to a fine point; chloroplast laminate (platelike), not extending the full length of the cell. Fig. 151 . . . . *Ourococcus*

217b Setae formed by a fine spine on the wall, extending from the narrowed tips of the cell. Fig. 152 . . . . *Schroederia*

(See also *Ankyra,* Fig. 153.)

218a (215) Cells many times (20 or more) longer than wide; the chloroplast with a row of pyrenoids. Fig. 126. . . . .
. . . . *Closteriopsis*

218b Cells less than 20 times the diameter long; slender or narrowly fusiform, with 1 pyrenoid sometimes evident. Fig. 125 . . . . *Ankistrodesmus*

219a **(213) Wtih a notch in the apices of the cell. Fig. 167** . . . . *Tetmemorus*

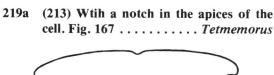

Figure 167

Figure 167   *Tetmemorus laevis* (Kütz.) Ralfs.

This genus belongs to the Placoderm desmids, having a wall in 2 sections that adjoin in the midregion. The median sinus is relatively shallow. The apical notch is prominent. Species that are found in the United States all seem to be confined to highly acid situations. Species are separated on the basis of semicell morphology; some tapering decidedly toward the apex, others more nearly cylindrical.

219b **Without a notch in the ends of the cell.** **220**

220a **Cells crescent-shaped, with an axial chloroplast bearing ridges in each horn. Fig. 149** . . . . *Closterium*

(See also *Spinoclosterium*, Fig. 148.)

220b **Cells not crescent-shaped, or with other types of chloroplasts** **221**

221a **Cells constricted in the midregion to form 2 semicells which are mirror images of one another** **222**

221b **Cells not constricted in the midregion to form semicells** **230**

222a **Cells furnished with whorls of protuberances which bear 1 or 2 spines; poles of the cell forked and spine-bearing. Fig. 168** . . . . *Triploceras*

Figure 168

Figure 168   *Triploceras gracile* Bailey.

Cells of this Placoderm desmid are highly ornate, possessing whorls of spines. They are practically cylindrical in general shape but have slightly enlarged, lobed and spine-bearing apices. Like many other desmids *Triploceras* occurs in acid or soft waters, especially in *Sphagnum* bogs.

222b **Cells not furnished with whorls of spiny protuberances** **223**

223a **Cells 10 or more times longer than broad, usually cylindrical or nearly so, with margins smooth or undulate; in some with walls bearing rectangular raised areas** **224**

223b **Cells less than 10 times their diameter in length; cylindric, fusiform or tumid, usually straight but sometimes slightly curved or crooked** **228**

224a Cells with a circle of folds or teeth at the base of the semicells where there is a shallow constriction. **225**

224b Cells without teeth or folds at the bases of the semicells **226**

225a Cells cylindrical, approximately the same diameter throughout, the margins usually undulate; poles truncate and smooth. Fig. 169 . . . . *Docidium*

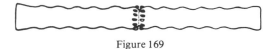

Figure 169

Figure 169 *Docidium undulatum* Bailey.

This genus is scarcely to be separated from *Pleurotaenium* (Fig. 171). In living cells the density of the chloroplast may obscure the characteristic creases in the wall at the bases of the semicells. These folds form marginal teeth as seen at the periphery. Some species have undulate walls, others smooth. The poles are always abruptly truncate and not decorated.

225b Cells subcylindrical, curved, somewhat inflated at the base of the semicell, the margins not undulate; poles truncate but bearing spines at the angles, often with a median notch; a circle of prominent teeth at the base of the semicells. Fig. 170. . . . . *Ichthyodontum*

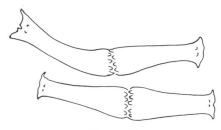

Figure 170

Figure 170 *Ichthyodontum sachlanii* Scott & Presc., one cell showing bipolarity.

This genus is practically cylindrical but the semicells are slightly enlarged at the base. The poles are abruptly truncate and in many instances (one species at least) show a dichotypical morphology; one pole is notched, with the angles spiniferous. There is a chloroplast in each semicell, longitudinally ridged and with a row of pyrenoids. The genus is known only from Java thus far.

226a (224) Cells cylindrical or subcylindric, sometimes slightly narrowed at the poles which are truncate, but lobed and with spines or teeth **227**

226b Cells cylindrical; poles truncate, smooth or with a circle of low granules, or teeth, but not lobed; the margins smooth or undulate, *or* with rectangular thickenings on the face of the semicells (one species with fine spines on the wall; sometimes hardened extrusions from mucilage pores appear as spines). Fig. 171 . . . . *Pleurotaenium*

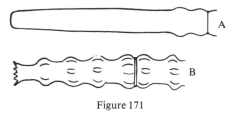

Figure 171

Figure 171 *Pleurotaenium.* (A) *P. trabecula* (Ehr.) Näg.; (B) *P. nodosum* Bailey.

There are more species and a greater variety of shapes in this genus than in *Docidium* (Fig. 169) and they are more widely distributed in types of habitats. Although all of them are elongate and usually have subparallel margins, there is considerable variation in details of the wall decoration. In some species the walls have nodes or are undulate; one spiny. Usually there is a circle of granules around the poles of the cell. Some species are not so restricted in their distribution as most desmids and may occur in basic or slightly alkaline waters as well as in acid or soft waters.

227a Poles of cells tri-lobed, the lobes bearing teeth. Fig. 172. . . . . *Triplastrum*

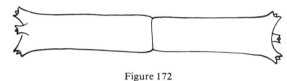

Figure 172

Figure 172 *Triplastrum indicum* Iyengar & Ram. (Redrawn from Iyengar and Ramanathan.)

The subcylindrical cells have a slight median constriction and truncate poles. At the apex are 3 short lobes, each bearing 2 to 4 spines. The chloroplast is longitudinally ridged and there are 3 or 4 pyrenoids. The genus is known only from India thus far.

227b Poles bi-lobed, each with 1 spine. Fig. 173 . . . . *Ichthyocercus*

Figure 173

Figure 173 *Ichthyocercus angolensis* W. & W.

This genus has subcylindrical cells and like *Triplastrum* (Fig. 172) is only slightly con-

stricted in the midregion. The poles are distinctly enlarged, however, and are slightly bi-lobed, and each bears a sharp, awl-like spine. There are 3 known species but they are rare. Conjugation in this genus has not been observed.

228a (223) Cells with 2 star-shaped, axial chloroplasts, 1 in each semicell. Fig. 82 . . . . *Cylindrocystis*

(Some species of *Cylindrocystis* have been transferred to *Actinotaenium,* a genus which also includes many species originally assigned to *Cosmarium.*

228b Cells with other types of chloroplasts 229

229a Cells short cylindric, or subcylindrical; chloroplasts 1 in each semicell (rarely with 3 or 4 chloroplasts in each semicell forming transverse zones in the protoplast), no terminal vacuoles with moving granules. Fig. 174 . . . . *Penium*

Figure 174

Figure 174 *Penium margaritaceum* (Ehr.) Bréb.

Some species of this genus are shaped somewhat like those in *Cylindrocystis* (Fig. 82), but *Penium* is a Placoderm desmid and has a wall in 2 sections that adjoin in the midregion. Also unlike *Cylindrocystis,* the walls may be punctate, poriferous, or bear rows of granules. In general *Penium* is more cylindrical than the latter genus, and because new wall sections are formed when cells divide they may become as long as some small species of *Pleurotaenium* (Fig. 171).

229b Cells slightly attenuated at the apices; chloroplasts with several pyrenoids; vacuoles with moving granules in the poles of the cell. Fig. 149 . . . *Closterium*

230a (221) Chloroplasts in the form of spiral ribbons 231

230b Chloroplasts some other shape. 232

231a Cells 'cigar'-shaped, the poles rounded, Fig. 175. . . . . *Spirotaenia*

Figure 175

Figure 175   *Spirotaenia condensata* Bréb.

The cylindrical cells in this genus (a Saccoderm desmid) may be slightly bent but usually are straight and 'cigar'-shaped with broadly rounded poles. A few have the poles tapering slightly. The coiled, ribbonlike chloroplast aids in making identification. Usually there are many pyrenoids.

231b Cells cylindrical with truncate poles. Fig. 176. . . . . *Genicularia*

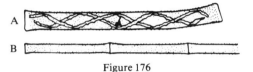

Figure 176

Figure 176 *Genicularia elegans* West. (A) single cell with spirally twisted chloroplasts; (B) filamentous arrangement.

In this genus the cylindrical cells may be solitary or adjoined end to end to form short, or relatively long filaments. Although the chloroplasts are spirally twisted and show a resemblance to *Spirogyra* (Fig. 229) this genus is usually identifiable by the cells being slightly enlarged at the poles, and also by the fact that the walls in most species are densely beset with short, sharp granules or short spines. *Genicularia* is a member of the Gonatozygonaceae, closely related to the true desmids and found associated with them in habitats. (See notes under *Gonatozygon* [Fig. 177].)

232a (230) Cells cylindrical, 10 or more times the diameter in length, the poles truncate; wall spiny or with sharp granules. Fig. 177 . . . . *Gonatozygon*

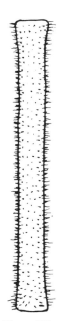

Figure 177

Figure 177   *Gonatozygon aculeatum* Hastings.

This genus has been placed in its own family, the Gonatozygonaceae, but in some systems of classification it has been assigned to the Mesotaeniaceae. Electronmicroscope studies, however, have shown that the wall possesses true pores and that it has the structure characteristic of the Placoderm desmids which requires that it be transferred from the Saccoderm class. The cells are solitary and often a bit crooked. The poles are slightly enlarged and truncate. Most species have short and stout, or long and fine spines. The chloroplast is a somewhat spiral or twisted ribbon that is axial rather than parietal as in *Genicularia* (Fig. 176).

232b   Cells fusiform, short-cylindric, less than 10 times the diameter in length; wall smooth 233

233a   Cells broadly fusiform or subcylindric; 2 chloroplasts, 1 in each half of the cell, bearing longitudinal ridges and with notched margins. Fig. 178 .... *Netrium*

Figure 178

Figure 178   *Netrium digitus* (Ehr.) Itz. & Roth, showing chloroplasts with fimbriate ridges.

These are 'watermelon'- or 'cucumber'-shaped cells which show the characteristics of the Saccoderm desmids. The wall is in one piece and only rarely is there a slight median invagination. Like other members of the Mesotaeniaceae, the cell-contents are divided into 2 portions, with 2 large, conspicuous and axial chloroplasts; the nucleus median. The chloro-

plasts are usually distinctly ridged with fimbriate margins. There are several pyrenoids in
each chloroplast.

**233b Cells cylindrical or narrowly fusiform,
with 1 chloroplast..............234**

**234a Cells 'cigar'-shaped, bowed; chloroplast
axial with from 4 to 6 pyrenoids.
Fig. 179.....................*Roya***

Figure 179

Figure 179   *Roya obtusa* (Bréb.) W. & W.

This rather rare Saccoderm desmid has slightly
curved, cylindrical cells in which there is but a
single chloroplast that is notched in the
midregion where the nucleus is located. There is
a row of pyrenoids. *Roya* often occurs in
subalpine regions; among wet mosses; also
found in mixtures of desmids from acid habitats occasionally. Four species have been reported from the United States.

**234b Cells elongate-ellipsoid or ovoid to
subcylindric; 1 parietal chloroplast, cell-
contents violet-colored. Fig. 83 ....
...................*Mesotaenium***

**235a (212) Cells constricted in the
midregion 236**

**235b Cells not constricted in the
midregion 247**

**236a Cells flat and nearly circular in
proportions; starlike in front view, the
median incision very deep; semicells also**

**deeply lobed or incised, in some spe-
cies with secondary lobes and lobules.
Fig. 180 ....*Micrasterias***

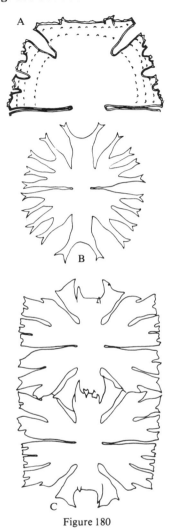

Figure 180

Figure 180 *Micrasterias*. (A) *M. americana*
var. *boldtii* Gutw.; (B) *M. radiata* Hass.;
(C) *M. foliacea* Bailey, portion of a filament.

This genus is a member of the true or
Placoderm desmids and includes some of the
most beautiful of microscopic objects. Al-

though the outline of the cell varies greatly among the 40 or more species reported from the United States, they can be identified by the flat, disclike shape. The sinus is deep and the semicells are variously lobed, sometimes with the lobes again incised so that arms are formed. The wall may be smooth or spiny and the lobes and lobules tipped with spines or not. When seen from the top or side the cells are much compressed. One species, *Micrasterias foliacea* Bailey (Fig. 180) has hooks on the polar lobes which enmesh with those of newly formed cells so that false filaments result. Most species are to be found in soft or acid-water ponds and *Sphagnum* bogs.

236b   Cells not flat and disclike 237

237a   Cells with shallow and broad, or a deep and narrow notch at the apex of the semicells. Fig. 181 . . . . *Euastrum*

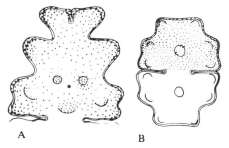

Figure 181

Figure 181 *Euastrum.* (A) *E. pinnatum* Ralfs; (B) *E. pectinatum* var. *inevolutum* W. & W.

There are numerous species of this Placoderm desmid, varying greatly in size, shape and type of wall oranamentation. Most of them, however, can be identified readily by the polar notch and by the protrusions or swellings on the face of the semicell. The cells are compressed when seen from the side, but show the facial swellings. The walls are often granular and, ac-

cording to species, may have definite mucilage pores (pore-organs). The generic as well as specific features are best seen in empty cells. Like *Micrasterias* (Fig. 180) most species prefer acid habitats.

237b   Cells without a notch in the apex of the semicell 238

238a   Apex of the cell extended into 3 or more lobes or arms, radiately extended into 3 or more planes so that the cells are star-shaped or triangular to 10-rayed in top view 239

238b   Apex of semicell not extended into arms, *or* if with arms only 2, and not extending in more than 1 plane 242

239a   Semicells various, transversely oval, pyramidal or urn-shaped, the apex only slightly if at all elevated; semicells and arms either smooth or variously ornamented with spines and verrucae; the two poles of the cell similar in morphology and arm arrangement. Fig. 182 . . . .
. . . . *Staurastrum*

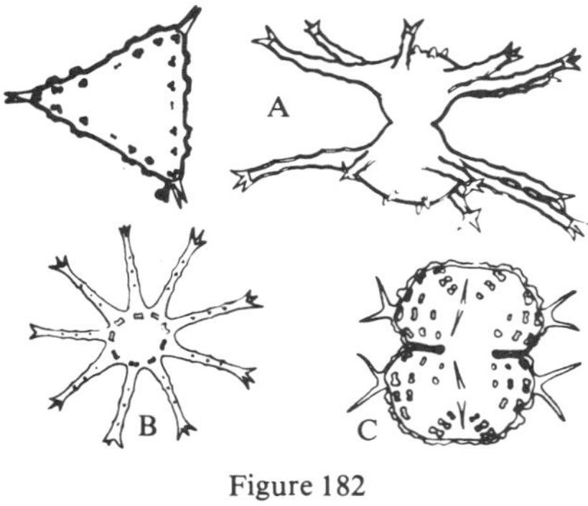

Figure 182

Figure 182 *Staurastrum.* (A) *St. rotula* Nordst. front view and (B) vertical view; (C) *St. cornutum* Arch. front view and (D) vertical view.

This genus of Placoderm desmids includes almost as many species as the prolific genus *Cosmarium* (Fig. 187). There is a tremendous variation in shape, size and ornamentation. The most distinctive feature is the extension of the semicell into 3 or more planes (radiating arms, lobes and processes) so that the cells are seen to be triangular in end view or to have as many as 10 rays. In one section of the genus the semicells have arms in one plane only (horizontal extensions of the semicell), but the shape of the arms and the type of spiny ornamentation identify such species with *Staurastrum*. Many species appear like *Cosmarium* (Fig 187) when seen in front view and one must focus carefully to see the arms or lobes of the semicell extending toward or away from the observer. Some species have long arms, especially among the euplanktonic species. Most are to be found in acid lakes and bogs, whereas a few are not selective and may occur in basic waters. (The smooth-walled species of *Staurastrum* and some species in *Arthrodesmus* [Fig. 188] have been assigned by some students to the genus *Staurodesmus*).

239b **Semicells variously shaped and variously lobed, but the apex or apical region distinctly elevated, the 2 poles of the cell not always the same in morphology and arm arrangement** **240**

240a **Semicells 3-lobed as seen in front view, the apical lobe elevated or protruding, with its lateral margins subparallel, the cell from 3- to 8-lobed (rarely 10-lobed as seen in end view which in outline is circular; poles of the cell similar in morphology; wall densely and uniformly granular. Fig. 183** . . . . *Euastridium*

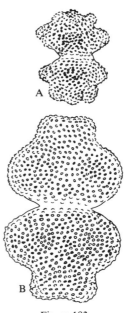

Figure 183

Figure 183 · *Euastridium*. (A) *E. verrucosum* Carter; (B) *E. prainii* W. & W.

This genus of desmids has features which resemble both *Euastrum* (Fig. 181) and *Staurastrum* (Fig. 182). The semicells are radiate, having lobes in several planes, and wall markings similar to *Euastrum*. The apex, however, is not notched as in the latter genus. There are only 2 or 3 species, widely distributed over the world, but especially in alpine regions.

240b **Cells shaped differently; wall not uniformly granular, the 2 poles of the cell different from one another morphologically** **241**

241a **Apex of one pole elevated and ornamented with a crown of granules;**

the other pole convex, smooth; semicells elliptic with 2 encircling series of long, spiny arms, one series near the base and one near the apex; end view circular in outline with 2 series of radiating arms, 10 in the lower series, 6 in the upper. Fig. 184. . . . . *Amscottia mira*

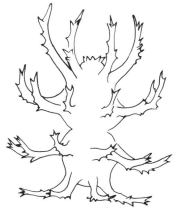

Figure 184

Figure 184   *Amscottia mira* Grönbl. & Kallio.

This is another example of dichotypical morphology (See *Ichthyocercus,* Fig. 173.), one semicell being different from the other, especially in the morphology of the polar lobe. The semicells are radiate with at least 10 spiny arms as seen in end view. The genus most nearly resembles *Staurastrum* (Fig. 182). Thus far only 1 species is known, from Africa, but it may be expected in other tropical or subtropical regions. The genus was named for the recent Arthur M. Scott.

241b   Semicells deeply 3-lobed, the apical lobe extended and furnished with a circle of spines; the lateral lobes terminating in

spines; the two poles different from one another in the arrangement of spines, one pole being bi-lobed; in end view irregularly radiate with the lobes divided and spiniferous at their tips. Fig. 185. . . . . *Allorgeia valiae*

Figure 185

Figure 185 *Allorgeia valiae* Gauth.-Liev. (Redrawn from Gauthier-Lievre.)

It is interesting that dichotypical desmids seem always to be collected from tropical regions. *Allorgeia,* another African genus, is distinctly dichotypical, with one semicell having lateral lobes at right angles to the long axis of the cell, those of the other semicell upwardly directed and bearing additional spines. The apical lobes of the two semicells also are different. The genus is named for the French phycologist Allorge.

242a   (238) Semicells with 2 extended arms at their apices as seen in front view, narrowly elliptic or fusiform when seen from the top. Fig. 186 . . . . *Staurastrum*

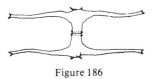

Figure 186

Figure 186 *Staurastrum leptocladum* Nordst., a species with biradiate semicells.

The species illustrated is an example of those few members of the genus *Staurastrum* which have arms of the semicell directed in 1 plane only. Most species are euplanktonic in soft-water lakes.

242b   **Cells without radiating arms; semicells compressed and broadly oval when seen from the top (sometimes semicells nearly round when seen from the side) 243**

243a   **Margins of semicells furnished with spines or processes 244**

243b   **Margin of cell without spines, smooth, or granular, the granules sometimes sharp or conical. Fig. 187 . . *Cosmarium***

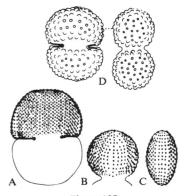

Figure 187

Figure 187 *Cosmarium*. (A) *C. panamense* Presc.; (B) lateral view and (C) vertical views; (D) *C. margaritatum* (Lund.) Roy & Biss., front and lateral views.

Like the genus *Staurastrum* (Fig. 182), *Cosmarium* includes thousands of species, more than any other genus in the Chlorophyta. There is accordingly considerable variation in shape, size and wall ornamentation. Whatever the shape of the semicell in face view, it usually is compressed, oval or elliptic when seen in vertical view. The sinus may be deep or only a slight invagination. Like all members of the Conjugales, the chloroplasts are few, large and showy, 1 or 2(4) in the semicell and each with a pyrenoid. One group of species is considered by some students to constitute the genus *Dysphinctium* (cells round in end view); others are grouped to form the genus *Actinotaenium* (cells round in end view with axial, radiate chloroplasts).

244a   **Face of semicell with protuberances, or with the wall thickened in the midregion, sometimes with knobs, best seen when the cell is rolled to a lateral view position 245**

244b   **Face of semicell without swellings or protuberances. Fig. 188. . *Arthrodesmus***

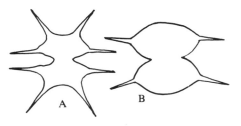

Figure 188

Figure 188 *Arthrodesmus*. (A) *A. octocornis* Ehr.; (B) *A. fuellebornei* Schm.

This genus of Placoderm desmids has compressed cells like *Cosmarium* (Fig. 187), narrower in side or end view than in front view. The angles of the semicells bear spines; the cell wall smooth, there being no granules, protuberances or pits. Most species occur in soft-water habitats, intermingled with other desmids. (See notes under *Staurastrum* [Fig. 186].)

245a **Semicells with lateral, either simple or bifurcated processes and with a pair of spines one on each angle of the apical margin; the face of the semicell with a median protrusion that is often granular; the cells compressed in side and vertical views. Fig. 189. . .** *Spinocosmarium*

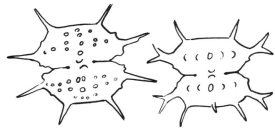

Figure 189

Figure 189 *Spinocosmarium quadridens* (Wood) Presc. & Scott, two forms.

This genus was erected to include those plants which have a combination of both *Cosmarium* and *Arthrodesmus* features plus some *Staurastrum* characteristics. The cells are compressed when seen from the side. The type species, *S. quadridens* (Wood) Presc. et Scott was formerly referred to *Arthrodesmus*.

245b **Cells otherwise, with a different type of ornamentation. 246**

246a **Apex of semicell furnished with prominent spines; facial protuberance (if any) one large, low swelling, the wall thickened here and often pitted or punctate. Fig. 190 . . . .** *Xanthidium*

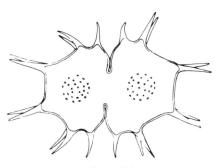

Figure 190

Figure 190 *Xanthidium cristatum* var. *uncinatum* Hass.

This genus, like *Arthrodesmus* (Fig. 188) has cells which are compressed (oval or elliptic in end or side view). There is usually a facial swelling of the semicell which may be pitted. The angles of the semicells bear 1 or a pair of spines. There is less variation in the shape of the semicell and wall ornamentation in this genus than in many other Placoderm desmids.

246b **Apex without spines, or with a short, toothlike spine on either side of the apical margin. Fig. 181. . . . . .** *Euastrum*

247a **(235) Cells spherical, enclosed by a spindle-shaped envelope which has longitudinal ridges. Fig. 166 . . . .**
. . . . *Desmatractum*

247b **Cells variously shaped but not enclosed in such an envelope 248**

248a **Cells oval, ovoid, spherical or ellipsoid. 249**

248b **Cells angular, pyramidal, trapeziform or polygonal 285**

249a Cells subcylindric, or ovoid, small, less than 4.5 μm in diameter, with a parietal, platelike chloroplast at one or both ends. Fig. 191 . . . .
. . . . *(Nannochloris) Diogenes*

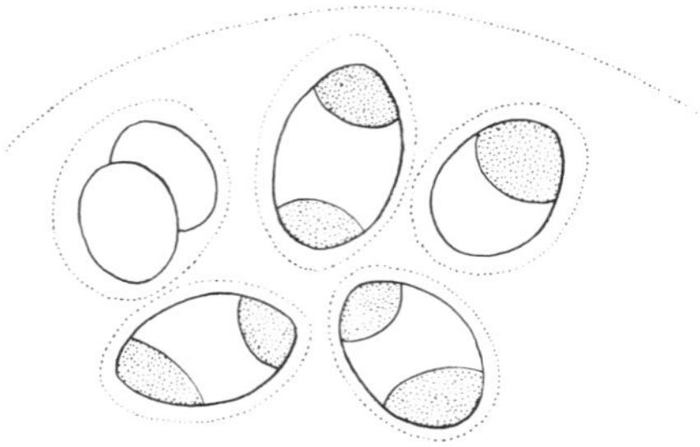

Figure 191

Figure 191 *Diogenes bacillaris* (Naum.) Bourr., a few cells from the margin of a gelatinous colony.

These small, bacilliform or oval cells usually occur in large numbers within a gelatinous mass. The cells, individually or in pairs, are enclosed by lamellate sheaths. The cells are able to undergo cell division in vegetative reproduction and hence are assignable to the Coccomyxaceae along with *Elakatothrix* (Fig. 88) and *Dactylothece* (Fig. 93). It is a frequenter of laboratory cultures. The genus name is regarded by some as synonymous with *Nannochloris*.

249b Cells different in size and shape, or with a different type of chloroplast 250

250a Cells bearing spines or ornamented with ridges, knobs, or granules 251

250b Cells without spines or decorations . . . .
. . . . 262

251a Cells oval, in pairs, bearing numerous, slender spines with thickened bases. Fig. 75. . . . . *Dicellula*

251b Cells otherwise, not so arranged or ornamented 252

252a Cells spherical, solitary or in clusters, free-floating; wall thick and marked with refringent alveolae. Fig. 143 . . . . .
. . . . *Keriochlamys*

252b Cells otherwise. 253

253a Cells with spines greater in length than the diameter. 254

253b Cells with spines shorter than the diameter of the cell; wall often decorated with a network of thickenings 260

254a Spines not tapering from base to apex (with parallel margins, but pointed at the tip), long and slender 257

254b Spines tapering from base to apex (thornlike), long and slender, or short and thick 255

255a Spines stout, broad at the base and tapering to a sharp point, evenly distributed over the cell. Fig. 192. . . . .
. . . . *Echinosphaerella*

Figure 192

Figure 192 *Echinosphaerella limnetica* G.M. Smith.

This is a relatively rare plant from the euplankton. In making identification care should be used to distinguish the single, parietal chloroplast by which the cell may be differentiated from some of the spiny zygospores of desmids (in which the cell content appears dark and massive, with no definitely shaped chloroplast distinguishable). Only 1 species is known, reported from several stations in the United States, including Alaska.

255b   Spines long and slender, thicker at the base and then abruptly narrowed . . . 256

256a   Cells oval, usually in pairs; spines numerous, needlelike but with a short section near the base thickened. Fig. 75. . . . . *Dicellula*

256b   Cells spherical, spines long and slender, with a distinct basal section thickened, then abruptly narrowed to a fine bristle. Fig. 193 . . . . *Acanthosphaera*

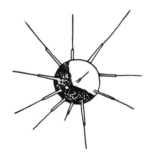

Figure 193

Figure 193 *Acanthosphaera zachariasii* Lemm.

This plant can be distinguished from *Echinosphaerella* (Fig. 192) because the spines are somewhat needlelike, arising from a base which is decidedly thicker than in the apical section. There is a parietal, platelike chloroplast. Euplanktonic.

257a   (254) Cells round 258

257b   Cells oval or ellipsoid 259

258a   Chloroplast with a distinctly reniform pyrenoid. Fig. 144 . . . . *Golenkinia*

258b   Chloroplast with a round pyrenoid. Fig. 145 . . . . *Golenkiniopsis*

259a   (257) Spines at the poles, or at the equator of the cell. Fig. 194 . . . . *(Lagerheimia) Chodatella*

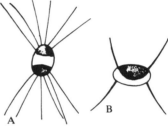

Figure 194

Figure 194 *Chodatella (Lagerheimia).* (A) *C. longiseta* (Lemm.) Printz; (B) *C. quadriseta* (Lemm.) G.M. Smith.

Unlike *Franceia* (Fig. 131) cells of this genus have long, needlelike spines confined to the poles, or to the poles and the equator. There are 13 species reported from the United States which are differentiated on the basis of cell shape and arrangement of spines. All are fairly common in the euplankton, and widely distributed. The genus should be compared with *Bohlinia* (Fig. 131A).

259b   Spines distributed over the cell wall. Fig. 131 . . . . *Franceia*

260a   (253) Cells round 261

260b Cells oval. Fig. 131A . . . . *Bohlinia*

Although sometimes found solitary, cells in this genus are mostly embedded in an amorphous mucilage. The wall bears fine, short spines which are more numerous near the poles of the ellipsoid cells than in the midregion. By some, *Bohlinia* is considered to be synonymous with *Franceia* in which the spines are evenly distributed. Thus far only *B. echidna* (Bohlin) Lemm. has been reported from this country.

261a Cells with bluntly rounded protuberances when mature (smooth-walled when young), pale green (sometimes almost colorless), inhabiting snow fields. Fig. 195. . . . . *Mycanthococcus*

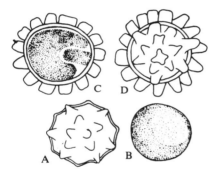

Figure 195

Figure 195 *Mycanthococcus antarcticus* Wille. (A) cell showing thick, spiny wall of cyst; (B) chloroplast; (C), (D) *M.* sp. from Olympic Mountain snowfields.

This is one of several genera of unicellular algae which occur in banks of permanent snow. There is a light green, parietal chloroplast. The cell wall is smooth when the cells are young, but becomes thickened and bears blunt protuberances in age, thus showing a resemblance to certain species of *Trochiscia* (Fig. 196). The species illustrated here has been found in the snows of Yellowstone National Park and in the

Olympic Mts. There are 3 or 4 species known, one of which has been collected from caves in Europe. Future studies may reveal that some records of this plant may have been confused with zygospore stages of *Chlamydomonas*.

261b Cells decorated otherwise, with ridges, reticulations, sharp spines, or with rounded protuberances; chloroplast definite, green (some species occurring in snow). Fig. 196 . . . . *Trochiscia*

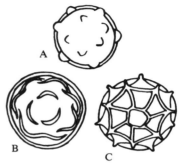

Figure 196

Figure 196 *Trochiscia*. (A) *T. granulata* (Reinsch) Hansg.; (B) *T. obtusa* (Reinsch) Hansg.; (C) *T. reticulata* (Reinsch) Hansg.

262a (250) Cells associated with fungi to form lichens. Fig. 197 . . . . *Trebouxia*

Figure 197

Figure 197 *Trebouxia cladoniae* (Chod.) G.M. Smith, showing axial chloroplast.

The species illustrated is one of the most common forms associated with fungi to form lichens. The cells are spherical and contain an axial rather than a parietal chloroplast that is characteristic of most members of the Chlorococcales.

262b Cells not associated with fungi in lichens 263

263a Chloroplast 1, axial and central, sometimes with radiating lobes extending to the wall 264

263b Chloroplast not axial, *or* more than 1 in a cell, without radiating lobes 267

264a Cells egg-shaped, with a flotation cap at the narrow end; cells neustonic, floating beneath the surface film of water; chloroplast axial. Fig. 198 . . . . . *Nautococcus*

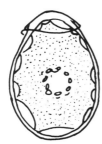

Figure 198

Figure 198 *Nautococcus piriformis* Korsch., showing flotation cap. (Redrawn from Starr.)

This unusual genus is one of the few neustonic algae, floating and suspended from the surface film. The cells are adapted by having a flotation cap at the smaller end of the elliptic or pyriform cell. There is 1 axial chloroplast and a central pyrenoid. Reproduction is by zoospores. Apparently there is but 1 species reported, rather rare but widely distributed.

264b Cells not neustonic, without a floatation cap 265

265a Chloroplast definitely star-shaped, with a central pyrenoid; cells spherical. Fig. 99 . . . . *Asterococcus*

265b Chloroplast not star-shaped; cells not spherical 266

266a Cells cylindrical or subcylindric; chloroplast an axial plate. Fig. 83 . . . . *Mesotaenium*

266b Cells spheroidal to pyriform, with lateral lobes; chloroplast irregularly lobed. Fig. 199. . . . . *Myrmecia*

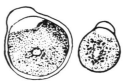

Figure 199

Figure 199 *Myrmecia aquatica* G.M. Smith. (Redrawn from G.M. Smith.)

These unicells are either spherical or somewhat pyriform and usually show a thickening of the wall at one side, giving them an unsymmetrical shape. Although the genus was originally described from subaerial situations, specimens in the United States have been collected from aquatic habitats.

267a (263) Cells spherical 268

267b Cells oval or ellipsoid (sometimes nearly round) 275

268a Cells large (up to 300 μm or more in diameter); wall thin; chloroplasts numerous, irregular in shape and lumpy with starch grains, arranged in radiating strands from the center of the cell, and also parietal. Fig. 200 . . . *Eremosphaera*

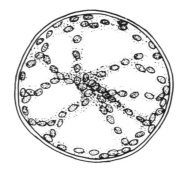

Figure 200

Figure 200 *Eremosphaera viridis* De Bary, showing radiately arranged and parietal chloroplasts.

This is one of the largest spherical cells (up to 300 μm, rarely 800 μm in diameter) encountered among unicellular algae. Although usually solitary, these green globes may appear in clusters of 4 enclosed by the old, mother-cell wall. There are 5 recognized species, one of which represents a transfer of *Oocystis eremosphaeria* G.M. Smith. The cells have numerous disclike chloroplasts in radiating cytoplasmic strands from a central region, as well as lining the wall. They are lumpy with attached starch grains. The cells occur mostly in soft-water habitats and are associated with desmids (at one time were thought to be a desmid).

268b   Cells not as above 269

269a   Cells enclosed in mucilage; individual cells usually with lamellate sheaths. . 270

269b   Cells not enclosed in mucilage 272

270a   Cells usually solitary (sometimes gregarious), enclosed eccentrically in a much-lamellated sheath; cell contents indefinite and chloroplast not clear; pyrenoid lacking; cells brownish or orange-colored. Fig. 18. . . . . *Urococcus*

270b   Plants otherwise 271

271a   Cells sometimes solitary but mostly many together in an irregularly globular, mucilaginous matrix which is lamellate; individual cells usually showing a lamellate sheath; cells mostly solitary within the colonial mucilage and irregularly scattered (not in pairs); cell with a cup-shaped, parietal chloroplast with a pyrenoid; contractile vacuoles present; colonies mostly aquatic (sometimes on moist subaerial strata). Fig. 86 . . . . *Gloeocystis*

271b   Cells globular or oval, many within an amorphous, gelatinous matrix which is homogeneous (not lamellate, but individual or paired cells showing thin, lamellate sheaths); cells without contractile vacuoles; plants terrestrial and the colony macroscopic. Fig. 12 . . . . *Palmella*

272a   (269) Chloroplast 1 (rarely 2); cells solitary, although sometimes gregarious but not colonial 273

272b   Chloroplasts several to many parietal, angular plates; pyrenoids lacking. Fig. 201. . . . . *Bracteacoccus*

Figure 201

Figure 201 *Bracteacoccus minor* (Chodt.) Petrova, with angular, parietal chloroplasts.

This genus resembles *Planktosphaeria* in the form of its flat, angular and parietal chloroplasts, but the cells occur singly rather than in colonies. The plants are easily confused with other 1-celled, spherical algae which grow in or on soil. There are about 7 species which have been isolated and cultured.*

273a **Cells associated to form an extended stratum on trees, wood and stones. Fig. 122** . . . . *(Protococcus. Pleurococcus) Desmococcus*

273b **Cells not forming such a stratum** . . . **274**

274a **Chloroplast a thin layer along the wall; pyrenoid usually lacking; free-living or in tissues of animals (sponges, *etc.*); reproduction by autospores (replicas of the adult cell. Fig. 156** . . . . *. . . Chlorella* including *Palmellococcus)*

274b **Chloroplast a massive cup with a pyrenoid; occurring in soil or on damp subaerial substrates (probably occurs in water also); reproducing by zoospores. Fig. 87** . . . . *Chlorococcum**

275a **(267) Cells oval, ovate, or irregularly globose, with thick-layered walls bearing knobs and prominent protrusions** . . **276**

275b **Cells without thick walls and knoblike protrusions** **277**

276a **Chloroplasts numerous, parietal, cone-shaped. Fig. 202** . . . . *Excentrosphaera*

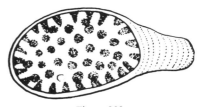

Figure 202

Figure 202 *Excentrosphaera viridis* G.T. Moore.

This is the only species reported for the genus. It is found both in the water and in subaerial habitats and is identified by its irregular shape produced by lamellated thickenings of the wall in 1 or more places. The chloroplasts are cone-shaped and are all directed inwardly from the parietal position along the wall.

276b **Chloroplast a massive, axial body with processes which are flattened against the wall. Fig. 203** . . . . *Kentrosphaera*

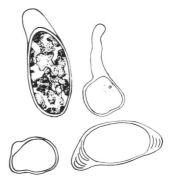

Figure 203

Figure 203 *Kentrosphaera bristolae* G.M. Smith.

There are 2 or 3 species of this genus (sometimes included in the genus *Chlorochytrium* [Fig. 160]). *Kentrosphaera bristolae* G.M. Smith and *K. facciolae* Borzi have been

*See Appendix for other unicellular green algae, p. 268.

found in the United States. The cells are similar in shape to *Chlorochytrium* but have a free-living habit, usually occurring on damp soil; *K. facciolae* has been found in Antarctica. Because of the processes of the chloroplast that are flattened against the wall the cells may appear to have several irregularly shaped chloroplasts.

277a **(275) Cells with spines 278**

277b **Cells without spines 280**

278a **Spines distributed over the wall 279**

278b **Spines localized at the poles, *or* at the midregion of the cell. Fig. 194 . . . . *(Lagerheimia) Chodatella***

279a **Spines shorter than the diameter of the cell. Fig. 131A . . . . *Bohlinia***

279b **Spines as long as or usually longer than the diameter of the cell, needlelike; cells solitary or in 2's. Fig. 131. . . . . *Franceia***

280a **(277) Cells with spiral longitudinal ribs on the wall. Fig. 204 . . . . *(Chloromonas) Scotiella****

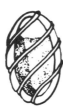

Figure 204

Figure 204 *Scotiella (Chloromonas ?) nivalis* (Chod.) Fritsch.

At least some species of this genus probably will be found to be the resting stage of *Chloromonas* (Fig. 46) and possibly other members of the Volvocales. A number of species have been described, most of which have been collected from red snow at high altitudes. Differences lie in the shape of the cell (usually oval) and the type of decorations on the wall. Occasionally *Scotiella* species are collected in the tychoplankton at low altitudes. Some authorities have suggested placing the genus in the Volvocales because of the type of chloroplast and the evidence of a basal-distal differentiation in the cell; whereas others include it in the Chlorococcales.

280b **Cells without longitudinal ribs 281**

281a **Cells small, less than 4.5 μm long, with a platelike chloroplast at one end. Fig. 191 . . . . . *(Nannochloris) Diogenes***

281b **Cells relatively large, without such a chloroplast arrangement 282**

282a **Two or more masses of dark mucilage appearing at either end, or making X-bands over the mother-cell wall of the previous generation; usually 2 or 4 cells in the old, mother-cell, but cells frequently found solitary. Fig. 142. . . . . *Gloeotaenium***

282b **Without such dark masses of mucilage about the cells 283**

283a **Cells enclosed in a mucilaginous sheath,**

*At least two species of *Scotiella* (*S. nivalis* [Chod.] Fritsch and *S. polyptera* Fritsch) have been shown to be the dormant stage of *Chloromonas* and *Cryocystis* in the Volvocales.

cells relatively small and numerous within a gelatinous investment; cells round or oval...................... **284**

283b Cells not enclosed in a mucilaginous sheath, but usually several together enclosed in the old, mother-cell wall which may be expanded and gelatinized (appearing upon first observation to be a sheath), sometimes solitary; cells oval, ellipsoid or lemon-shaped; chloroplasts 1, or 2, or in a few species many parietal plates or irregular discs; reproduction by autospores. Fig. 146......... *Oocystis*

284a Cells oval, often solitary but usually gregarious, forming a mucilaginous expanse on moist, subaerial substrates. Fig. 156 . . . . *(Palmellococcus) Chlorella*

284b Cells solitary, aquatic, ellipsoid, enclosed in a very wide, gelatinous, brownish sheath which is impregnated with iron. Fig. 205 . . . . *Sphaerellocystis*

Figure 205

Figure 205 *Sphaerellocystis ellipsoidea* Ettl. (Redrawn from Ettl.)

The cells appear like a non-motile *Chlamydomonas,* enclosed in a wide, ample, gelatinous sheath. The sheath bears nodules of iron deposits. There are 4 species, not yet reported from the United States but are expected.

285a **(248) Cell body actually spherical but**

with 4 long, narrow, brown armlike appendages radiating from it. Fig. 206 . . . . *Pachycladon*

Figure 206

Figure 206 *Pachycladon umbrinus* G.M. Smith. (Redrawn from G.M. Smith.)

This rare plant, 1 species only known for the genus, occurs in the euplankton of lakes. The long, darkly colored appendages from a relatively small, subspherical cell body make identification certain.

285b Cells differently shaped, or not with such appendages **286**

286a With several long spines forming a tuft at the angles of the cell. Fig. 207 . . . . *Polyedriopsis*

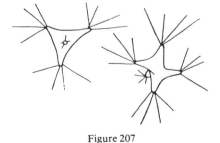

Figure 207

Figure 207 *Polyedriopsis spinulosa* (Schm.) Schmidle.

There are 3 species in this genus, all of which are euplanktonic. They are rectangular or polyhedral in shape, with from 1 to 4 long spines at each angle. *Polyedriopsis quadrispina* G.M. Smith has but 1 stout spine at the angles; is quadrate in shape. Some species of *Tetraedron* (Fig. 209) have been transferred to *Polyedriopsis*.

286b   With 1, 2 or 3 spines at the angles, *or* with papillate mucros at the tips of the angles **287**

287a   Without long spines, but with teeth or papillate mucros at the tips of angles or at the apex of arms. **288**

287b   With 1 spine at each angle of the cell **289**

288a   Body of the cell polyhedral, gradually narrowed at the angles to form hornlike, twisted processes tipped with teeth. Fig. 208 . . . . *Cerasterias*

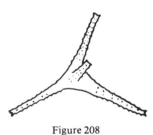

Figure 208

Figure 208   *Cerasterias irregulare* G.M. Smith.

Apparently there is only 1 good species in this genus although 3 or 4 have been recognized. *Cerasterias irregulare* G.M. Smith is characterized by being irregularly pyramidal, with twisted processes or arms at the angles. Thus far the plant has been reported only from lakes in mid-America. It should be compared with *Pseudostaurastrum* (Fig. 343), a genus

which contains pyramidal cells but belonging to the Chrysophyta.

288b   Body of semicell not narrowed to form arms, but polyhedral or pyramidal with lobes and broad protrusions. Fig. 209 . . . . *Tetraedron* (p.p.)

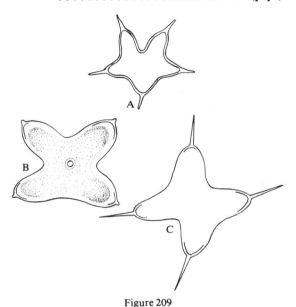

Figure 209

Figure 209   *Tetraedron.* (A) *T. caudatum* (Corda) Hansg.; (B) *T. minimum* (A. Br.) Hansg.; (C) *T. incus* (Teil.) G.M. Smith.

Species of *Tetraedron* are pentaedrical, pyramidal or polyhedral, the angles tipped with a spine or mucro. Morphologically they cannot be differentiated in a key to genera from the genus *Polyedriopsis*. This genus is established to group those species which reproduce by zoospores; whereas *Tetraedron* reproduces by non-motile autospores. *Tetraedron* displays many geometric shapes as treated in many reference works, and has had many species referred to it, as many as 95 according to a recent summary. But many of these have been shown to belong to the genus *Pseudostaurastrum*

(Fig. 343) in the Xanthophyceae. In that genus there are several to many disclike, yellow-green chloroplasts, rather than a single parietal chloroplast with a central pyrenoid as in *Tetraedron*.

289a Cells symmetrically triangular or pyramidal, with a stout spine at each angle, in some the spines with an inflated base, length greater than the diameter of the cell. Fig. 210 . . . .
. . . . *Treubaria*

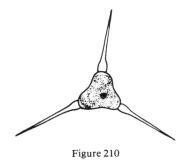

Figure 210

Figure 210   *Treubaria crassispina* G.M. Smith.

This genus should be compared with *Pachycladon* (Fig. 206) which has long, arm-like, brown-colored extensions of the angles. *Treubaria* cells are similar to *Tetraedron,* in the pyramidal or polyhedral shapes, but in that genus the spines are relatively short or only papillate mucros at the tips of the lobes. There is 1 parietal chloroplast with a central pyrenoid. Four species have been described from the United States.

289b Cells irregularly pyramidal; spines not stout 290

290a Cells irregularly lobed, polyhedral, the lobes relatively broad, the angles tipped with short spines that taper from base to apex. Fig. 209. . . . . . . *Tetraedron* (p.p.)

290b Cells polyhedral, with relatively narrow lobes; spines (in *P. spinulosa*) needle-like. Fig. 207. . . . . . *Polyedriopsis* (p.p.)

291a (81) Plant a microscopic, *unbranched* filament (structure of the plant not visible without a microscope); attached or free-floating, *or* if macroscopic (form of the plant determinable without a microscope) the thallus in the form of an expanded sheet, a tube; or a *branched,* arbuscular (treelike), gelatinous and beaded growth. (See Figs. 235, 297.) . . .
. . . . 292

291b Plant a *branched* filament; either a coenocytic tube without cross walls, microscopic with cells in 1 series, sometimes in a wide gelatinous sheath; *or* an attached prostrate filament or cushion; *or* a disclike thallus in which the branching habit is obscure because of closely appressed branches 359

292a Cells constricted in the midregion, with 2 large chloroplasts, one in either part of the cell. 293

292b Cells not constricted in the midregion 301

293a Cells adjoined by the interlocking of short, straight, hornlike or hooked processes, or by buttonlike knobs . . 294

293b Cells adjoined by their end walls, either along the entire apical surface or by armlike extensions of the cell apex . . 296

294a Cells adjoined by overlapping, slender horns or by hooked processes. 295

294b Cells adjoined by buttonlike processes at the apex. Fig. 211 . . . . *Teilingia*

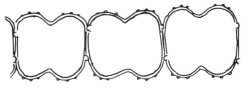

Figure 211

Figure 211 *Teilingia granulata* (Roy & Biss.) Bourr.

This genus of filamentous desmids is composed of those forms in which there are short, buttonlike apical projections on the apical wall. These adjoin similar projections of other cells to form filaments. Formerly the genus was included in *Sphaerozosma* (Fig. 212) in which there are longer, more prominent and hornlike processes which interlock to form filaments. The cells have about the same shape and proportions of some *Cosmarium*.

295a Interlocking processes in the form of forked lobes which bear recurved spines. Fig. 180 . . . . *Micrasterias foliacea*

295b Interlocking processes slender, straight and hornlike. Fig. 212. . . *Sphaerozosma*

Figure 212

Figure 212 *Sphaerozosma laeve* var. *latum* (W. & W.) Förster.

Cells in this filamentous desmid have relatively long apical processes which interlock with those of adjoined cells. The processes may bear knobs at their tips. The cells are similar to *Cosmarium* in shape and proportions. Included in this genus according to one classification scheme is *Onychonema* of other authors.

296a **(293)** Semicells transversely elliptic or oval, the median incision of the cell relatively deep. Fig. 213 . . . . *Spondylosium*

Figure 213

Figure 213 *Spondylosium planum* (Wolle) W. & W.

Although this genus can have cells that are triangular in end view, most species have cells that are compressed and are much like some species of *Cosmarium* (Fig. 187), arranged in a filament. One species which is rather rare is *S. pulchrum* (Bail.) Archer. It has semicells which are much extended laterally so that the cell is much wider than long. The apices of the cells in this species are furnished with a protrusion which adjoins that of adjacent cells in the filament. The walls are smooth and undecorated in this genus.

296b Cells not transversely elliptic; median incision not deep, sometimes only a slight concavity in the lateral wall. . . 297

297a Cells cylindrical, subcylindric, or barrel-shaped 298

297b Cells quadrate or angular, usually with the margins conspicuously lobed. (See Fig. 217 *Desmidium baileyi*, however.) 300

298a Cells barrel-shaped or subcylindric,

enlarged at the base of the semicells where there is a notch or sinus, the lateral margins either smooth or bearing a circle of spines on each semicell; wall with a ring of creases or folds near the apices of the cell; chloroplast axial, stellate as seen in end view. Fig. 214 . . . . . .
. . . . *(Gymnozyga) Bambusina*

Figure 214

Figure 214 *Bambusina brebissonii* Kütz. (*Gymnozyga moniliformis* Ehr.)

298b Cells other shapes, without a circle of creases near the apices; chloroplast stellate 299

299a Cells very slightly if at all inflated in the midregion, rectangular or subcylindric; median incision very shallow; often with rings of pores in the wall near the apices; chloroplast stellate. Fig. 215 . . . .
. . . . *Hyalotheca*

Figure 215

Figure 215 *Hyalotheca dissiliens* (Smith) Bréb.

In this filamentous desmid the median incision is merely a slight invagination and sometimes is obscure. The cell contents are divided, with an axial chloroplast in each semicell. The most common species have somewhat rectangular cells and one, *Hyalotheca mucosa* (Dill.) Ehr. is identified by having a wide, gelatinous sheath. In *H. dissiliens* (Smith) Bréb. there are transverse rings of puncta near the poles.

299b Cells cylindrical, slightly constricted in a median enlargement; chloroplast an axial plate or ribbon. Fig. 216 . . . .
. . . . *Groenbladia*

Figure 216

Figure 216 *Groenbladia neglecta* (Racib.) Teiling.

The long, cylindrical cells of this filamentous desmid are somewhat swollen in the midregion where there is a slight notch. The chloroplast, 1 in each semicell is an axial plate with 2 pyrenoids. One species placed in this genus has a whorl of spines on each semicell.

300a (297) Cells wider than long or as wide as long, without a median incision or with but a slight median notch; walls at the poles of young semicells infolded or replicated. Fig. 217 . . . . *Desmidium*

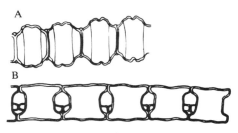

Figure 217

Figure 217 *Desmidium.* (A) *D. grevillii* (Kütz.) De Bary; (B) *D. baileyi* (Ralfs) Nordst.

Cells of this filamentous desmid vary much in shape. Some are oval and moniliform when seen in end view, some are triangular and some are quadrangular. When seen in front view (in filamentous arrangement) one must focus carefully upward and downward to detect the presence of lobes in the vertical plane. A characteristic habit of some species is to show a spiral twisting of the cell arrangement so that in any one view they do not have their lobes or processes in the same plane throughout the length of the filament. *Desmidium baileyi* (Ralfs) Nordst. has 3 or 4 polar processes which adjoin those of other cells in the filament so that there are fenestrations between the individuals. In such a species the cells are triangular or quadrangular in end view.

300b Cells a little longer than wide; rectangular with a shallow median incision; 4-lobed in end view; margins of the cells parallel; poles of young cells infolded. Fig. 218. . . . . *Phymatodocis*

Figure 218

Figure 218 *Phymatodocis nordestedtiana* Wolle, a section of a filament and vertical view of a cell to show lobing.

Although rarely found, this filamentous desmid may be the dominant form in some soft-water habitats which are especially favorable for these plants. The cells appear somewhat quadrangular when seen in front view as they occur in the filament, but are quadrilaterally symmetrical and are 4 lobed as seen in end view, the lobes sometimes curved.

301a (292) Chloroplast parietal, of various shapes, netlike, ringlike, or platelike, with pads and thin areas (Fig. 219); *or if* axial, plants in the form of a macroscopic thallus as in Fig. 236 316

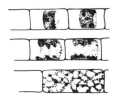

Figure 219

Figure 219 Three types of parietal chloroplasts. A complete parietal ring (above); an incomplete ring; a parietal network (below).

301b Chloroplast axial, an irregularly shaped plate, a broad band, or star-shaped; *or if* parietal, in the form of a ribbon as in Fig. 220; microscopic. 302

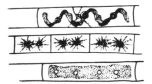

Figure 220

Figure 220 Three types of axial chloroplasts. A parietal ribbon as in *Spirogyra* (above); axial stellate as in *Zygnema;* an axial band or plate as in *Mougeotia* (below).

302a Chloroplast 1, or 2 in a cell; stellate, with radiating processes from a central core which includes a pyrenoid. 303

302b Chloroplasts other shapes 306

303a Cells quadrate, with 1 star-shaped chloroplast containing a single pyrenoid. Fig. 221. . . . .*Schizogonium*

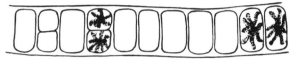

Figure 221

Figure 221 *Schizogonium murale* Kütz.

The species illustrated and 2 others are found in the United States, growing on dripping rocks or wet soil. *Schizogonium crenulatum* (Kütz.) Gay has short, crinkly filaments. The basically filamentous habit may become expanded so that a frondlike thallus is produced. This genus together with *Prasiola* (Fig. 236) have sufficient structural and reproductive characteristics to warrant placing them in a separate family (Schizogoniaceae), and the order Schizogoniales. The star-shaped chloroplast is helpful in making determinations.

303b Cells mostly longer than wide, with 2 or more chloroplasts 304

304a Chloroplasts from 2 to 6, relatively small and biscuit-shaped or somewhat

star-shaped, connected in the midregion of the cell by a strand of cytoplasm that encloses the nucleus; conjugating cells becoming filled with layers of pectic substances; zygospores cushion-shaped, compressed-spheroid or subquadrangular. Fig. 222 . . . .*Zygnemopsis*

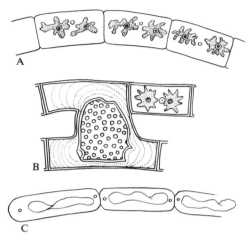

Figure 222

Figure 222 *Zygnemopsis*. (A) *Z. decussata* (Trans.) Trans. showing semistellate chloroplasts; (B) *Z. decussata* conjugation showing types of zygospore; (C) *Z. desmidioides* (W. & W.) Trans. with an axial platelike chloroplast.

The differences between this genus and *Zygnema* (Fig. 223) are to be found mostly within the reproductive habit; hence determination of the plants in the vegetative condition cannot be certain. In *Zygnema* the chloroplasts are definitely star-shaped, often with long rays, whereas in *Zygnemopsis* the chloroplasts are more padlike with a few, short radiating processes. The latter genus is much less common than *Zygnema;* has fewer species. In *Zygnemopsis desmidioides* (W. & W.) Trans. the chloroplast is an axial plate and not at all stellate.

304b Chloroplasts different in shape, *or* if star-shaped, larger than above and always 2; conjugating cells not becoming filled with pectic substances; zygospores globose, compressed-globose or oval 305

305a Chloroplasts 2, definitely star-shaped, each containing a large, central pyrenoid; aquatic. Fig. 223 . . . . . . *Zygnema*

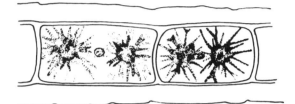

Figure 223

Figure 223 *Zygnema pectinatum* (Vauch.) C.A. Agardh, vegetative cells showing axial stellate chloroplasts and a central nucleus.

There are numerous species of *Zygnema,* differentiated on the basis of zygospore morphology (shape, size, wall markings). The paired, definitely star-shaped chloroplasts, each with a conspicuous pyrenoid, make identification possible. Frequently the cells are so densely packed with starch grains and cytoplasmic granules (waste material) that the shape of the chloroplasts is obscured. Application of an iodine solution often facilitates observations, or if one examines several lengths of filaments under low magnification the stellate form of the chloroplasts will become apparent. A few species have a conspicuous gelatinous sheath and a very thick wall. The filaments of *Zygnema* form green clumps and floating mats, but not the large, balloonlike or 'cloud' masses in the water as does *Spirogyra* (especially when submersed); usually are not as densely green as in that genus.

305b Chloroplast axial as above, but with radiating processes much reduced, the two chloroplasts sometimes bridged so as to form a dumb-bell shaped mass; terrestrial or subaerial. Fig. 224 . . . . *Zygogonium*

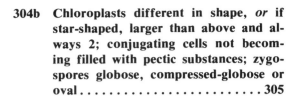

Figure 224

Figure 224 *Zygogonium ericetorum* Kütz.

These filaments are somewhat irregular because the cell walls are unevenly thickened and usually are invested by a layer of mucilage. The cells have the habit of putting out rhizoidal protrusions. Sometimes the protrusions are branched. Occasionally the conjugating tubes,when they fail to meet another tube, will continue to grow as rhizoidal processes. The plants are most commonly found in subaerial habitats; sometimes in water on submerged stumps, *etc.*

306a (302) Cell sap or chloroplast purplish to violet-green 307

306b Cell sap not purplish 309

307a With 2 disclike chloroplasts. Fig. 225 . . . . *Pleurodiscus*

Figure 225

Figure 225 *Pleurodiscus purpureus* (Wolle) Lag., showing disclike chloroplasts.

There is but 1 species of this genus reported thus far from the United States. It is a plant readily identified by the unique curved, plate-like chloroplast and purplish cell sap. In conjugation an oval or ellipsoid zygospore is formed within the tubes and enclosed by a collar. It is to be sought in subaerial habitats, seeping rocks.

307b   With 1 bandlike chloroplast 308

308a   Cells elongate-oval to subcylindric, in short interrupted filaments; pyrenoids 1 or 2; chloroplast often purplish or violet-green. Fig. 226 . . . . . *Ancylonema*

Figure 226

Figure 226 *Ancylonema nordenskioldii* Bergg.

The cylindrical cells with broadly rounded poles form short filaments, either continuous or interrupted. The 1 species is known from snow and glaciers; forms what is called "black snow" or "black ice." The cell contents are often violet-green, sometimes reddish. It is referred to the Desmidiaceae although conjugation has not been observed for complete confirmation. There is 1 twisted, bandlike chloroplast containing a pyrenoid.

308b   Cells long-cylindric, occurring in long filaments; chloroplast axial, bandlike, containing a row of conspicuous pyrenoids. Fig. 227 . . . . *Mougeotia*

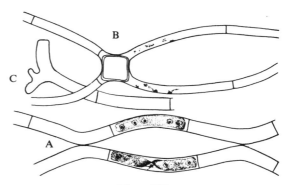

Figure 227

Figure 227 *Mougeotia*. (A) *M. genuflexa* (Dillw.) C.A. Agardh, showing geniculate or 'knee-bending' type of conjugation and the platelike, axial chloroplast; (B) *M. elegantula* Wittr., zygospore, with residues in the conjugating cells; (C) *M.* sp., showing rhizoidal branches.

309a   (306) Chloroplast in the form of a parietal, spiral or twisted ribbon, with many pyrenoids 310

309b   Chloroplasts axial bands or plates, pyrenoids none, or few 312

310a   Cell wall densely and minutely granular. Fig. 176 . . . . *Genicularia*

310b   Cell wall smooth 311

311a   Chloroplasts nearly parallel, only slightly twisted; conjugation without the formation of tubes between the gametangia. Fig. 228 . . . . *Sirogonium*

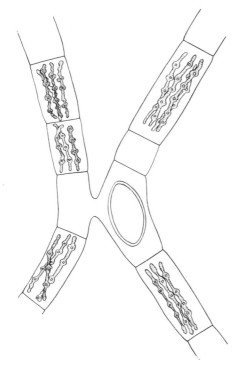

Figure 228

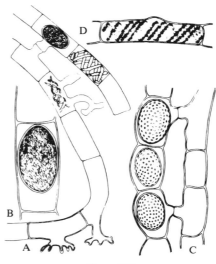

Figure 229.

Figure 228 *Sirogonium sticticum* (Engl. Bot.) Kütz., showing semiparallel chloroplasts and conjugation to form a zygospore.

This genus is differentiated from *Spirogyra* (Fig. 229) by the arrangement of the chloroplasts which are practically straight or only slightly twisted in the cell. Further, in *Zygogonium* conjugation is accomplished by geniculations of the filament rather than by tubes. The reproducing filaments are brought into juxtaposition by a bending of the reproductive plants.

311b Chloroplasts definitely spiralled; conjugation by the formation of tubes from one or both gametangia, either between two different filaments (scalariform conjugation) or between adjacent cells in the same filament (lateral conjugation). Fig. 229 . . . . *Spirogyra*

Figure 229 *Spirogyra*. (A) *Sp. rhizobrachialis* Jao, showing rhizoidal branches and conjugation; (B) *Sp. aequinoctialis* G.S. West; (D) single cell showing chloroplasts with pyrenoids in the nodes.

This is the most frequently encountered member of the order Zygnematales (Conjugales). There are more than 300 species, differentiated by a combination of vegetative characters and reproductive details, mostly in reference to the morphology of the zygospore and its wall markings. Some species have replicate (infolded) cross walls. Identification of *Spirogyra* species is not possible without mature zygospores. These plants form green 'clouds' of filaments below surface; usually appear as floating mats when entering upon the reproductive state at which time the plant mass become yellowish-brown. Conjugation may be either lateral or scalariform.

312a (309) Chloroplast a folded plate or short band, without a pyrenoid; conjugation scalariform; zygospores formed in the conjugation tube and extending into the gametangia. Fig. 230 . . . *Mougeotiopsis*

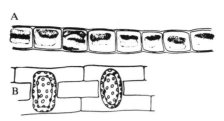

Figure 230

Figure 230 *Mougeotiopsis calospora* Palla. (A) vegetative cells with platelike chloroplasts (lacking pyrenoids); (B) mature zygospores.

This is a relatively rare genus, only 2 species of which are known for the United States. It is possible to make tentative identification because the chloroplast lacks a pyrenoid. The cells are characteristically short-cylindric. In reproduction it is similar to *Debarya* (Fig. 232) in that the entire contents of the conjugating cells (gametangia) become fused to form the zygospore.

312b **Chloroplasts with from 2 to several pyrenoids.** **313**

313a **Conjugating cells (gametangia) shortened by division of vegetative cells; plants relatively rare. Fig. 231** . . . .
. . . . *Temnogametum*

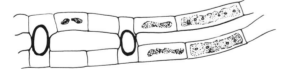

Figure 231

Figure 231 *Temnogametum* sp., showing platelike chloroplasts with pyrenoids, the gametangial cells cut off from the vegetative cells, and 2 zygospores formed within the conjugating tubes.

At the time of reproduction vegetative cells in this genus become divided, cutting off a small portion which serves as a gametangium. There is scarcely a tube formed between conjugating cells. The resulting zygospores are formed between the conjugating filaments, but extend into the two gametangia. The chloroplast is an axial band with pyrenoids either in a linear series, or scattered. Compare with *Debarya* (Fig. 232) which also has a chloroplast in the form of an axial plate (usually twisted), but in which the empty gametangial cells become filled with lamellated substance.

313b **Conjugating cells not shortened by division of vegetative cell** **314**

314a **Filaments composed of cylindrical cells with 1 or more axial ribbons and many pyrenoids; filaments formed by incidental adherence of cells following divisions; wall with short spines or sharp granules. Fig. 177** . . . . *Gonatozygon*

A recent study has shown that the cell wall in this genus contains pores; hence must be transferred from the Saccoderm to the Placoderm desmids.

314b **Filaments long and composed of permanently adjoined cells; walls smooth** **315**

315a **Filaments of slender cells, mostly under 12 μm in diameter (rarely as much as 30 to 42 μm); chloroplast an axial plate, usually not filling the cell; conjugating cells becoming filled with pectic substances; granular residues not found in the emptied reproductive cells; plants relatively rare. Fig. 232** . . . . . . *Debarya*

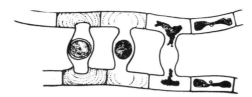

Figure 232

Figure 232 *Debarya* sp., showing formation of zygospores and the lamellated substance deposited in the conjugating cells.

This genus, named for the famous botanist De Bary, has filaments that resemble slender species of *Mougeotia* (Fig. 227) or *Temnogametum,* (Fig. 231) and cannot be positively identified in the vegetative condition. The cells are more delicate and the chloroplast is a relatively small, twisted plate. *Debarya* is much less frequently collected than *Mougeotia*. In reproduction all of the contents of the conjugating cells enter into the formation of the zygospore, and the space once occupied by the protoplasts becomes filled with lamellated pectic material which is light refractive.

315b **Filaments wider; cells long-cylindric or rarely short-cylindric; chloroplast a broad, axial band with conspicuous pyrenoids, usually filling the diameter of the cell, but not always in length; conjugating cells not filled with pectic substances; granular residues present in the emptied reproductive cells; plants common. Fig. 227 . . . . *Mougeotia***

316a **(301) Plant a tuft of short, erect filaments, usually branched but sometimes appear unbranched when young; some species forming prostrate, attached discs. Fig. 233 . . . . . *Coleochaete***

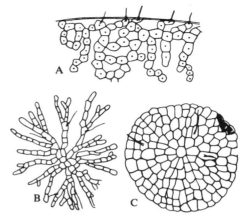

Figure 233

Figure 233 *Coleochaete.* (A) *C. nitellarum* Jost endophytic in *Nitella;* (B) *C. soluta* (Bréb.) Pringsh.; (C) *C. orbicularis* Pringsh.

There are 9 species of this genus reported from the United States. They are differentiated by the habit of growth (prostrate and disclike, or in erect tufts), and by the size of cells. One species, *Coleochaete nitellarum* Jost occurs only in the walls of *Nitella* (Fig. 3) which is nearly always found with the endophyte. *Coleochaete* appears conspicuously when *Nitella* is allowed to age in the laboratory. The sheathed setae which characterize *Coleochaete* arise from a granule (blepharoplast) within the cell and emerge through a pore in the wall. The disclike thallus formed by some species is frequently found on the sides of glass aquaria. In nature such colonies occur on other algae or on submersed stems of cattail, or on glass and crockery in the water.

316b **Thallus not in the form of a cushion, nor as a tuft of filaments 317**

317a **Thallus a macroscopic, expanded sheet, 1 cell in thickness, principally in salt water but also in cold streams; attached. Fig. 234. . . . . *Monostroma***

Figure 234

Figure 234 *Monostroma latissimum* (Kütz.) Wittr., habit of thallus, attached to rock substrate.

317b **Thallus not a large, flat, crinkled sheet; attached.** **318**

318a **Thallus an intestiniform, hollow tube, with the wall 1 cell in thickness. Fig. 235** . . . . *Enteromorpha*

Figure 235

Figure 235 *Enteromorpha intestinalis* (L.) Grev. (A) habit of branched, tubular thallus; (B) a few cells from the wall showing parietal position of chloroplasts.

Like *Monostroma* (Fig. 234) *Enteromorpha* is primarily a marine alga but becomes adapted rather successfully in freshwater habitats. The long, hollow tubes are frequently branched, forming slender threads, or crinkled, elongate sacks. The plants are always attached (at least when young) to submersed aquatic plants, sticks and stones, especially in flowing water and nearly always in hard (calcareous) or saline waters. One species is common in far western United States where it often occurs in troublesome abundance in irrigation ditches.

318b **Thallus not an intestiniform tube** . . . **319**

319a **Plant a lobed or ruffled, stalked disc of cells, 2 to 3 or up to 10 cm across; chloroplast axial and somewhat radiate; attached by a central, short stipe; usually on rocks in alpine and polar habitats. Fig. 236** . . . . *Prasiola*

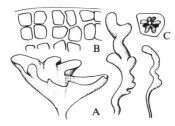

Figure 236

Figure 236 *Prasiola crispa* (Lightf.) Menegh. (A) representative forms of the thallus; (B) cells from the margin of the thallus showing tendency to be arranged in 4's; (C) a single cell showing stellate, lobed chloroplast.

Eight species of *Prasiola* have been reported from the United States, mostly from alpine or subalpine habitats. In the Arctic and Antarctic the plant is a conspicuous element of the subaerial flora, growing on bones, boards, or on the ground especially where there has been an enrichment from nitrogenous matter. The thalli are foliose when fully developed but may be filamentous and ribbonlike when young. The genus and *Schizogonium* (Fig. 221) comprise the Schizogoniaceae, a family with star-shaped chloroplasts.

319b **Plant otherwise; a filament or a gelatinous strand** **320**

320a **A filament of cells** **328**

320b **A gelatinous strand, or tube, or a plant which includes a gelatinous tube that may or may not have cross partitions.** **321**

321a **Thallus a short, tubelike strand, sometimes forked, containing many transverse lamellations (layers); cells at the tips of the strands. Fig. 237** . . . . *Hormotila*

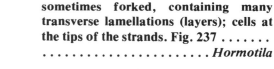

Figure 237

⤻Figure 237 *Hormotila mucigena* Borzi.

This curious plant is a branched, colonial form by virtue of the fact that as the cells divide they secrete mucilage and construct gelatinous strands that branch and rebranch, the cells always occurring at the distal end of the strands. Zoospores produced in asexual reproduction are biflagellate. *Hormotila* is usually regarded as a member of the Tetrasporales near *Gloeocystis* (Fig. 86). It has been suggested that *Hormotila* is related to *Urococcus* (Fig. 18).

321b **Thallus not as above** **322**

322a **Cells located at the ends of undivided tubes, the cells bearing a seta with a sheathed base. Fig. 117** . . . . *Chaetosphaeridium*

322b **Cells without setae** **323**

323a **Cells constricted in the midregion; occurring at the ends of parallel tubes which are united in colonies and are impregnated with lime. Fig. 238** . . . . *Oocardium*

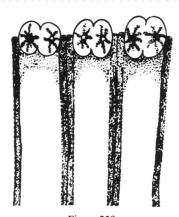

Figure 238

Figure 238 *Oocardium stratum* Näg., a few cells of a colony in calcareous tubes.

This is a very rare desmid, or at least it has been reported but few times, probably because it is easily overlooked by collectors since the tubes are enclosed in calcareous deposits; occurs on sticks and stones in hard water. It has been found in calcareous deposits on larger algae. Each cell is located at the end of a tube with the long axis of the cell at right angles. Each semicell has a star-shaped chloroplast with a central pyrenoid. *Oocardium* is similar to *Cosmarium* in general proportions.

323b **Cells not constricted in the midregion; not arranged as above** **324**

324a Cells in tubes which are attached to microfauna. Fig. 120 . . . . *Colacium*

324b Cells not attached to microfauna . . . 325

325a Cells elongate-oval; brackish, or freshwater 326

325b Cells round; not in brackish water . . 327

326a Cells at the ends of jointed, branching tubes; protoplast with an eye-spot; plants of brackish water, or marine. Fig. 239. . . . . *Prasinocladus*

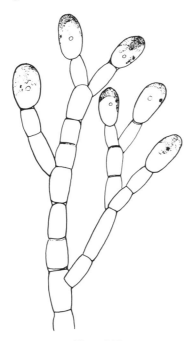

Figure 239

Figure 239 *Prasinocladus lubricus* Kuckuck.

Although essentially marine, *Prasinocladus lubricus* Kuck. has been found in brackish water and may occur in freshwater in coastal regions.

It is an attached, branching strand composed of a series of compartments, forming a treelike thallus in which the oval protoplasts occur only at the tips of the branches. There is 1 chloroplast at the forward end of the cell which actually is the posterior pole because, like some of its relatives (*Malleochloris,* Fig. 162), the cells are in an inverted position with the anterior end downward; have a red eye-spot. The genus belongs to the Chlorangiaceae in the Tetrasporales.

326b Cells at the ends of a series of membranous cups, telescoping one within another, formed from the remains of old, mother-cell walls, the chain of cups often branching. Fig. 118 . . . .
. . . . *Ecballocystis*

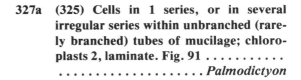

327a (325) Cells in 1 series, or in several irregular series within unbranched (rarely branched) tubes of mucilage; chloroplasts 2, laminate. Fig. 91 . . . .
. . . . *Palmodictyon*

Palmodictyon viride Kütz. and *P. varium* (Näg.) Lemm. are fairly common in mixtures of algae from shallow pools or the tychoplankton of lakes, but never appear in any abundance. The former species has cells enclosed in individual sheaths, whereas the latter is without such sheaths. Some strands of the colonial mucilage may be simple, others irregularly branched and sometimes anastomosing.

327b Cells in many series, *i.e.,* arranged in clusters of 4 throughout an elongated strand of mucilage; chloroplast 1, a parietal cup; (colonies of some species saccate or balloonlike). Fig. 79 . . . .
. . . . *Tetraspora*

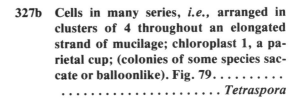

328a (320) Filaments of more than 1 series throughout; (axes with corticating cells); either microscopic or macroscopic, branched thallus encased in mucilage 356

328b Plants otherwise; either uniseriate throughout, or uniseriate in the basal portion and multiseriate above; microscopic 329

329a Filaments uniseriate below, becoming multiseriate in the upper portions of old plants 330

329b Filaments uniseriate throughout when mature. 332

330a Filaments of cylindrical cells below, soon multiseriate with bricklike, angular cells; walls thick. Fig. 240 . . *Schizomeris*

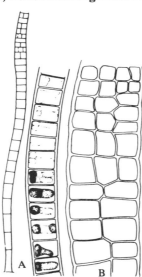

Figure 240

Figure 240 *Schizomeris leibleinii* Kütz. (A) base of filament, and a portion of the uniseriate filament enlarged; (B) section of upper portion of filament.

There are only 2 species in this genus, 1 of which is widely distributed over the world. The filaments are relatively large when fully developed and rather coarse. They occur in dark green clumps in standing water (sometimes on mud just above the water line), and have the macroscopic appearance of *Spirogyra* (Fig. 229) or of some large *Ulothrix* (Fig 257). But unlike those genera, *Schizomeris* filaments separate readily and can be seen individually within a clump of collected material. There is some evidence that the plant favors habitats rich in nitrogenous matter and is to be looked for in shallow water of lakes near the entrance of drains, effluent of sewage treatment plants, *etc.*

330b Filaments otherwise, upper cells not bricklike 331

331a Filaments of cylindrical cells, uniseriate throughout most of the plant's length, finally developing globular cells in more than 1 series anteriorly (becoming palmelloid). Fig. 241. . . *Pseudoschizomeris*

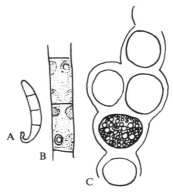

Figure 241

Figure 241 *Pseudoschizomeris caudata* Deason & Bold. (A) young plant; (B) cells showing chloroplasts; (C) palmelloid phase of old plant (in culture). (Redrawn from Deason & Bold.)

The plants are filamentous and ulotrichoid when young but become palmelloid and form multiseriate strands when older. The chloroplast is parietal and all but covers the wall, with 1 pyrenoid. The only species has been cultured from Texas soil.

331b Filaments of cylindrical cells at first in 1 series, becoming multiseriate and forming packets of sarcinalike, angular cells above. Fig. 242 . . . . *Trichosarcina*

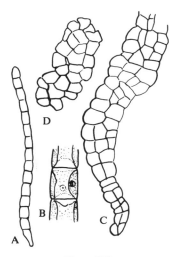

Figure 242

Figure 242 *Trichosarcina polymorpha* Nichols & Bold. (A) young plant with uniseriate arrangement of cells; (B) cells with chloroplasts; (C), (D) multiseriate arrangement of cells in older plants. (Redrawn from Nichols & Bold.)

This genus was established for plants which in their early development are uniseriate, but which eventually become multiseriate with cubical packets of cells. It is sometimes regarded as synonymous with *Pseudendoclonium*. *Trichosarcina* is monospecific; has been cultured from Texas soils.

332a **(329) Filaments rarely solitary, mostly in tufts from a prostrate cluster of cells, the cells of the filament embedded in a wide, firm, gelatinous sheath and separated from one another; plants gray or olive-green with several chloroplasts. Fig. 243 . . . . *Kyliniella***

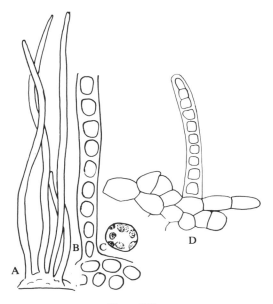

Figure 243

Figure 243 *Kyliniella latvica* Skuja. (A) habit of thallus with upright filaments from a prostrate growth; (B) cell arrangement; (C) one cell showing chloroplasts; (D) prostrate portion of thallus with a young, erect filament.

In this genus of fresh-water red algae (Rhodophyta) there are horizontal plates of cells that give rise to erect tufts of unbranched filaments. These are relatively short and inconspicuous in their early development but become several centimeters long when mature. The cell contents are pink at times although the plant mass is gray-green. The cells may be adjoined or the protoplasts may be widely separated in a relatively wide sheath. Plants are to be sought in swiftly running water. As far as known *K. latvica* Skuja is the only species and has been found in eastern United States.

332b   Plants otherwise **333**

333a   Chloroplast a parietal network, usually close and dense, covering the entire wall; pyrenoids many and conspicuous. . . 334

333b   Chloroplast otherwise; pyrenoids few or lacking. **335**

334a   Cells cylindrical, usually many times their diameter in length (may be only 3 times longer than wide); wall thick, distinctly lamellate in most species. Fig. 244. . . . . *Rhizoclonium*

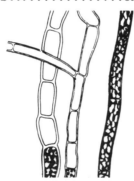

Figure 244

Figure 244   *Rhizoclonium hookeri* Kütz., left; *R. hieroglyphicum* (C.A. Agardh) Kütz., right.

The species belonging to this genus are all coarse, wiry and but very little, if at all branched. The filaments are composed of relatively long, cylindrical, coenocytic cells, or if somewhat shorter cells with lateral walls slightly convex and irregular. The walls are frequently lamellate, a character which shows especially at the joints. The chloroplast is a loose or dense meshwork composed of many small, interconnected discs, each with a pyrenoid. In some instances the network is disrupted and there ap-

pear to be many individual chloroplasts. The branches of the filament (when present) are short, one or a few cells in length, and arise at right angles to the main axis; extend back rather than toward the apex of the filament. It does not show the arbuscular, treelike plan of branching found in *Cladophora, e.g.* When the branches are long, as in a few species, the plants intergrade with some forms of *Cladophora* (Fig. 294) in which the regular plan of branchings has been reduced or lost. *Rhizoclonium* forms dense, tangled mats in standing water, or long, stringy, sometimes ropelike strands in flowing water. *R. hieroglyphicum* (Ag.) Kütz. is the most common species, one which has rather uniformly cylindrical cells with relatively thin walls; does not branch.

334b   Filaments of cells which are not quite cylindrical but are slightly larger at the anterior end (practically cylindrical in a few of the many species); plants with a hold-fast and basal-distal differentiation; cells with 1 or more ringlike scars on the wall immediately below the cross wall (rings formed during cell division); when mature showing enlarged female cells (oogonia) and small boxlike cells with spermatozoids (antheridia). Fig. 245. . . . . *Oedogonium*

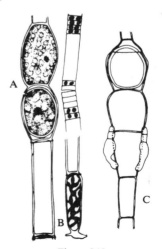

Figure 245

Figure 245 *Oedogonium.* (A) *Oe. crispum* Kütz., portion of filament with 1 fertilized and 1 unfertilized egg; (B) basal hold-fast cell and portion of a filament containing antheridia with antherozoids; (C) *Oe. westii* Tiff., showing dwarf male filaments epiphytic on female plant.

There are over 300, possibly more than 400 species of this genus which belongs to the Oedogoniaceae, a family which includes only 2 other genera (*Bulbochaete,* Fig. 296 and *Oedocladium,* Fig. 268). Species are differentiated and are identified when in the reproductive condition by the size, shape and morphology of the sex organs and their location, and the details of the mature oospore. Whereas some species have the antheridia on filaments the same size as those which bear the oogonia, others possess dwarf male plants which grow as epiphytes on or near the oogonia. *Oedogonium* plants begin as attached filaments and may remain so throughout life, or they may become free-floating and form cottony masses near or on the surface of the water, usually becoming pale, yellow-green or cream-colored in age. Often these masses are so dense that if left to dry by the evaporation of habitat water they may form what is known as "algal paper." Common habitats are on overhanging grass leaves or the culms of rushes, old cattail stalks, *etc.* Bewildering to the student, several species may grow together in close proximity and some patience is required to differentiate them.

335a (333) **Chloroplast a parietal plate, a ring, or a band which incompletely encircles the cell (but see *Ulothrix zonata,* Fig. 257 in which the ring is complete)** 336

335b **Chloroplast massive and dense, difficult of determination, *or* a parietal sheet of thick and thin areas (padded appearance), *or* a branched, beaded thread (See *Microspora,* Figs. 246, 262.)** 353

Figure 246

Figure 246 *Microspora.* Types of chloroplasts. (A) beaded; (B) padded parietal.

Types of chloroplasts found in *Microspora.*

336a **Filaments of cylindrical cells with a broad parietal chloroplast that has a fringed margin (fingerlike extensions), with several pyrenoids. Fig. 247.** . . . . . .
. . . . *Entransia*

Figure 247

Figure 247 *Entransia dichloroplastes* Presc.

This plant is enigmatic, having cells which are ulotrichoid in some respects, but with zygnemataceous pyrenoids. Conjugation has never been observed in the 2 known species. Apparently the plant is to be sought only in soft-water habitats.

336b Filaments and chloroplasts otherwise 337

337a Filaments composed of long, cylindrical multinucleate units; chloroplast in the form of several rings in each unit. Fig. 248 . . . . *Sphaeroplea*

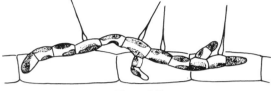

Figure 248

Figure 248 *Sphaeroplea annulina* (Roth) Ag. (A) vegetative cell with ringlike chloroplasts; (B) one cell containing fertilized eggs.

Four species of this genus have been reported from the United States, but only *S. annulina* (Roth) Ag. is at all common. Whenever it occurs it is found in abundance, often coloring the water red when the oospores are mature. The characteristic long, cylindrical cells may be mistaken for species of *Rhizoclonium* (Fig. 244), especially when preserved material is examined. The annular, bandlike chloroplasts aid in identification. In the species illustrated eggs and motile antherozoids are produced within cells of different filaments apparently, at least in western United States.

337b Filaments not composed of long, multinucleate units; chloroplasts otherwise, 1, sometimes 2 in each cell 338

338a Filaments prostrate, 'creeping' on or in other algae; cells bearing 1 or 2 bristles, some genera with branched hairs . . . 339

338b Filaments not creeping on other algae; floating, *or* if prostrate with cells in discontinuous series 340

339a Filaments epiphytic, 'creeping' over walls of larger, filamentous algae in a 'wormlike' fashion; some cells bearing a bulbous-based hair. Fig. 249 . . . . *Aphanochaete*

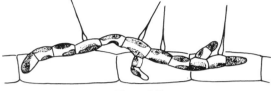

Figure 249

Figure 249 *Aphanochaete repens* A. Braun, growing on a filament of *Rhizoclonium*.

There are 6 species of *Aphanochaete* reported from the United States, 3 of which are very common but are easily overlooked because of their small size and their epiphytic habit on other filaments. The simple setae, with swollen bases, extending from the cell wall are helpful in making identification. *Aphanochaete polychaete* (Hansg.) Fritsch is characterized by having several such setae borne on a single cell. *A. repens* A. Braun often has the appearance of a minute measuring worm.

339b Filaments endophytic, 'creeping' beneath the wall of other algae, upper outer wall of some cells bearing long, tapering bristles. Fig. 250 . . . *Ectochaete*

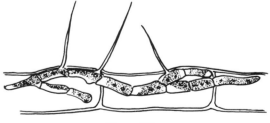

Figure 250

Figure 250 *Ectochaete endophytum* (Möb.) Wille, growing on a filament of *Rhizoclonium*.

(See also *Coleochaete nitellarum*, Fig. 233.)

This genus is somewhat like *Aphanochaete* (Fig. 249) which is growing beneath the wall of algae rather than on them. The hairs on the wall do not have swollen bases. Such algae as *Cladophora* (Fig. 294) and *Rhizoclonium* (Fig. 244) are frequently endophytized by *Ectochaete*.

340a   (338) Filaments very short (up to 20 cells) often interrupted or discontinuous 341

340b   Filaments longer, in a continuous series 343

341a   Chloroplast a parietal, folded plate or merely a disc extended over but a small part of the wall; cells short-cylindric, with rounded poles. Fig. 251 . . . . *Stichococcus*

Figure 251

Figure 251   *Stichococcus bacillaris* Näg.

The difference between *Stichococcus* and the small filaments characteristic of *Chlorhormidium* (Fig. 259) is difficult to define. In the former genus the filaments are relatively short (10 to 20, sometimes 40 cells) and have a tendency to break into short sections intermittently. Of the 11 species which occur in the United States, most are found on the bark of trees, old boards, or damp soil. The species illustrated is the most common, often occurring with *Desmococcus* (Fig. 122) in subaerial habitats, the short filament twisted, contorted, or coiled in 1 plane.

341b   Chloroplast broader and cells of different shape. 342

342a   Cells long-cylindric, or if short-cylindric with rounded poles, oval; chloroplast ribbonlike or a broad plate covering most of the wall; filaments sometimes coiled. Fig. 252 . . . . *Gloeotila*

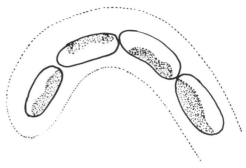

Figure 252

Figure 252   *Gloeotila contorta* Chodat, portion of coiled filament in a gelatinous sheath.

The free-floating filaments of *Gloeotila* have oval-cylindric cells in short or long series. In one species the cells have rounded poles and the filaments are coiled and twisted, enclosed in a wide, gelatinous sheath. The plants appear much like a *Stichococcus* with the cells longer and more convex at the poles. There is no basal-distal differentiation. The parietal chloroplast lacks a pyrenoid.

342b   Cells transversely oval, in short, interrupted series, 2 or 4 cells enclosed in a separate gelatinous sheath within the series. Fig. 253 . . . . *Hormidiopsis*

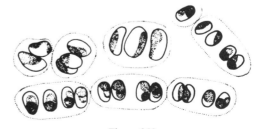

Figure 253

Figure 253   *Hormidiopsis ellipsoideum* Presc.

The species illustrated is the only one reported from the United States and possibly cannot be differentiated from *Chlorhormidium* (Fig. 259) except that the filaments are frequently interrupted and are decidedly constricted at the cross walls, the cells being oblong or oval rather than cylindrical. Characteristically, the chloroplast extends only part way around the cell wall.

343a (340) Filaments composed of units which include 2 oval or subspherical protoplasts; the space between the protoplasts and the wall filled with layered (lamellose) material. Fig. 254 . . . . *Binuclearia*

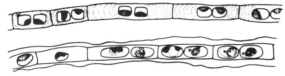

Figure 254

Figure 254 *Binuclearia tatrana* Wittr., one filament showing a gelatinous sheath.

There are 3 species of this genus reported from the United States, but the one illustrated is most frequently found. The plants occur intermingled among other filamentous algae, especially in mixtures taken from bogs. The paired protoplasts within each cylindrical unit of the filament make identification certain.

343b Filaments of cells without paired protoplasts as above 344

344a Filaments with a gelatinous sheath . . 345

344b Filaments without a gelatinous sheath 350

345a Cells cylindric-quadrate, globose or ellipsoid; adjoined at the end walls . . 346

345b Cells oblong, not adjoined at the end walls. Fig. 255 . . . . *Geminella*

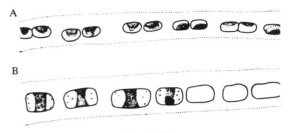

Figure 255

Figure 255 *Geminella*. (A) *G. interrupta* (Turp.) Lag.; (B) *G. mutabilis* (Breb.) Wille.

These are filamentous plants that have cylindrical or broadly oval cells encased in a wide sheath of mucilage. The cells may be adjoined, or rather evenly spaced one-half to 2 cell lengths apart. Like *Chlorhormidium* (Fig. 259) the chloroplast covers but a small portion of the wall.

346a Cells quadrate, cylindrical, with rounded poles; wall in 1 piece 347

346b Cells globose, subglobose, or ellipsoid; wall usually of 2 overlapping pieces that meet in the midregion of the cell and which form short, lateral projections, resulting from a rim around the cell in the midregion. Fig. 256 . . . . *Radiofilum*

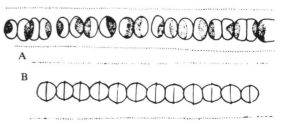

Figure 256

Figure 256 *Radiofilum*. (A) *R. flavescens* G.S. West; (B) *R. conjunctivum* Schmidle.

The globose or subglobose cells of *Radiofilum* filaments help to separate it from *Geminella* (Fig. 255) which also possesses a gelatinous sheath. Some species, *Radiofilum conjunctivum* Schm., *e.g.,* have the walls in 2 sections which form a conspicuous overlapping in the midregion. There are 10 species reported from the United States, differentiated by shape, size and morphology of the cells.

347a **Filament of cylindrical cells with a lobed, parietal chloroplast and 1 pyrenoid; filaments becoming multiseriate and the cells globular in age. Fig. 241 .. .. *Pseudoschizomeris***

347b **Filaments otherwise............ 348**

348a **Filaments composed of slender, cylindrical cells, truncate or sometimes rounded at the apices, enclosed in a wide, gelatinous sheath; chloroplast a broad plate or a twisted band. Fig. 252 .... *Gloeotila***

348b **Filaments otherwise............ 349**

349a **Cells short-cylindric with rounded poles; chloroplast a small plate or a narrow band covering only about one-third of the cell wall; filament enclosed in a wide, gelatinous sheath; cells usually in interrupted series. Fig. 255 ...... *Geminella***

349b **Cells quadrate to long-cylindric; chloroplast a broad, parietal band covering most of the cell wall, or a median band completely encircling the cell; pyrenoids 1 to several; at least 1 species with basal-distal differentiation from a hold-fast**

cell; without a wide gelatinous sheath, but may have a close, firm sheath immediately external to the wall. Fig. 257 ... . *Ulothrix*

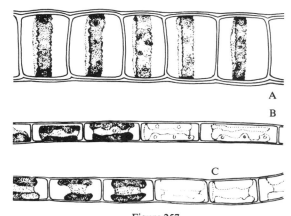

Figure 257

Figure 257 *Ulothrix*. (A) *U. zonata* (Weber & Mohr) Kütz., with ringlike chloroplasts; (B) *U. cylindricum* Presc.; (C) *U. aequalis* Kütz.

Species of this genus have short, rectangular cells, or in a few species long, cylindrical cells. There is considerable variation in cell diameter, some species having cells wider than long. The most familiar species and the largest is *Ulothrix zonata* (Weber *et* Mohr) Kütz. It has a basal hold-fast and shows basal-distal differentiation. The chloroplast in this species, at least in mature cells, forms a complete, parietal band, with 1 or more pyrenoids. Other species have chloroplasts which encircle the wall two-thirds or three-fourths of a circle. Whereas most species occur in standing water, *U. zonata* prefers rapidly flowing streams, or wave-washed shores where it forms bright green, streaming clumps or fringes on rocks and submersed wood. Reproduction is by quadriflagellate zoospores, formed 4 or 8 in a cell, or by biflagellate isogametes produced from 16 to 64 in a cell.

350a **(344) Filaments not showing basal-distal differentiation** **351**

350b **Filaments with a basal hold-fast.** **352**

351a **Filaments short, from 2 to 6 (rarely up to 32) cells; filaments tapered at both ends. Fig. 258** .... *Raphidonema*

Figure 258

Figure 258   *Raphidonema nivale* Lag.

Some species of this genus are scarcely filaments, the plants consisting of only a few cells adjoined. They are more fusiform at times

than cylindrical and the poles are tapering so that the cross walls are oblique. Very frequently this genus is found in snow fields at high altitudes. The genus *Koliella* (Fig. 258A) has similar cells, occurring singly or in linear pairs, decidedly pointed at the apices. This genus is known to have oogamous reproduction. *Raphidonema,* according to the above interpretations includes species with filaments up to 32 cells long; reproduces only by cell division as far as known.

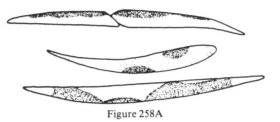

Figure 258A

Figure 258A   *Koliella* sp. from snow banks in Olympic Mountains.

The species of *Koliella* illustrated is an example of the cell arrangement that characterizes the genus and which, according to some students, separates it from *Raphidonema* (Fig. 258).

351b **Filaments indefinitely long; cells not tapered at the apex of the filament. Fig. 259** .... *(Hormidium)* *Chlorhormidium*

Figure 259

Figure   259   *Chlorhormidium   dissectum* (Chod.) Fott.

The name *Hormidium* is now a synonym of *Chlorhormidium*. The genus includes several species of simple, unbranched filaments of cylindrical cells (or somewhat quadrangular cells) which have chloroplasts that extend only part way around the wall and which are only about one-half the cell's length. Species are both aquatic, intermingled with other filamentous algae, and subaerial, occurring on wet soil and on dripping rocks, or in seeps.

352a **Cells elongate-cylindric, the apical cell unsymmetrically pointed, Fig. 260.....** **.......................** ***Uronema*****

Figure 260

Figure 260 *Uronema elongatum* Hodgetts, anterior portion of a filament.

Five species of this genus have been described, but these are in questionable position inasmuch as there is justification for considering the plants as growth forms of *Ulothrix*. The cells are long and cylindrical with a ulotrichoid chloroplast. Usually the filament is only a few cells in length. The unsymmetrically pointed apical cell is the chief identifying character. Younger stages in the development of *Stigeoclonium* (Fig. 272) should be kept in mind when identification of *Uronema* is made.

352b **Cells short-cylindric, apical cell not tapering. Fig. 257...........** ***Ulothrix*****

353a **(335) Cells quadrate or oval to subglobose, enclosed in a wide, stratified, gelatinous sheath. Fig. 261 ......** **....................** ***Cylindrocapsa*****

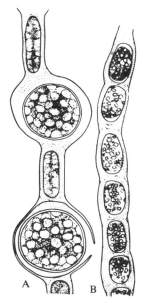

Figure 261

Figure 261 *Cylindrocapsa geminella* var. *minor* Hansg. (A) portion of filament with oogonia; (B) vegetative cells with massive chloroplasts.

Although filaments of this genus begin as attached plants they soon become free-floating and are found intermingled with other filamentous algae, especially in soft-water lakes and bogs. The chloroplasts are so dense and the contents include so much stored food material that few structural characteristics can be determined. The female reproductive organ (oogonium) is globular and greatly swollen to several times the diameter of the vegetative cells. The contents (female gamete) are red, especially after fertilization, at which time an oospore wall is formed. The antheridia occur as a single or double row of small, somewhat quadrangular cells and also are red. Each antheridial cell produces 4 quadriflagellate antherozoids.

353b **Filaments formed otherwise.......354**

*Sometimes assigned to *Ulothrix*, Fig. 257.

354a Cells often orange or yellowish-red with carotenoids (haematochrome); epiphytic or subaerial on wood and stones . . . **355**

354b Cells never with haematochrome; plants aquatic; chloroplast a perforated and padded sheet or a branched, beaded ribbon. Figs. 246, 262 . . . . *Microspora*

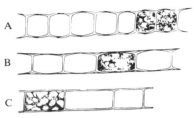

Figure 262

Figure 262 *Microspora.* (A) *M. loefgrenii* (Nordst.) Lag.; (B) *M. willeana* Lag.; (C) *M. floccosa* (Vauch.) Thur.

In this genus the simple, unbranched filaments have quadrangular to short-cylindric cells. The chloroplast varies greatly in appearance, even within the same filament, being either a parietal, folded, discontinuous plate, or a meshwork of strands (Fig. 246). There are 15 or 16 species, differentiated by size and proportions of the cells and by the wall characters. The wall is in 2 sections which overlap in the midregion. The overlapping is conspicuous in thick-walled species, but at least can be determined in the thin-walled species by examining the broken ends of filaments where the sections of the terminal cell protrude. When the filaments fragment and the cells dissociate H-shaped sections are found. The plants produce akinetes rather frequently and often an entire filament may appear as a chain of globular spores. Because of the wall features and the nature of the chloroplast *Microspora* is placed in its own family and order.

355a Filaments prostrate, forming a disclike expanse on the epidermis of leaves of terrestrial plants. Fig. 4 . . . . *Phycopeltis*

355b Filaments erect, forming orange, fuzzy growths on trees, rocks and timbers in humid situations. Fig. 7. . . *Trentepohlia*

356a (328) Plants macroscopic, gelatinous, arbuscular thalli, appearing beaded to the unaided eye **357**

356b Plants microscopic, *or* if macroscopic not appearing beaded. **358**

357a Branches whorled (beaded effect); when mature with dense clusters of spores (carpospores) formed around the base of the female sex organ (carpogonium) as a result of fertilization. Fig. 263 . . . .
. . . . *Batrachospermum*

(**B. dillenii Bory is not beaded.**)

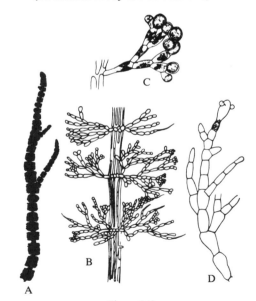

Figure 263

Figure 263 *Batrachospermum.* (A) *B. moniliforme* Roth, habit of plant; (B) portion of thallus showing small spermatia cells at tips of branches; (C) *B. vagum* (Roth) Ag., detail of branch with spermatia; (D) *B. boryanum* Sirod., carpogonial branch with 2 spermatia attached to the trichogyne of the carpogonium (female organ).

This genus belongs to the Rhodophyta and is one of the most common of freshwater red algae. Although it does not show a red color, the pigments do contain phycoerythrin and others peculiar to the Rhodophyta. The macroscopic thalli, highly branched and beaded are encased in a copious, soft mucilage. They form arbuscular growths on stones and sticks in flowing water, although a few species occur in pools of soft water, ususally growing on vertical walls of *Sphagnum* bog pools. The thallus may be gray-green, bluish-green, or olive to tawny. *Batrachospermum vagum* (Roth) Ag. is perhaps the most common species in the United States, occurring in large patches on stones in flowing water. It is characterized by the trichogyne (tip of the female organ) being spatula-shaped. This and most other species of the genus prefer shaded areas. Microscopically, *Batracospermum* is one of the more elegant of freshwater algae. Species are differentiated by the morphology of the female organ, by the habit of branching, and by the location of the sex organs; some species monoecious, others dioecious. The spermatia (male) cells occur as small, spherical units at the tips of the ultimate cells of the branches. The female organ is a flask-shaped cell at the tip of a special branch. It consists of a swollen, female gamete-containing cell which extends into a variously shaped elongation, the trichogyne. The spermatia are non-motile and reach the trichogyne by drifting to initiate fertilization. When mature, and after fertilization dense heads of carpospores are formed around the base of the carpogonium as a result of zygote activity. These can be detected easily under low magnification, appearing as dark clumps scattered about in the whorls of branches.

357b **Thallus similar to above; carpospores solitary or in small clumps of 2 or 3, the spores produced on wandering filaments (gonimoblasts) which develop from the zygote in the carpogonium. Fig. 264 . . .**
. . . . *Sirodotia*

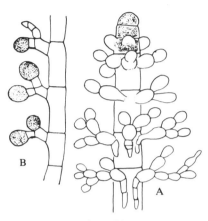

Figure 264

Figure 264 *Sirodotia suecica* Kylin. (A) apex of filament with young branches developing at nodes; (B) carpospores borne on a wandering gonimoblast filament.

These plants occur in the same types of habitats as *Batrachospermum* (Fig. 263) and appear much like that genus macroscopically. The chief difference is that the zygote in *Sirodotia* develops gonimoblast filaments which grow through the thallus, cutting off clusters of spores 2 or 3 together or singly here and there. In *Batrachospermum* the spores form in dense clumps in the immediate vicinity of the carpogonium where the egg has been fertilized.

358a **(356) Filaments uniseriate below, becoming multiseriate in the upper section, with the cells bricklike in shape and arrangement; cells adjoined. Fig. 240 . . . .**
. . . . *Schizomeris*

358b Filaments multiseriate throughout; cells not adjoined but arranged in irregular, linear series within a gelatinous strand to form a false filament. Fig. 91 . . . . *Palmodictyon*

359a (291) Plants composed of cellular units; cross walls present 360

359b Plants multinucleate threads (coenocytic) without cross walls (except when reproductive structures are developed and separated by a cross partition from the main filament) 415

360a Filaments prostrate, epizoic, forming a disc on the carapace of turtles. Fig. 265 . . . . *Dermatophyton**

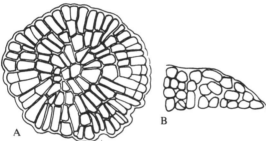

Figure 265

Figure 265 *Dermatophyton radians* Peter. (A) surface view of epizoic thallus; (B) portion of thallus in cross section (semidiagrammatic).

The thallus in this genus is a disc, 1-celled thick, formed by closely appressed, branched filaments radiating from a common center. The cells are thick-walled and multinucleate; have a single chloroplast with several pyrenoids. Apparently the plants grow only on the carapaces of turtles.

Figure 266

Figure 266 *Trichophilus welcheri* Weber van Bosse, short-branched filaments on the hair scales of sloth. (Redrawn from Bourrelly.)

360b Plants otherwise 361

361a Plants growing on moist soil 362

361b Plants aquatic, *or* epiphytic or parasitic on higher plants, or if terrestrial without rhizoidal branches 363

362a Filaments subterranean with tufts of branches appearing above the surface; apices of branches pointed but without setae. Fig. 267 . . . . *Fritschiella*

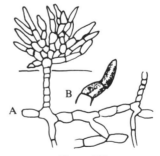

Figure 267

Trichophilus welcheri Web. v. Bosse (Fig. 266) is sparsely branched, semipalmelloid species which grows among the hair scales of the sloth along with *Cyanoderma* (Fig. 19).

Figure 267 *Fritschiella tuberosa* Iyengar. (A) portion of thallus showing horizontally growing subterranean filaments and an erect tuft of branches; (B) tip of branch to show parietal, folded chloroplast.

This thallus has tufts of branches arising above the soil. The exposed cells have a ulotrichoid chloroplast and 1 or 2 pyrenoids. The subterranean filament also gives rise to downward-growing rhizoidal branches. First found in India, this genus is now known from southwest United States. The prostrate portion produces zoospores and gametes on diploid and haploid individuals respectively, thus exhibiting an alternation of isomorphic generations.

362b   **Plants branched filaments growing on moist soil, with rhizoidal extensions; cells cylindrical, the apical cell with a cap (thimble); reproduction by antherozoids and non-motile egg borne in an enlarged cell (oogonium). Fig. 268** . . . . . .
. . . . *Oedocladium**

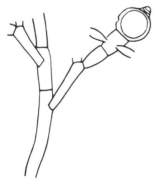

Figure 268

Figure 268 *Oedocladium hazenii* Lewis, portion of a branched filament with an oogonium and 2 epiphytic, 1-celled male plants; growing on soil.

363a   **(361) Plants mainly *prostrate,* growing horizontally; mostly epiphytic or endophytic and forming discs or flat expanses; (but see some species of *Coleochaete* which have upright branches from a prostrate base)** 364

363b   **Plants mainly *erect,* or free, not growing horizontally on a substrate, but sometimes a thallus with some horizontal growth at the base of the erect portions; free-floating or attached; or epiphytic or parasitic on higher plants; in some forms perforating wood or shells** 379

364a   **Filaments short and incidentally formed as a series of cells arising from encrusting patches of globular cells growing on trees and boards. Fig. 122** . . . . . .
. . . . *(Pleurococcus) Desmococcus*

364b   **Filaments otherwise** 365

365a   **Thallus a freely and openly branched filament; cells usually bearing setae or spine-shaped hairs** 366

365b   **Thallus not freely branched but forming a disc, cushion (pseudoparenchymatous), or a flat expanse of cells** 373

366a   **Endophytic in the walls of other algae** 367

366b   **Not endophytic in the walls of other algae** 368

*In humid, subtropical regions *Oedogonium* also grows on moist soil and subaerial substrates; is unbranched.

367a Thallus irregularly branched; angular cells lying within or under the walls of *Nitella;* cells bearing setae (sometimes very scarce) which are sheathed at the base. Fig. 233 . . *Coleochaete nitellarum*

367b Cells not angular; branches short; cells never bearing setae. Fig. 269 . . . .
. . . . *Entocladia*

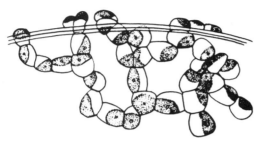

Figure 269

Figure 269 *Entocladia polymorpha* (G.S. West) G.M. Smith, growing within walls of *Pithophora.*

This genus includes only 3 species in the United States but there are others in Europe and they are probably widely distributed over the world; are easily overlooked. The thalli are small and grow inconspicuously within the walls of larger algae. The filaments are short and irregularly branched; have a parietal, ulotrichoid chloroplast. Unlike somewhat similar plants the cells do not bear setae.

368a (366) Some cell walls bearing setae with a sheathed base; terminal cells of branches not tapering to form setae. Fig. 233 . . . . *Coleochaete*

368b Cells with setae not sheathed at the base, *or* if without setae, ends of branches tapering to hairs or spinelike extensions; sometimes endophytic 369

369a Setae and terminal hairs multicellular. 370

369b Setae one-celled 372

370a Multicellular setae in the form of especially slender branches arising from lateral walls of some cells. Fig. 270 . . . .
. . . . *Pseudochaete*

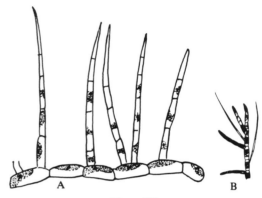

Figure 270

Figure 270 *Pseudochaete.* (A) *P. crassisetum* G.S. West; (B) *P. gracilis* W. & W.

The species illustrated is rarely found. It grows partly erect on submersed plants, logs, old cattail stems. The branched filaments taper at both ends. Some of the cells have long, narrow and finely tapering hairs. Some students of the algae regard *Pseudochaete* as a growth form of *Stigeoclonium* (Fig. 272).

370b Plants with multicellular hairs resulting from the apical tapering of branches . . .
. . . . 371

371a Filaments horizontally spreading in and on duckweed thalli; with erect branches that taper to fine points. Fig. 271. . . . . .
. . . . *Endoclonium*

Figure 271

Figure 271 *Endoclonium polymorphum* Franke, habit on and in tissues of *Lemna.*

The branched filaments 'creep' among the cells of *Lemna,* especially *L. minor.* There are both prostrate and erect portions of the thallus. As a genus *Endoclonium* is not well understood. Some regard it as a growth form or a juvenile stage of some species of *Stigeoclonium.*

371b Filaments growing as erect, branched tufts or plumes; branching alternate or opposite, the branches thornlike and short or long and tapering to fine points. Fig. 272. . . . . *Stigeoclonium*

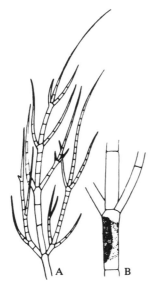

Figure 272

Figure 272 *Stigeoclonium flagelliferum* Kütz. (A) upper portion of thallus showing tapering branches; (B) cell showing laminate chloroplast and branch habit.

There are several species of this genus differentiated by size, by plan of branching, and by the general morphology of the thallus as a whole. Some species form long, graceful tufts; others are bunched growths with part of the thallus prostrate. It should be compared with *Cloniophora* (Fig. 289).

372a (369) Growing in the mucilage of other algae. Fig. 273 . . . . *Chaetonema*

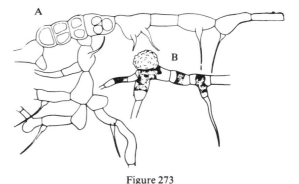

Figure 273

Figure 273 *Chaetonema irregulare* Nowak. (A) branched filament containing antheridial cells; (B) portion of a filament with an oogonium.

Apparently there are only 2 species described for this genus. It is rather rarely seen because its habitat is the gelatinous matrix of highly branched algae such as *Chaetophora* (Fig. 297) and *Batrachospermum* (Fig. 263) where it is well-camouflaged.

372b Growing in a 'creeping' fashion on the walls of larger algae. Fig. 249 . . . .
. . . . *Aphanochaete*

373a (365) Cells bearing setae with sheathed bases. Fig. 233 . . . . *Coleochaete*

373b Cells without setae, *or* if setae present without a sheath base 374

374a Plants endophytic or epiphytic. 375

374b Not epiphytic or endophytic in the walls of other algae. 376

375a Filaments 'creeping' beneath the walls of other algae. Fig. 269 . . . . *(Endoderma) Entocladia*

375b Filaments growing on and among the cells of *Lemna* thalli. Fig. 271 . . . . *Endoclonium*

376a (374) Some cells with setae on the walls. Fig. 274 . . . . *Chaetopeltis*

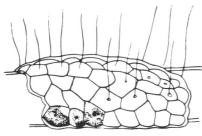

Figure 274

Figure 274   *Chaetopeltis orbicularis* Berth.

This plant forms relatively small, circular discs composed of indistinctly radiate 'filaments' which grow closely side by side. It should be compared with *Coleochaete* (Fig. 233). It is rather oddly placed in the Tetrasporales because the cells bear pseudocilia which are very

long and erect. These have a different morphology than the hairs or setae produced by other somewhat similar prostrate plants. *Chaetopeltis orbicularis* Berth. is to be found as an epiphyte on larger filamentous algae and on the stems of aquatic plants.

376b Setae lacking 377

377a Thallus a thin expanse, 1 cell in thickness; a circular disc, sometimes with irregular margins. Fig. 275. . . . . *Protoderma*

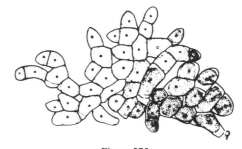

Figure 275

Figure 275   *Protoderma viride* Kütz.

This prostrate plant forms a cushionlike thallus 1 cell in thickness at the margin and one which shows irregular branching of short filaments. It is to be found growing on the stems of submersed aquatic plants; cells have a single, parietal chloroplast and 1 pyrenoid.

377b Thallus cushionlike, several cells in thickness in the midregion 378

378a Cells with several chloroplasts; thallus enclosed in a mucilage; disc several cells thick in the midregion, 1 cell thick at the margin. Fig. 276 . . . . *Pseudulvella*

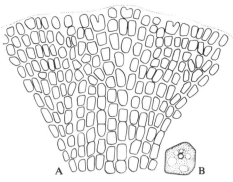

Figure 276

Figure 276 *Pseudulvella americana* (Snow) Wille.

The prostrate, disclike thalli of this plant are relatively large; often are visible to the unaided eye, growing on cattail stems, rushes, and on *Chara*. The entire thallus is enclosed by a gelatinous sheath which is evident at the margin. The sheath often bears gelatinous bristles which are deciduous and frequently the hair-bearing character does not show. The thallus is 1-celled thick at the margin, several cells thick in the midregion. Reproduction is by quadriflagellate zoospores; sexual reproduction seems never to have been reported. The one species is known from western and midwestern United States.

378b   Cells with but 1 chloroplast; thallus not enclosed in a sheath. Fig. 277 . . . *Ulvella*

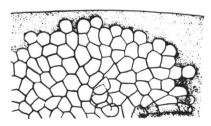

Figure 277

Figure 277 *Ulvella involvens* (Savi) Schm., diagram of attached colony showing arrangement of cells.

There is some confusion between this genus and *Dermatophyton* (Fig. 265), one species of which has been referred to *Ulvella* (but apparently incorrectly). Both of these disclike thalli may occur on turtle carapaces, but *Ulvella* also grows on aquatic plants and on stones; seems to have thinner cell walls and biflagellate zoospores. The cells are uninucleate. There are both marine and freshwater species of *Ulvella;* the one illustrated having been found on turtles.

379a   (363) Plant a sparsely branched filament growing erect from the walls of other algae; cells bearing 1, 2 or 3 setae with inflated bases. Fig. 278 . . . .
. . . . *Thamniochaete*

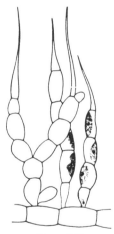

Figure 278

Figure 278 *Thamniochaete huberi* Gay, filaments growing erect from algal host.

This genus has both prostrate filaments and erect branches; occurs as an epiphyte on filamentous algae. The cells are either subcylindric, barrel-shaped or nearly globular. The terminal cells of the branches especially, but others as well bear long, colorless setae. Thus far one species has been reported from the United States in California.

379b   Plants not as above 380

380a   **Filaments irregularly branched, composed of moniliform cells, mostly erect but sprawling and prostrate in part, on tree trunks and leaves of higher plants. Fig. 279** . . . . *Physolinum*

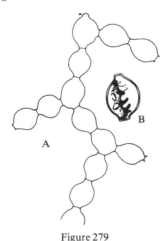

Figure 279

Figure 279 *Physolinum monilia* (de Wild.) Printz. (A) prostrate and erect filaments of moniliform cells; (B) cell with chloroplast.

This plant consists of lemon-shaped cells adjoined to form branched chains without any definite central axis and apparently without basal-distal differentiation. The thallus is mostly prostrate as an epiphyte on higher plants (especially leaves in humid situations). The genus is sometimes included in *Trentepohlia*

(Fig. 7). The cells divide by producing a 'bud' which enlarges and constricts, after which a cross wall is formed separating the new cell from the old. There is 1 parietal chloroplast, often with projections, or the chloroplast may become divided into 2 or 3 discs, but always without pyrenoids. Like *Trentepohlia* the cells are yellow or orange with carotenoid type of haematochrome.

380b   **Plants not as above** 381

381a   **Plants parasitic and epiphytic under the epidermis of leaves and fruits.** 382

381b   **Plants not parasitizing leaves although some may be surface growing and epiphytic** 383

382a   **Plants parasitic, living beneath the epidermis of leaves but with erect, multicellular branches bearing sporangia externally. (On *Magnolia*, *Thea* (tea), *Musa* (banana). Fig. 6** . . . . *Cephaleuros*

382b   **Thallus consisting of a bulbous, subepidermal cell which gives rise to an erect, few-celled stalk and, or hair bearing a sporangium. Fig. 5** . . . . *Stomatochroon*

383a   **Cells without setae, filaments not tapering to hairlike tips** 384

383b   **Cells bearing setae or with branches tapering to hairlike tips or fine points** 409

384a   **Branches short, 1 or 2 cells, in some, irregular and rhizoidal, often formed near one end of the filament.** 385

384b Branches longer, multicellular, usually forming a definite pattern of growth, opposite or alternate on the main axis, or branches both long and short on the same filament 389

385a Chloroplast a spiral ribbon. Fig. 229 . . . . *Spirogyra*

385b Chloroplast not a spiral ribbon 386

386a Chloroplast a parietal plate 387

386b Chloroplast not a parietal plate 388

387a Filament of oval or nearly globular cells, sometimes bearing short branches; free-floating, in a wide, gelatinous sheath. Fig. 280 . . . . *Hazenia*

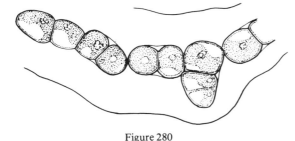

Figure 280

Figure 280 *Hazenia mirabilis* Bold. (Redrawn from Bold.)

The branched filaments of this genus are enclosed in a soft, mucilaginous, tubelike sheath, the sheaths sometimes anastomosing. The subquadrate to subglobose cells are in chains and are not always in contact with one another. At times there is vertical as well as transverse division of the cells. The chloroplast is ulotrichoid. As far as known there is no basal-distal differentiation, but the cell at the terminus of a filament is definitely conical. Only one species is known, from soil in Tennessee.

387b Filament irregularly twisted, composed of irregularly shaped, vermiform or subcylindrical cells, enclosed in an oval or globular sheath; branching indefinite. Fig. 281 . . . . *(Heterodictyon)* *Helicodictyon*

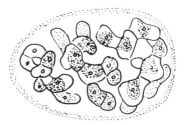

Figure 281

Figure 281 *Helicodictyon planctonicum* Whitford. (Redrawn from Bourrelly.)

This genus, like *Hazenia* (Fig. 280), has a filament enclosed in a wide, gelatinous sheath. The cells are short, bacilliform or irregularly lobed and often are arranged in a twisted, rather indefinite branched filament. There is a ulotrichoid chloroplast with a pyrenoid. The name is synonymous with *Heterodictyon* Whitford.

388a (386) Chloroplast an axial plate or band with conspicuous pyrenoids; wall relatively thin. Fig. 227. . . . . *Mougeotia*

388b Chloroplast a parietal network of thickenings and thin strands, with many scattered pyrenoids in the meshwork; walls thick, sometimes lamellate. Fig. 244. . . . . *Rhizoclonium*

389a (384) Growing in wood, shells or within limestone. Fig. 282 . . . . *Gomontia*

Figure 282

Figure 282 *Gomontia holdenii* Collins, habit of the thallus showing erect branches.

These plants must be sought in old wood, shells or in limy concretions, or on the surface in part. The thallus occurs as a cushionlike, irregularly tangled mass of short filaments from which some elements grow downward to form rhizoidal, penetrating threads. Reproductive structures (sporangia) are borne on the upper parts of the thallus at the ends of short, erect branches. Akinetes are also produced. In germination these often form aplanospores. Cell walls, especially in the prostrate portion, are thick-walled and lamellate; there is a dense, laminate chloroplast with several pyrenoids. Most species are marine.

389b **Not growing *in* wood nor in shells, but may occur *on* wood and old shells; sometimes on tree trunks** **390**

390a **Plants containing reddish carotenoid haematochrome.** **391**

390b **Plants without carotenoids** **392**

391a **Plants erect filaments, forming orange-colored, cottony growths on tree trunks, wood and rocks in moist situations. Fig. 7** . . . . *Trentepohlia*

391b **Plants consisting of horizontal filaments growing prostrate, epiphytic on leaves of higher plants. Fig. 4** . . . . . . *Phycopeltis*

392a **Thallus encrusted with lime** **393**

392b **Thallus not encrusted with lime** **394**

393a **Thallus in the form of a cushion, giving rise to compactly arranged, upright branches; cells broadest near the tip of the filaments; growing on wood or shells (sometimes on other plants in water). Fig. 283** . . . . *Gongrosira*

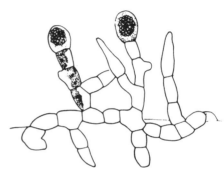

Figure 283

Figure 283 *Gongrosira debaryana* Rab., horizontal and erect branches with terminal sporangia.

Like *Gomontia* (Fig. 282) these plants grow on shells and submersed wood, or on aquatic plants, but form entirely external thalli rather than penetrating the substrate. Plants are sometimes lime-encrusted. The erect branched portion of the thallus is more extensively developed than in *Gomontia* (some filaments are very long). The chloroplast is parietal and more definite than in the latter genus and less of a dense meshwork. Some authors include the genus *Ctenocladus* (Fig. 286) in *Gongrosira*.

393b **Thallus composed of loosely branched filaments, the branches arising unilaterally. Fig. 284** . . . . *Chlorotylium*

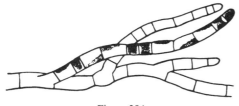

Figure 284

Figure 284 *Chlorotylium cataractum* Kütz., portion of thallus showing characteristic habit of branch development.

The attached, lime-encrusted thalli of this branched filamentous plant are usually found in flowing water. The filaments present a distinctive appearance when seen microscopically because pairs of short, green cells with a parietal chloroplast (often with a reddish tinge) alternate with a more elongate and sometimes nearly colorless cell. Of the 5 known species 2 occur in the United States.

394a **(392) Thallus a tuft of dichotomously branched, radiating, yellow-green filaments. Fig. 285. . . . .*Leptosira***

Figure 285

Figure 285 *Leptosira mediciana* Borzi, portion of plant showing horizontal and erect branching systems.

Filaments occur in yellowish (light green) tufts and are usually attached to substrates. The ir-

regularly branched filaments of beadlike or barrel-shaped cells arising from a prostrate portion of the thallus help in making identification. The chloroplast is a dense, parietal plate (usually). Of the 3 known species 1 only is known from the United States, rather rare but widely distributed.

394b **Thallus not a dense tuft of yellow-green filaments 395**

395a **Bearing enlarged, thick-walled akinetes (vegetative spores), interspersed among the cylindrical cells of the filament, or with such spores at the ends of branches 396**

395b **Without akinetes, or if rarely present, scattered, not arranged as above . . . 397**

396a **Akinetes globular, in chainlike series. Fig. 286 . . . .*Ctenocladus***

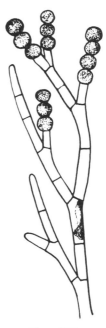

Figure 286

Figure 286 *Ctenocladus circinnatus* Borzi, showing terminal series of akinetes.

This is a branched filamentous plant with cylindrical cells which seems to prefer hard water or alkaline situations. They grow epiphytically on angiospermous plant stems and exposed roots, sometimes forming extensive growths throughout an entire lake. The branches terminate in chains of globular akinetes. *Ctenocladus* is sometimes included under *Gongrosira* (Fig. 283).

396b   Akinetes barrel-shaped or oval, solitary Fig. 287 . . . . *Pithophora*

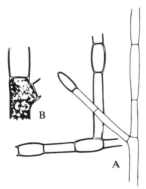

Figure 287

Figure 287 *Pithophora*. (A) *P. mooreana* Collins; (B) *P. oedogonia* (Mont.) Wittr., showing a sample of the chloroplast.

There are 8 species of this irregularly branched, filamentous genus in the United States, differentiated by dimensions of the filament and by the size and shape of the much swollen akinetes that are formed intermittently throughout the plant. The cells are coenocytic and have the cladophoraceous type of chloroplast. When occurring in laboratory aquaria, usually having been brought in on material from biological supply houses, the filaments often fail to develop the characteristic akinetes, the cells becoming exceedingly long and losing some of the appearance by which they are usually identified. Branches arise mostly at right angles to the main axis; frequently an akinete will bear a branch.

397a   (395) Growing on submersed wood and stones, with a prostrate; cushionlike mass of branches from which vertical branches arise. Fig. 283 . . . . *Gongrosira*

397b   Thallus formed differently 398

398a   Plants growing on the shells of turtles (rarely on submersed bricks or rough stones); branches arising only from the extreme base of the main filament. Fig. 288 . . . . *Basicladia*

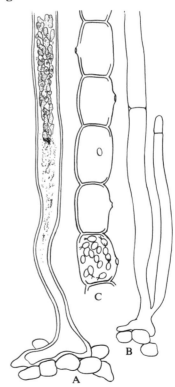

Figure 288

Figure 288 *Basicladia chelonum* (Collins) Hoffman & Tilden. (A) cells at the base of the filament; (B) basal branching habit; (C) series of sporangia formed in the upper section of a filament.

The species illustrated and 3 others (differentiated mostly by size) comprise the genus which is distinctive because of the very long, cylindrical cells and the habit of branching. The genus is invariably found on the backs of snapping (and a few other) turtles. Occasionally plants will be found on bricks in water, or on other especially rough surfaces. The upper cells become zoosporangia and develop lateral pores in the midregion for the release of zoospores.

398b   Plant not growing on turtles, or with other types of branching 399

399a   Plants showing basal-distal differentiation, usually attached (floating in age); branching usually arbuscular (treelike) or bushlike. 400

399b   Plants not forming bushy tufts. 407

400a   Filaments erect, on trees and wood; cells green when growing in dense shade, brick-red or orange-yellow when growing in lighted habitats. Fig. 7 . . . . *Trentepohlia*

400b   Cells without haematochrome; green. 401

401a   Plants stout; chloroplast a dense reticulum; walls thick and lamellate; filaments with arbuscular branching . . . 405

401b   Plants delicate; walls thin, *or* if thick only in the basal portion of the plant; chloroplast a parietal plate or band 402

402a   Filaments either free-floating or attached, branching with cross walls above the level of the branch origin 404

402b   Filaments not as above; branch septations at the point of origin 403

403a   Filaments with arbuscular branching; branches thornlike and short and also long and slightly tapering on the same axis, the cells of the branches smaller than those of the main axis. Fig. 289 . . . . *Cloniophora*

Figure 289

Figure 289 *Cloniophora spicata* (Schm.) Islam, upper portion of thallus to show habit of branching.

These erect, branched filaments have a profuse growth of downward-directed rhizoidal branches and fascicled, arbuscular branches above. The cells of the main axis are larger than those of the branches, but not the same variation in diameter found in *Draparnaldia* (Fig. 299). Unlike *Stigeoclonium* (Fig. 272) which it resembles, the branches do not end in tapering chaetae. The cells contain a chaetophoraceous chloroplast. Three species are recognized, mostly from tropical and subtropical streams.

403b **Filaments irregularly branching, partly prostrate and partly erect, with all cells approximately the same size; chloroplast a parietal plate; plants showing a tendency to form a palmelloid stage. Fig. 290** . . . . *Pseudendoclonium*

(Possibly synonymous with *Trichosarcina*, Fig. 242.)

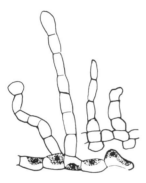

Figure 290

Figure 290 *Pseudendoclonium submarinum* Wille, prostrate and erect branches.

The filaments in this genus grow horizontally on moist substrates but have irregular, short, erect branches. The filaments frequently have dissociated cells which form packets and show the characteristics described for the genus *Trichosarcina* (Fig. 242). Sometimes the erect branches occur as tufts. Plants are known to produce zoospores but sexual reproduction has not been observed. Of the 2 known species, 1 is marine.

404a **(402)** **Thallus arbuscular, minute, sometimes densely tufted, attached (but readily broken from the substrate and floating freely). Fig. 291** . . . . *Microthamnion*

Figure 291

Figure 291 *Microthamnion strictissimum* Rab.

This genus is frequently overlooked because of the small size of the branched, epiphytic growths. Although beginning as attached thalli they frequently float freely and are found in the tychoplankton, intermingled with other filamentous algae. A characteristic which helps in identification is that the septation of the branch occurs some distance above the level of origin. The branches may taper slightly but do not terminate in hairs. The taxonomic position of the genus is under question.

404b **Thallus gelatinous, free-floating or entangled among other algae, with radiating series of irregularly dissociated cells; branching profuse, but irregular. Fig. 292** . . . . *Heterococcus*

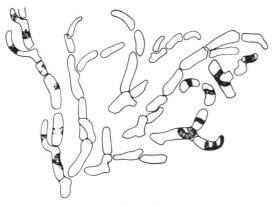

Figure 292

Figure 292 *Heterococcus arcticus* Presc., habit of branching and radiate arrangement of short, pseudofilaments within a gelatinous matrix.

The thallus of this genus consists of irregularly radiating series of (often) dissociated cells. Branches form near the apex of a cell, the new cell soon becoming dissociated. The apical cells are tapered or thornlike and curved in the species illustrated. The chloroplast is a parietal, chaetophoraceous plate covering part of the cell wall. Although lightly green in color, the genus belongs to the yellow-green algae (Tribonematales in the Xanthophyceae).

405a (401) Branching open and spreading; not enclosed in mucilage; apices of branches not tapering to form setae 406

405b Branching close and entangled, resulting in a spongy thallus. Fig. 293 . . . .
. . . . *Aegagropila*

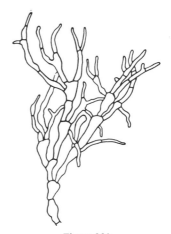

Figure 293

Figure 293 *Aegagropila profunda* (Brand) Nordst.

This genus is more irregularly branched than *Cladophora* (Fig. 294); is sometimes included in that genus. It has a profuse development of downward-directed, rhizoidal branches. The upper filaments are densely entangled and the cells are more irregular in shape than in *Cladophora*. The species illustrated is well-named for it is found growing on the bottom of lakes at depths up to 200 feet, especially in clear water. The chloroplast is cladophoraceous and the cells are coenocytic.

406a Branching forming fasicles, both short and long branches from the same axis; walls thick in the lower part of the filament but not lamellate; chloroplast a parietal band with 1 or a few pyrenoids. Fig. 289 . . . . *Cloniophora*

406b Branching not tufted, arising singly (or in pairs), arbuscular the branching becoming irregular in old plants or in water-washed thalli; cell walls thick and usually lamellate; chloroplast a parietal network with numerous pyrenoids; coenocytic. Fig. 294 . . . . *Cladophora*

Figure 294

Figure 294 *Cladophora glomerata* (A) cell showing parietal, netlike chloroplast (which often appears broken into small plates); (B) habit of branching.

There are numerous species of this genus in both fresh and salt water. Many of the names, however, have been shown to be given to growth-forms or habitat variations of the same species. They are differentiated by size, habit of branching and over-all thallus form. The cell walls are often thick and lamellate. Chloroplasts are disclike thickenings interconnected to form a meshwork and there are many pyrenoids. Although the general habit is arbuscular, plants which winter over or which become wave-washed often develop irregular habits of branching and in many instances intergrade with branched species of *Rhizoclonium* (Fig. 244). Perhaps the most characteristic habit of freshwater *Cladophora* is flowing water, especially on dams and waterfalls.

407a   (399) Vegetative cells very long and cylindrical, somewhat regularly interrupted by swollen, thick-walled spores (akinetes). Fig. 287 . . . . *Pithophora*

407b   Akinetes lacking; cells all cylindrical or nearly so 408

408a   Branches scarce and short, or wanting altogether; if branches present, without repeated branching. Fig. 244 . . . .
. . . . *Rhizoclonium*

408b   Branches many-celled, bearing secondary branches which arise irregularly so that the arbuscular habit is almost lost; wave-washed and winter forms. Fig. 294 . . . . *Cladophora*

409a   (383) Setae without cross walls at the base, formed by lateral extensions of cells immediately below the anterior cross partition of the cell. Fig. 295 . . . . .
. . . . *Fridaea*

Figure 295

Figure 295 *Fridaea torrenticola* Schmidle.

The species illustrated is very rare (the only one known for the genus) but occurs in abundance in particular habitats. It is readily identified by the long, cylindrical cells which bear threadlike extensions that are given off laterally below the anterior cross wall of the cell. The filaments occur in compact tufts and usually are yellow-green. The chloroplast is parietal and laminate, but often so dense that the exact organization is difficult of determination. Zoosporangia are elongate and saclike, produced laterally on the filaments.

409b   **Setae formed otherwise** **410**

410a   **Setae bulblike at the base. Fig. 296** . . . .
. . . . *Bulbochaete*

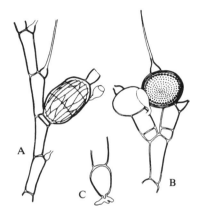

Figure 296

Figure 296 *Bulbochaete*. (A) *B. insignis* Pringsh., showing oogonium, oospore and an attached male plant (1-celled); (B) *B. congener* Hirn; (C) basal hold-fast cell.

Like its close relative *Oedogonium* (Fig. 245) this genus contains numerous species which are likewise differentiated by dimensions and characteristics of the sex organs and the oospore. They cannot be identified to species in the vegetative condition. The branched filaments are attached (at least when young) and are readily recognized by the bulbous- based, unicellular setae that develop at the anterior end of the cell (but lateral). The plants can be sought on overhanging grass, or on the culms of rushes, submersed plants, *etc.* Most species have dwarf male plants that grow epiphytically on the female sex organ (oogonium).

410b   **Setae shaped and formed otherwise** . . . .
. . . . **411**

411a   **Setae sheathed at the base. Fig. 233** . . . .
. . . . *Coleochaete*

411b   **Setae not sheathed at the base** **412**

412a   **Thallus not embedded in mucilage, or if so, enclosed in a very soft, watery mucilage without definite shape.** **413**

412b   **Thallus enclosed in a firm mucilaginous matrix of definite shape, globular or somewhat elongate and irregularly arbuscular (sometimes strands are from 4 to 14 cm long). Fig. 297.** . . . *Chaetophora*

Figure 297

Figure 297 *Chaetophora*. (A) *C. elegans* (Roth) C.A. Agardh, habit of thallus; (B) *C. incrassata* (Huds.) Hazen, habit of thallus; (C) portion of thallus showing longitudinal filaments with out-turned branches.

Microscopically, species of this genus are delicately and gracefully branched filaments that occur in macroscopic tufts, or as gelatinous, small balls. One, *Chaetophora incrassata* (Huds.) Hazen is composed of cables of elongate cells which give off laterally dense tufts of dichotomous branches. The resulting growth produces a somewhat arbuscular, gelatinous strand which may be as much as 14 cm long. Other species form spherical or irregularly globose balls, 1 or 2 mm in diameter on submersed leaves and wood (especially in cold water), or on cattail stems. Within the globular colonies the filaments are radiate and dichotomously branched. Colonies are often gregarious and may form extensive patches. The firmness of the mucilage in which the plants are encased macroscopically differentiates *Chaetophora* from *Draparnaldia* which has very soft, amorphous mucilage, and from *Stigeoclonium* which exhibits no mucilage.

413a   **Thallus composed of slender, repeatedly branched filaments; cells all about the same size but in the branches tapering to fine points or setae. Fig. 272** . . . .
. . . . *Stigeoclonium*

413b   **Thallus consisting of main filaments of larger cells from which arise tufts and plumes of branches composed of much smaller cells** **414**

414a   **Main axis consisting of cells of 2 sizes, cylindrical or barrel-shaped, and short, rectangular; the tufts of branches composed of much smaller cells arising opposite one another only from the short cells. Fig. 298** . . . . *Draparnaldiopsis*

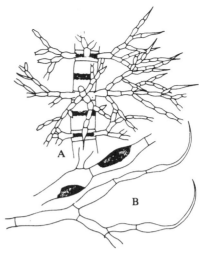

Figure 298

Figure 298 *Draparnaldiopsis salishensis* Presc. (A) portion of main axis with branch tufts; (B) sample of branch showing shape of cells.

This genus is similar to *Draparnaldia* (Fig. 299) but the main axis cells are of 2 sizes, the longer cells alternating with the shorter which bear the lateral branches. The branchings are dichotomous and rather rigid in appearance in contrast to the graceful plumes found in most species of *Draparnaldia*. Two species are known for the United States, both western in distribution.

414b   **Main axis composed of cylindrical or barrel-shaped cells all the same length; branches arising opposite, alternate, or in whorls. Fig. 299** . . . . . . *Draparnaldia*

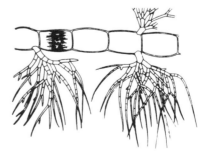

Figure 299

Figure 299 *Draparnaldia glomerata* (Vauch.) Ag., showing tufts of lateral branches, with barrel-shaped axial cells containing bandlike chloroplasts.

The genus is strikingly characterized by having a filament of large cells forming an axis from which tufted plumes of branches of small cell size arise. Different species vary in size and shape of the branched tufts. The thallus is enclosed in amorphous, soft and watery mucilage; occur mostly in cold, flowing water and trickles from springs. Macroscopically the plants appear as pale yellow-green gelatinous strands that easily slip through the fingers when gathered.

415a   **(359) Parasitic in higher plants such as *Arisaema* (Indian Turnip). Fig. 159. . . .** . . . . *Phyllosiphon*

415b   **Not parasitic on higher plants 416**

416a   **Filaments repeatedly, dichotomously branched, regularly constricted at the base of the forkings. Fig. 300 . . . .** . . . . *Dichotomosiphon*

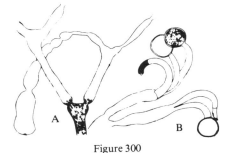

Figure 300

Figure 300 *Dichotomosiphon tuberosus* (Braun) Ernst. (A) habit of branching showing constrictions; (B) tip of a reproductive branch showing globular oogonia and hooklike antheridia.

The species illustrated is the only one in the genus. It occurs in dense, tangled tufts and mats, usually on the bottom of lakes, although occasionally on damp soil and seeps about springs. There are downward-growing, rhizoidal branches and upward-directed vegetative and sex organ-bearing branches. The oogonia when mature are so large that they can be discerned with the naked eye. They are yellowish and globular. Curved, antheridial organs are borne immediately below the oogonia which are terminal on the branch. The plants seem to reproduce sexually only when growing in relatively shallow water (up to 4 feet) suggesting a light factor relationship. Plants dredged from 60 feet or more never seem to have reproductive organs. The filaments are siphonous, coenocytic tubes. The chloroplasts are numerous, parietal discs or ovals, and starch is produced as a food reserve. Now placed in its own family, *Dichotomisiphon* was at one time associated with *Vaucheria* (Fig. 301) which now is in the Chrysophyta.

416b   **Filaments not dichotomously branched; without constrictions; iodine test for starch negative. Fig. 301 . . . . *Vaucheria***

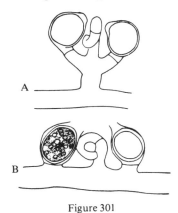

Figure 301

Figure 301 *Vaucheria*. (A) *V. geminata* (Vauch.) De Cand., sex organs on a short pedicel; (B) *V. sessilis* (Vauch.) De Cand.

This genus usually forms dark green, velvety

mats on damp soil, on rocks in flowing water, or occasionally woolly strands streaming from stones. Plants may break away from the substrate and form floating, often 'dirty' mats at the surface. The siphonous filaments are large enough to be seen individually with the unaided eye. Several species are common in freshwater, but there are also many marine and brackish water forms. They are differentiated by shape and position of the sex organs, and by size of the filaments. The mats harbor a veritable zoological garden of small animals. Long considered as a member of the Chlorophyta, this genus actually belongs to the Xanthophyceae (Chrysophyta) in spite of its densely green color. The food reserve is oil rather than starch and the multiflagellate zoospores have pairs of flagella unequal in length.

417a (6) **Plants macroscopic, *or* if microscopic, with pseudopodia or flagella; chloroplast violet-green, gray-green, bluish, rarely reddish. The thallus macroscopically may appear tawny, brownish, or gray, sometimes dark purple to black. Look here for freshwater members of the Rhodophyta** **418**

417b **Plants microscopic; chloroplasts some other color than above: yellow-green, golden-yellow, or brownish, with xanthophylls and carotenes predominating; rarely some yellowish chloroplasts or the protoplasm will show bluish-green tinges.** **431**

418a **Plant a protoplasmic meshwork over and in *Sphagnum;* parasitic, with bluish (sometimes brownish) color, with pseudopodia in the form of fine strands, radiating and anastomosing. Fig. 157** . . . . *Chlamydomyxa*

418b **Plant not a protoplasmic meshwork** **419**

419a **Plant a motile, reniform cell with 2 laterally attached flagella; chloroplast bluish (although sometimes described as brownish), a parietal lobed plate covering most of the cell wall. Fig. 302.** . . . . *Protochrysis*

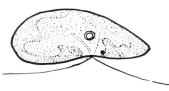

Figure 302

Figure 302 *Protochrysis phaeophycearum* Pascher. (Redrawn from Pascher.)

This reniform cell is much like *Heteromastrix (Nephroselmis* [Fig. 54]) in shape but has no eye-spot or gullet. It is included here because of the brownish or bluish chloroplast. Although the classification is not certain these genera comprise the Nephroselmidaceae in the Cryptophyta. Both genera are similarly pigmented. Apparently only 1 species each reported from the United States.

419b **Plants otherwise** **420**

420a **Plants microscopic.** **429**

420b **Plant macroscopic** **421**

421a **Plant a gelatinous, tubular sack, attached in flowing water to stones and snail shells; pinkish or reddish tan; wall of sack one-celled thick. Fig. 303.** . . . . *Boldia*

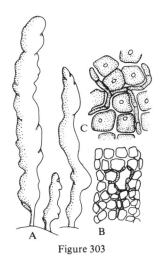
Figure 303

Figure 303 *Boldia erythrosiphon* Hern. (A) habit of tubular thallus; (B) peripheral wall cells of the tube showing filaments growing among the larger corticating cells; (C) a few cells of the cortex and the wandering, filaments.

Boldia plants are hollow sacks, the walls of the tube composed of large and irregularly shaped cells, among which, externally, small filaments 'creep'. The latter form inward-protruding cells from which sexual reproductive elements are produced. There appear to be 2 species occurring in southeastern United States, but also in Missouri.

421b   Plant not formed as above **422**

422a   Plant an attached, much-branched, soft threadlike growth, up to 50 cm in length; pink or reddish; in age becoming expanded in the upper portions; branches curved or S-shaped; thallus consisting of longitudinal filaments giving rise laterally to a cortex of short, branched filaments, the margin of the thallus smooth. Fig. 304 . . . . *Nemalionopsis*

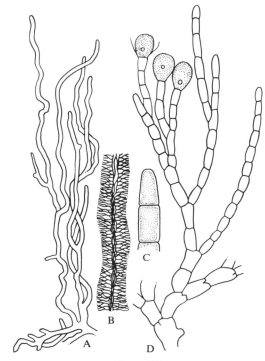
Figure 304

Figure 304 *Nemalionopsis shawi* Skuja. (A) habit of branched, cordlike thallus; (B) diagram of longitudinal section of thallus to show vertical cable of filaments and outturned branches; (C) cells from a branch to show platelike chloroplast; (D) a branch bearing terminal sporangia.

This genus, for which there is only 1 species known, occurs in flowing water, attached to stones. The plants when mature are cordlike and freely and irregularly branched. Microscopically the structure resembles that of *Thorea* (Fig. 306) but the outturned branches are very short and equal in length so that the periphery is smooth. Thus far *Nemalionopsis* has been found only in southeast United States.

422b   Plant not a soft, branched threadlike growth **423**

423a Thallus spinelike or spurlike, (2 to 20 cm) high, with intermittent nodelike swellings; relatively stiff and cartilaginous; thallus simple and mostly unbranched, but microscopically composed of a complex of longitudinal filaments and a thick cortex of cells; plants of swift water and waterfalls. Fig. 305 .... *Lemanea*

Figure 305

Figure 305 *Lemanea annulata* Kütz., habit of plant.

This genus is a member of the Rhodophyta, but like other freshwater red algae it is some other color, being gray- or olive-green or tawny. The thalli are cartilaginous and stand erect from an attached base—but are bent flat in swiftly flowing water over dams and rapids. The slender, spinelike growths (up to 20 cm in length), devoid of practically any macroscopic branching make the plant readily identified. The 'spurs' have intermittent swellings or nodes. The thallus is composed of a complex of branched filaments which have apical growth. Carpogonia are internal, in the nodal region, and the spermatia are borne externally in clusters near the pores in the nodes. At least 13 species have been reported from the United States; are widely distributed.

423b Plants not stiff and cartilaginous . . . 424

424a Plants embedded in soft mucilage; thallus consisting of an axial filament bearing whorls of short branches, giving a beaded effect. 425

424b Plants not embedded in soft mucilage 426

425a Mature plants bearing a dense cluster of carpospores which develop at the base of the female organ (carpogonium) after fertilization, the clusters showing as dark masses here and there throughout the thallus. Fig. 263. . . . . *Batrachospermum*

425b Mature plant with carpspores borne singly or in clusters of 2 or 3 along filaments (gonimoblasts) which have developed from the fertilized egg and which wander throughout the thallus; dense heads of spores not evident as in *Batrachospermum*. Fig. 264 . . . . *Sirodotia*

426a (424) Thallus consisting of multiaxial filaments (strands or filaments) with numerous, feathery branchings; without definite nodes and internodes, and without whorls of branches. Fig. 306 . . . . *Thorea*

Figure 306

Figure 306 *Thorea ramosissima* Bory, habit of plant showing numerous, fine branches.

This is a feathery thallus macroscopically; microscopically composed of a multiaxial cable of filaments with short, compactly arranged, outturned branches. They may be as much as 50 cm in length. The genus is of infrequent occurrence but abundant in habitats where it does appear. Sexual reproduction has not been observed as yet. Four species have been reported from the United States; widely distributed.

426b   Thallus otherwise. 427

427a   Thallus cartilaginous and irregularly dichotomous in branching; composed of a complex of central filaments from which a thick cortex of superficial cells develops (examine slightly crushed tips of branches to determine the structural plan of the thallus). Fig. 307 . . . .
. . . . *Tuomeya*

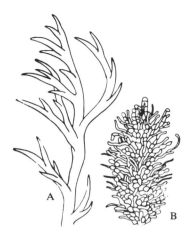

Figure 307

Figure 307 *Tuomeya fluviatilis* Harvey. (A) habit of thallus; (B) apical portion of branch.

This rather rigid and cartilaginous member of the Rhodophyta is identified by the complex, dichotomous or antlerlike habit of thallus branching. The thallus is composed of multiaxial series of filaments and corticating cells. There are 2 species from the United States but only in the eastern half.

427b   Thallus constructed otherwise; not cartilaginous 428

428a   Thallus rather regularly branched, arbuscular, composed of a monaxial filament which is enclosed and surrounded by a cortex of compactly arranged and appressed, polygonal cells which form and develop immediately behind the apex (examine tips of branches) plants of fresh, often hard water. Fig. 308 . . . .
. . . . *Compsopogon*

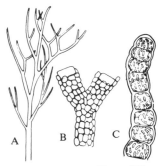

Figure 308

Figure 308 *Compsopogon coeruleus* (Baĺbis) Mont. (A) habit of thallus; (B) portion of axis showing corticating cells; (C) apical portion of uncorticated branch showing chloroplasts.

This is a member of the Rhodophyta which occurs more often in hard or brackish water, attached to stems of woody plants, weeds and stones. *Compsopogon coeruleus* (Balb.) Mont. is the most common species; has been found in irrigation ditches in far western United States. Although essentially filamentous, the thallus becomes macroscopic in proportion and appears as a rather delicately branched, tufted or bushy growth, blue-green to gray- or violet-green in color. The main axial filament becomes encased in a cortication of small, angular cells. Sexual reproduction is unknown. It is primarily a tropical and subtropical genus.

428b Thallus regularly dichotomously branched; multiaxial (several associated filaments), and corticated; thallus partly prostrate and somewhat dorsiventrally differentiated, with the concave side down, the branches tending to curl at the tips; somewhat tufted, dark gray to blackish-green in color; plants of brackish situations, or marine. Fig. 309 . . . . .
. . . . . *Bostrychia*

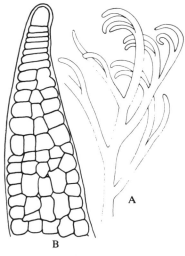

Figure 309

Figure 309 *Bostrychia scorpioides* (Gmel.) Mont. (A) habit of branching in upper portion of thallus; (B) apex of branch showing development of cortex.

This genus of the red algae is mostly marine, growing on mangrove, but at least 1 species occurs in brackish water and in estuaries. *Bostrychia scorpioides* (Gmel.) Mont. is found on the walls of locks in the Panama Canal and on *Typha* in the adjacent lakes. The thallus is dichotomously branched, somewhat flattened with a tendency to have enrolled margins; the branches are curled at the tips.

429a (420) Filaments violet or brownish-green, unbranched except for rhizoidal proliferations at the base 430

429b Plants branched, violet or gray-green, usually growing in tufts; branches not tapering at the apex; cells with 1 disclike or ribbonlike chloroplast which covers only part of the wall. Fig. 310 . . . .

.... *(Audouinella, Acrochaetium)*
Rhodochorton

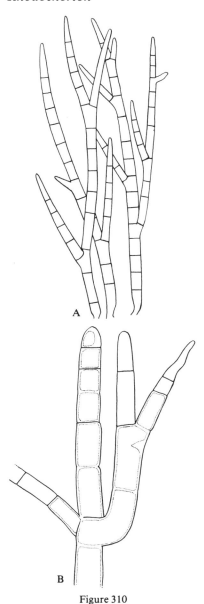

Figure 310

Figure 310 *Rhodochorton* sp. (A) habit of thallus; (B) branches at apex of a filament.

This member of the Rhodophyta has been known previously as *Audouinella*. The erect, branched (and usually tufted) filaments are violet- or gray-green, at least when occurring in macroscopic clumps. The main axial cells have disclike or short, ribbonlike chloroplasts. In size and form and habit of growth these plants may be mistaken for juvenile stages of *Batrachospermum,* and if that genus is present in the habitat, it may be assumed that what appear at first observation to be *Rhodochorton* is the so-called "Chantransia" stage of *Batrachospermum.* The chloroplast of the juvenile stage of that genus is platelike and extends the full length and width of the cell (at least in young, alive specimens).

430a   Filaments relatively stout, mucilaginous or somewhat cartilaginous, with a firm gelatinous sheath enclosing cells separated from one another; chloroplast stellate, microscopically reddish; plants mostly gregarious, sometimes solitary. Fig. 311 . . . . *Bangia*

Figure 311

Figure 311 *Bangia atropurpurea* (Roth) Ag., habit of filament showing downward proliferations of basal cells.

This is a simple, unbranched member of the Rhodophyta, either greenish or light purplish in color. Filaments occur in tufts, sometimes forming dense fringes on rocks at water level in lakes or in the splash zone. Although most species of *Bangia* occur in marine habitats, *B. atropurpurea* (Roth) Ag. is found inland. Possibly the plant known as *B. fuscopurpurea* (Dillw.) Lyngb. from marine and estuarian habitats is the same. The filaments are uniseriate below but become multiseriate above, the protoplasts separate from one another in a firm, gelatinous sheath. The lowermost cells send down rhizoidal elongations as hold-fast structures. The chloroplast is stellate. Thus far, *B. atropurpurea* is known in the United States only from the Great Lakes, and possibly from Pennsylvania.

430b   Filaments gray-green in colonial mass (sometimes brownish), but cells pink microscopically; tufted, the filaments arising from a prostrate patch of cells; chloroplasts several parietal plates. Fig. 243 . . . . *Kyliniella*

431a   (417) Chloroplasts, yellow-brown, or dark golden-brown; plants motile or non-motile. 501

431b   Plants not yellow- or golden-brown. 432

432a   Chloroplasts yellow, pale yellow-green, with xanthophyll predominating; iodine test for starch negative; cell contents with oil or with leucosin showing a metallic lustre. (See *Botrydium,* [Fig. 326] and *Vaucheria* [Fig. 301], members of the Chrysophyta but green, chlorophyll predominating the carotenoid and yellow pigments which are present.) Look here for members of the Xanthophyceae in the Phylum Chrysophyta 445

This group of the algae is difficult of separation on the basis of the chloroplast color alone inasmuch as the shades of green, and the relative abundance of carotenoid and xanthophyll pigments cannot be clearly distinguished from the shades of green in the Chlorophyta. In addition to the iodine test for starch (which when negative is confirmative that the plant does not belong to the Chlorophyta), heating specimens in concentrated sulphuric acid (when the specimens lend themselves to such treatment) provides a helpful differentiation. The yellow-green algae (Xanthophyceae or Heterokontae) become blue-green in acid; whereas Chlorophyta remain unchanged in color.

432b   Chloroplasts not yellow-green 433

433a Cells with blue or bluish protoplasts **434**

433b Cells with chloroplasts or protoplasts not blue **439**

434a Organisms with 2 flagella **435**

434b Organisms not flagellated **436**

435a Cells ellipsoid or oval, or slipper-shaped, usually bilobed anteriorly with the flagella somewhat lateral; chloroplasts 3, blue or bluish, parietal plates. Fig. 312 . . . . *Chroomonas*

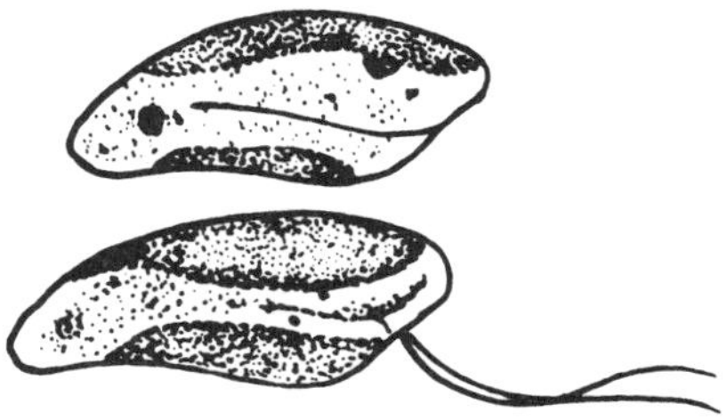

Figure 312

Figure 312 *Chroomonas nordstedtii* Hansg., dorsal and ventral surfaces.

These minute, slipper-shaped organisms have 2 parietal, blue or bluish-brown chloroplasts and 2 flagella that are attached immediately below the cell apex. They move rapidly and determinations cannot be made unless some medium is introduced to the mount to retard action. Use 5% glycerin. *Chroomonas* should be compared with *Cryptomonas* (Fig. 16) which has yellowish-green chloroplasts or at times reddish. The latter genus has a gullet whereas *Chroomonas* does not.

435b Cells reniform, with laterally attached flagella and 2 blue or bluish-green chloroplasts. Fig. 302 . . . . *Protochrysis*

436a (434) Organism an anastomosing, protoplasmic meshwork, 'creeping' by pseudopodia over *Sphagnum*, with blue-green chloroplasts. Fig. 157 . . . . *Chlamydomyxa*

436b Organisms otherwise **437**

437a One to 4 spherical or oval cells in a mucilaginous sheath which bears a gelatinous bristle; protoplast cup-shaped with oval, blue 'chloroplasts' (endophytic blue-green cells). Fig. 313 . . . . *Gloeochaete*

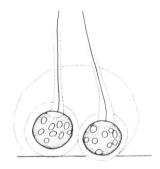

Figure 313

Figure 313 *Gloeochaete wittrockiana* Lag., two cells showing blue-green 'cells' within a cup-shaped protoplast.

This is an anomalous organism, the morphology and physiology of which are interpreted as that of symbiotism involving a colorless member of the Tetrasporaceae (Chlorophyta) and a blue-green (Cyanophyta) endophyte. The protoplast is a parietal cup similar to the chloroplast of many of the green algae but contains several oval blue-green bodies. The long, slender, gelatinous hairs make this genus readily identified. The cells occur in clumps, rarely solitary, attached to filamentous algae.

437b Plants not as above; sheaths present or absent although cells may be enclosed in old, mother-cell walls. **438**

438a A linear series of globular or oblong cells within a mucilaginous sheath; chloroplasts star-shaped. Fig. 314 . . . . *Asterocystis*

Figure 314

Figure 314 *Asterocystis smaragdina* (Reinsch) Forti.

The bright blue-green, star-shaped chloroplasts of this genus help in making identification. The cells are oval or oblong, arranged in mostly uniseriate, false filaments. The gelatinous strands may branch. Plants occur attached as an epiphyte, usually on members of the Cladophoraceae, but may appear in mixtures of free-floating algae in the tychoplankton. The genus is regarded as a 'lower' member of the Rhodophyta, based on the pigmentation and the nature of the food reserve.

438b Two to 4 or 8 globose or oval cells contained within an enlarged mother-cell wall; chloroplastlike bodies vermiform (few and long, or many and short). Fig. 315 . . . . *Glaucocystis*

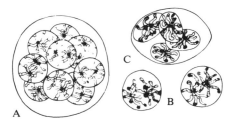

Figure 315

Figure 315 *Glaucocystis.* (A), (B) *G. duplex* Presc.; (C) *G. nostochinearum* Itz.

There are about 5 species of this genus which, like *Gloeochaete* (Fig. 314) involve an endophytic blue-green alga and a colorless member of the Chlorophyta, an oocystaceous host. The protoplasts are bright blue-green and occur in different shapes and arrangements within either globular or oval cells, according to species. The plants are free-floating, solitary or in families, enclosed by the old, mother-cell wall. They seem to prefer soft-water habitats.

439a (433) Cells in compact, irregularly shaped colonies, appearing brown or orange-colored because of dark mucilage. Fig. 95 . . . . *Botryococcus*

439b Cells not in opaque or orange-colored mucilage as above; contents red, violet-red or green with a red tinge **440**

440a Living in snow and alpine regions (red snow). Fig. 17 . . . . *Chlamydomonas*

440b Not living in snow **441**

441a A colony of oval or globose cells enclosed in a layered, colored sheath. Fig. 86. . . . . *Gloeocystis*

441b Cells not enclosed in a layered sheath **442**

442a Cells irregular in shape; living as an endophyte in higher plants. Fig. 20 . . . . *Rhodochytrium*

442b Cells not endophytic **443**

443a Cells spherical, solitary or gregarious, in thin, mucilaginous strands; terrestrial, forming dark red or purplish patches on damp soil (common in greenhouses). Fig. 9 . . . . *Porphyridium*

443b Cells round, ellipsoid, or fusiform, not arranged as above; not terrestrial (or rarely so) 444

444a Cells fusiform; orange-red carotenoid coloring the green cell; with 1 flagellum. Fig. 13 . . . . *Euglena*

444b Cells round or ellipsoid; protoplast with a wide wall which appears as a gelatinous sheath; flagella 2, but usually not showing when cells are encysted at which time they exhibit a red color. Fig. 14 . . . . *Haematococcus*

445a (432) Plants filamentous 446

445b Plants not filamentous. 454

446a False filaments; cells not adjoined in continuous series 447

446b Plant a true filament; cells in continuous series 449

447a Thallus an attached, dichotomously branched stalk with transverse striations; cells pyriform, thick-walled, solitary or in pairs at the end of the stalk branches. Fig. 316 . . . . . *Malleodendron*

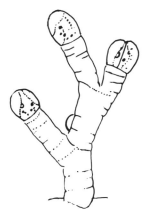

Figure 316

Figure 316 *Malleodendron gloeopus* Pascher. (Redrawn from Pascher.)

The branched, gelatinous stalks are relatively thick, being about the same diameter as the cells they bear. The stalks have transverse lamellations. *Malleodendron* grows epiphytically on filamentous algae. The cells divide vertically and the daughter cells add to the gelatinous stalk from their lower side. The genus is known from both the United States and Europe but is relatively rare.

447b Thallus otherwise. 448

448a Thallus a series of branched, gelatinous tubes arranged in a chain with 1 or 2 globular cells in the apex of each joint of the branching, tubelike strands. Fig. 317 . . . . *Mischococcus*

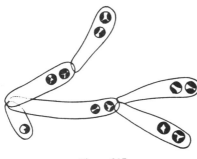

Figure 317

Figure 317   *Mischococcus confervicola* Näg.

The globose, yellow-green cells of this species occur at the ends of repeatedly branched, gelatinous stalks, attached to filamentous algae. Only 1 species seems to be known and this is widely distributed in the United States.

448b   Thallus consisting of a floating, gelatinous mass in which short, interrupted series of cells radiate as short, branched, filaments. Fig. 292 . . . . *Heterococcus*

(Regarded by some to be synonymous with *Monocilia* [Fig. 319].)

449a   (446)   Filaments prostrate, laterally adjoined to form a plate or an attached disc. Fig. 318 . . . . *Phaeodermatium*

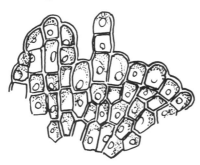

Figure 318

Figure 318   *Phaeodermatium rivulare* Hansg. (Redrawn from Pascher.)

This plant forms a dislike, encrusting growth, several cells thick in the center; 1-celled thick and showing radiating filaments at the margin. It has been suggested that this is a stage in the life cycle of *Hydrurus* (Fig. 353), which grows in the same habitat—cold mountain streams.

449b   Plants otherwise 450

450a   Filaments branched, erect. Fig. 319 . . . . *Monocilia*

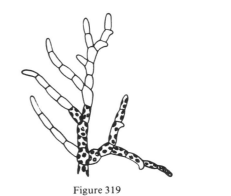

Figure 319

Figure 319   *Monocilia viridis* Gerneck.

This and 1 other species, *M. flavescens* Gerneck are the only ones reported from the United States, although if included with the genus *Heterococcus* (Fig. 292) as recommended by some authors, there are many species. *Monocilia* occurs as an irregularly branched filament in soil. A member of the Chrysophyta, the disclike chloroplasts are yellow-green or bright yellow and the food reserve is oil; starch test negative. See *Heterococcus,* Fig. 292.

450b   Filaments not branched 451

451a Filaments short, of from 2 to 8 barrel-shaped cells; chloroplasts brownish. Fig. 320 . . . . *Stichochrysis*

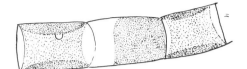

Figure 320

Figure 320 *Stichochrysis immobilis* Pringsh. (Redrawn from Pringsheim.)

This odd genus has very short filaments of cylindrical or barrel-shaped cells. There are usually 2 laminate chloroplasts (or only 1) along the lateral walls; pale yellow to brownish. Reproduction occurs only by cell division as far as known. It is known from the United States.

451b Filaments otherwise 452

452a Cells long-cylindric; wall in 2 pieces which overlap at the midregion, the 2 sections usually evident when cells are empty; filaments showing H-shaped pieces upon fragmentation; cells with parallel or slightly convex lateral margins; chloroplasts 2 to many parietal discs. Fig. 321 . . . . *Tribonema*

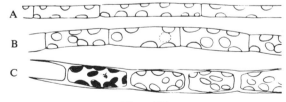

Figure 321

Figure 321 *Tribonema*. (A) *T. bombycinum* var. *tenue* Hazen; (B) *T. bombycinum* (Ag.) Derbes & Solier; (C) *T. utriculosum* (Kütz.) Hazen.

There are many species of this simple, unbranched, pale-colored member of the Xanthophyceae (Heterokontae), differentiated by the size and proportions of the cells and by the shape and number of the chloroplasts. Like *Microspora* (Fig. 262) in the Chlorophyta, the cell walls are composed of 2 sections which adjoin and overlap in the midregion. Hence, when the filaments fragment, the cells disjoin at the midsection rather than at the cross walls, resulting in the characteristic H-shaped sections. By careful focusing the overlapping of the wall sections can be seen in the unfragmented portions of the filament, especially in those few species which have a thick wall. The cells are rectangular or cylindrical although some species have somewhat convex lateral walls. The chloroplasts are pale green parietal discs or folded plates. In northern latitudes *Tribonema* is one of the first of the algae to appear in ditches and swamps after the ice thaw.

452b Cells otherwise 453

453a Cells quadrate, the sides decidedly parallel; the wall in 2 sections but overlapping not at all evident except at the broken ends of the filaments; occasionally and intermittently with brownish collars about the cells where the 2 sections overlap. Fig. 322 . . . . . . *Bumilleria*

Figure 322

Figure 322 *Bumilleria sicula* Borzi.

There are 2 species of the 3 that are known for this genus in the United States. *Bumilleria exilis* Klebs is common and much the smaller (6 μm in diameter). The unbranched filaments are similar to *Tribonema* (Fig. 322) but the cells are more nearly rectangular in optical section, with decidedly parallel lateral walls. Sometimes external overlapping H-shaped sections of thicker wall layers, brownish in color occur intermittently along the filament.

453b **Cells oval with truncate apices, the filaments distinctly constricted at the cross walls; cell walls in 1 section only; chloroplasts 1 or 2 parietal, folded plates. Fig. 323 . . . . *Heterothrix***

Figure 323

Figure 323   *Heterothrix ulothricoides* Pascher. (Redrawn from Bourrelly.)

In this genus the filaments are composed of cylindrical or subcylindrical cells which may have broadly rounded poles. The filaments are usually relatively short and some species have a tendency for the cells to dissociate. Occasionally there is also a tendency to produce branches in the dissociation, as in *Heterothrix stichococcoides* Pascher. There is no basal-distal differentiation. The wall, unlike *Tribonema* (Fig. 321) is in 1 section. Chloroplasts are 1 or 2 parietal, folded plates; reproduction by zoospores. Of the 13 described species only 1 has been reported from the United States, Massachusetts.

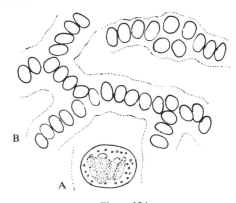

Figure 324

Figure   324   *Sphaeridiothrix   compressa* Pascher & V1. (A) single cell showing chloroplast; (B) portion of thallus. (Redrawn from Pascher.)

454a **(445) Plant a small (1-2 mm diameter) green vesicle, balloon-shaped; terrestrial. Fig. 325 . . . . *Botrydium***

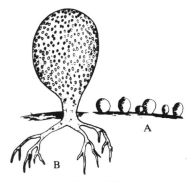

Figure 325

Figure 325   *Botrydium granulatum* (L.) Grev. (A) habit of thalli on soil; (B) vesicle enlarged (semidiagrammatic) to show subterranean rhizoidal extensions and numerous oval chloroplasts.

These tiny, 1-celled, green balloonlike algae appear on damp soil in greenhouses or on mud where water has receded. *Botrydium* is siphonaceous and coenocytic; has underground, rhizoidal extensions. Here the cell contents may concentrate, followed by rounding up of segments to form a special kind of spore known as

*See *Sphaeridiothrix* (Fig. 324) in which oval, unadjoined cells are uniseriately arranged as a filament within a gelatinous sheath, sometimes forming anastomosing branches and becoming uniseriate, resembling some expressions of a filamenous relative of *Heterothrix,* Fig. 323.

a hypnospore. Chloroplasts are numerous, green discs. Oil rather than starch is produced as a food reserve and zoospores have 2 flagella of different length, characteristic of the Xanthophyceae. Besides the species illustrated there is another which is less frequently found, *B. wallrothii* Kütz. in which the vesicle has a thick, wrinkled and lamellate wall.

454b Plant not a green vesicle; aquatic . . . 455

455a Cells solitary or incidentally clustered 456

455b Cells in colonies, definite or indefinite in shape and arrangement; sometimes forming stalked colonies 489

456a Cells attached, sessile, *or* seriately arranged on a stalk 457

456b Cells free-floating or swimming 466

457a Cells sessile 458

457b Cells on a long or short stalk 461

458a Cells in an attached, colorless, thin-walled lorica; appressed on its side (prostrate). Fig. 326 . . . . *Kybotion*

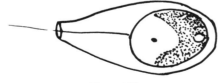

Figure 326

Figure 326 *Kybotion eremita* Pascher.

This curious genus has a colorless, vaselike lorica that lies on its side on aquatic substrates (especially leaves of aquatic plants). The protoplast within has a yellowish chloroplast and a rhizoidal thread extending toward or through the lorica aperture. There are 3 species known, differentiated by the shape of the lorica.

458b Cells otherwise. 459

459a Cell membrane in 2 sections, the upper lifting off at maturity to allow escape of aplanospores (small globular spores); cells oval or short cylindric. Fig. 327 . . . . *Chlorothecium*

Figure 327

Figure 327 *Chlorothecium pirottae* Borzi.

This cylindrical plant with parietal, yellow-green chloroplasts is attached by a short stalk and a disc to submersed plants, including larger algae. It is rather rare in the United States; only 1 of the 12 species being reported. It is easily overlooked because it occurs in dense mixtures of algae from bogs.

459b Cell membrane not in 2 sections 460

460a Cells globose or subglobose; cytoplasm reticulate. Fig. 328 . . . . *Perone*

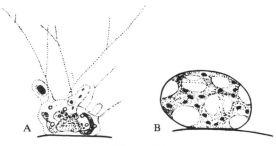

Figure 328

Figure 328 *Perone dimorpha* Pascher. (A) rhizoidal stage; (B) vegetative cell or secondary stage.

There is a freely moving, amoeboid stage and an attached or epiphytic, encysted stage in the life history of this organism. In the resting stage the cell is to be found in *Sphagnum* or other moss leaves, and in this state shows a highly reticulated, faintly pigmented protoplast. The chloroplasts are several small discs. In the amoeboid phase the cell throws out long, fine, threadlike and branched pseudopodia. Heterokont zoospores are formed in reproduction. The species illustrated is the only one reported thus far.

460b Cells shaped otherwise; cytoplasm not highly reticulated nor alveolar. Fig. 329. . . . . *Characiopsis*

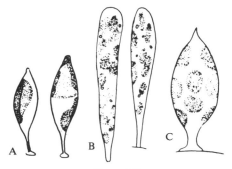

Figure 329

Figure 329 *Characiopsis.* (A) *C. acuta* (Braun) Borzi; (B) *C. cylindrica* (Lambert) Lemm.; (C) *C. spinifer* Printz.

There are several fairly common species in this genus which occur as epiphytes on filamentous algae. They vary in shape and in length of the attaching stalk. Unlike the genus *Characium* (Fig. 150), some species of which are very similar in shape, oil is formed rather than starch as a food reserve, and the chloroplasts are pale yellow-green. The starch-iodide test must be used to differentiate the 2 genera.

461a (457) Cells with a vaselike, pitcher-shaped lorica (envelope) with a neck and a terminal opening. Fig. 330 . . . . *Stipitococcus*

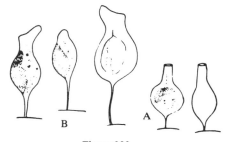

Figure 330

Figure 330 *Stipitococcus.* (A) *S. vasiformis* Tiff.; (B) *S. urceolatus* W. & W.

There are 5 or 6 species of this genus of rather uncommon occurrence, although in particular habitats the various species are relatively abundant on filamentous algae. *Stipitococcus urceolatus* West & West is perhaps more frequently seen than others and is readily identified by the pitcher-shaped lorica. There is a pale, yellow-green, platelike chloroplast and a rhizoidal thread which often is obscure.

461b Cells without a lorica **462**

462a Cells cylindrical, straight or curved, sometimes with a spine at one or both ends. Fig. 331 . . . . *Ophiocytium*

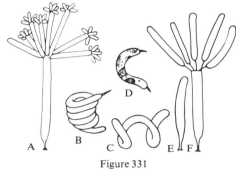

Figure 331

Figure 331 *Ophiocytium.* (A) *O. arbuscula* Rab.; (B) *O. cochleare* A. Braun; (C) *O. parvulum* (Perty) Braun; (D) *O. capitatum* Wolle; (E) *O. desertum* fa. *minor* Presc.; (F) *O. gracillipes* (Braun) Rab.

The factors which determine the distribution of this genus are unknown, but there is good evidence that highly refined chemical qualities of water are critical. Several species always occur in the same habitat, along with other genera of the Xanthophyceae. Species are differentiated by presence or absence of polar spines and whether free-floating or stalked and epiphytic. They usually occur intermingled with miscellaneous algae in bogs or swamps which are acid. Most species are solitary but *Ophiocytium arbuscula* Rab. is colonial because of the habit of zoospores to germinate at the apex of the parent cell to which the new cells become attached.

462b Cells shaped differently **463**

463a Cells club-shaped or somewhat pyriform; walls in 2 sections, the upper lifting away to permit escape of spores. Fig. 327 . . . . *Chlorothecium*

463b Cells spherical, angular in front view, fusiform or oval; wall in 1 piece **464**

464a Stipe slender, threadlike, longer than the cell body. Fig. 332 . . . . *Peroniella*

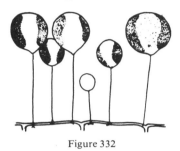

Figure 332

Figure 332 *Peroniella hyalothecae* Gobi.

Species of this genus occur solitarily or in gregarious clusters on other algae, or are attached in the mucilage of colonial forms. The one illustrated seems to occur no other place but on the filaments of the desmid *Hyalotheca* (Fig. 215). Like other members of the Chrysophyta, the chloroplasts are not grass-green but a pale, apple-green, and there are usually refractive globules of oil or some other food reserve than starch. The chloroplasts are parietal plates.

464b Stipe stout, shorter than the cell body in length (or rarely equalling it) **465**

465a Cells transversely fusiform, attached from the side by a short stalk. Fig. 333 . . . . *Dioxys*

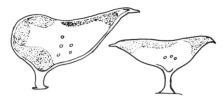

Figure 333

Figure 333 *Dioxys inermis* Thompson. (Redrawn from Thompson.)

The attached cells in this genus vary in shape but mostly are triangular and somewhat flattened. At least 1 species has 3 horns. The attaching stipe is relatively slender but short. The yellowish chloroplasts (2 to 4) are disclike plates folded around the wall. Two of the 6 known species occur in the United States, but are very rare.

465b **Cells elongate-fusiform or tubular, attached by a broad stipe at the posterior end. Fig. 329** . . . . *Characiopsis*

466a **(456) Cells with 2 flagella of unequal length*** **467**

466b **Cells without flagella** **468**

467a **Cells metabolic, changing shape in swimming, sometimes with pseudopodia, chloroplast yellowish-green. Fig. 334** . . . . *Chloramoeba*

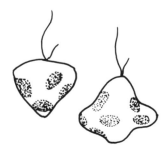

Figure 334

Figure 334 *Chloramoeba* sp., showing two cell shapes and the heterokont flagella.

The shapeless, almost colorless members of this genus have 1 very short and 1 very long flagellum that is directed forward. There are several faintly pigmented chloroplasts. The cells are metabolic, readily changing shape in motion; often found on aquatic plants. It is to be expected in the United States but so far has not been reported.

467b **Cells firm, oval or pyriform, usually not metabolic. Fig. 335** . . . . *(Chlorochromonas) Ochromonas*

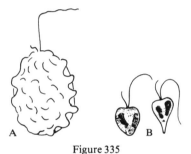

Figure 335

Figure 335 *Ochromonas.* (A) *O. verrucosa* Skuja; (B) *O. (Chlorochromonas) minuta* (Lewis). (A, redrawn from Popovsky.)

The cells in this genus are broadly or narrowly pyriform, but may be metabolic and change shape in motion. The flagellation is heterokont. There are 2 parietal chloroplasts, contractile vacuoles and a red eye-spot has been reported to occur. Non-motile, pear-shaped cells with yellowish chloroplasts which occur in the microscope field should be examined carefully for flagella or traces of them. Frequently the cells come to rest and attach themselves to a substrate at the posterior end. Eight species have

*See *Phaeaster* (Fig. 357). This organism exists for the most part as a globular, palmelloid colony, but it also has a solitary, uniflagellate, swimming stage, regarded by some as being a zoospore produced from the cells of the colony.

been reported from the United States of the 15 that are known, differentiated by the presence or absence of a pyrenoid, the color of the chloroplast and the morphology of the cysts that are formed as a resting stage. Some species have passed under the name *Chlorochromonas.*

468a (466) Cell wall smooth 469

468b Cell wall sculptured or decorated, sometimes spiny. 480

469a Cells spherical, subglobose, or broadly ovate to pyriform. 470

469b Cells rectangular, pyramidal, polyhedric, cylindric or crescent-shaped 474

470a Cells contained in a gelatinous envelope, solitary or 2 together. Fig. 336. . . . .
. . . . *Chlorobotrys*

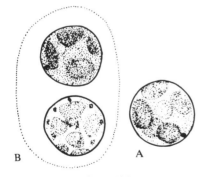

Figure 336

Figure 336 *Chlorobotrys regularis* (West) Bohlin. (A) single cell; (B) 2 cells enclosed in common mucilage.

The identifying character of the species illustrated is the paired arrangement of the spherical cells enclosed in a mucilaginous

sheath. There are several parietal chloroplasts and invariably a red spot which is an oil globule. Even though preserved, when some identifying characteristics are lost, the cells retain the dark-colored spot. *Chlorobotrys* is both eu- and tychoplanktonic. Six species have been reported from the United States, differentiated by cell shape, number of cells in the colony and number of chloroplasts.

470b Cells not enclosed in a gelatinous envelope 471

471a Cell wall in 2 sections, separating and persisting as membranous sections near the liberated autospores. Fig. 337 . . . . .
. . . . *Diachros*

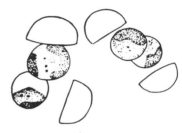

Figure 337

Figure 337 *Diachros simplex* Pascher.

The species illustrated is the only one reported from the United States. It is somewhat like *Schizochlamys* (Fig. 94) in the Chlorophyta in that the mother-cell wall fragments are retained after new cells (autospores) are released and the pieces persist as hemispherical, transparent cups near the daughter cells. There are 1 or 2 parietal chloroplasts.

471b Cell wall in 1 piece; mother cell not forming persisting sections but disintegrating to liberate spores 472

472a Cells spherical 473

472b Cells broadly ovoid or pyriform. Fig. 338. . . . . *Leuvenia*

Figure 338

Figure 338 *Leuvenia natans* Gardner. (Redrawn from G.M. Smith.)

Although essentially unicellular, this genus may occur as a film at the water surface. Young cells are spherical and have 1 or 2 chloroplasts, whereas older cells become pyriform or ovate and have numerous yellow-green chloroplasts. Only 1 species is known for the genus and in the United States is reported from only 2 localities.

473a Cells solitary (rarely clumped), with 1 or 2 parietal, golden-yellow chloroplasts, attached or free-floating. Fig. 339 . . . . (*Epichrysis*) *Chrysosphaera*

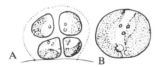

Figure 339

Figure 339 *Chrysosphaera paludosa* (Korsch.) Bourr. (A) cluster of cells enclosed by mucilage; (B) single cell showing chloroplasts and characteristic chrysophycean globules.

The cells are solitary or clumped; have 2 bright yellowish chloroplasts. They may be free-floating but often form aggregations of indefinite numbers on stems of aquatic plants, appearing as epiphytes. Reproduction is by autospores or by uniflagellate zoospores.

473b Cells solitary or in clusters, with a parietal reticulation of yellow-green chloroplasts; organisms attached but usually free-floating, often forming "blooms." Fig. 340 . . . . . . *Botrydiopsis*

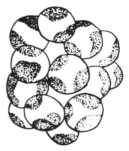

Figure 340

Figure 340 *Botrydiopsis arhiza* Borzi.

In the same habitats where *Ophiocytium* (Fig. 332) occurs one may find *Botrydiopsis,* spherical unicells or sometimes in clumps and colonial. Small cells, when young may contain a single, parietal chloroplast, but in age there are many yellowish-green bodies. *B. eriensis* Snow is larger than the species illustrated and is euplanktonic whereas the latter is tychoplanktonic.

474a (469) Cells rectangular, boxlike in 1 view, narrowly elliptic when seen from the side, with a spine at each corner. Fig. 341 . . . . *Pseudotetraedron*

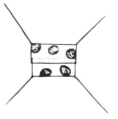

Figure 341

Figure 341 *Pseudotetraedron neglectum* Pascher.

The rectangular cells in this genus show their chrysophyte relationship by the fact that the wall is in 2 sections, a feature that can be seen when the cells are turned to show their quadrangular shape. There are several yellow-green chloroplasts; oil occurs as a food reserve. There is 1 species reported for the genus and this is very rare; one record from the United States.

474b   Cells other shapes **475**

475a   Cells pyramidal, polyhedral or triangular, with arms or with a spine or a mucro at each angle. Fig. 342 . . . .
. . . . *Pseudostaurastrum*

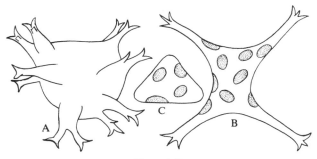

Figure 342

Figure 342 *Pseudostaurastrum.* (A) *Ps. enorme* (Hansg.) Chod.; (B) *Ps. hastatum* (Reinsch) Chod.; (C) *Ps. muticum* (A.Br.) Bourr.

Cells in this genus vary greatly in shape. They may be simple polygons with a spine or a mucro at each angle or they may be variously lobed with branched arms and spine-tipped elongations from the angles. Some species are smooth-walled; others are coarsely punctate or scrobiculate. The chloroplasts are several parietal bands, or numerous discs, always without a pyrenoid. They are golden-yellow to yellowish-green in color. Several earlier described species of *Tetraedron* (Fig. 209) are properly referred to *Pseudostaurastrum.* Like

Tetraedron species in this genus may reproduce by autospores, but unlike that genus zoospores are also produced.

475b   Cells cylindrical or crescent-shaped . . . .
. . . . **476**

476a   Cells elongate-cylindric, coiled or S-shaped, equally rounded at both poles. Fig. 331 . . . . *Ophiocytium*

476b   Cells oblong, sides convex, short-cylindric or fusiform, sometimes not equally rounded at both poles; curved but not coiled or twisted. **477**

477a   Cells short-cylindric, 1 1/2-2 times as long as broad; poles symmetrically rounded. Fig. 343 . . . .
. . . . *(Monallantus) Ellipsoidon*

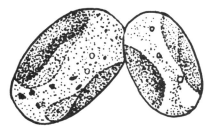

Figure 343

Figure 343 *Ellipsoidon brevicylindrus* (Pascher) Bourr.

This genus occurs in the same habitats as *Ophiocytium* (Fig. 332) and *Bumilleriopsis* (Fig. 345). There are 2 thin, folded parietal, platelike chloroplasts. Two species have been reported from the United States (1 under the name of *Monallantus* which is regarded as being synonymous).

477b   Cells fusiform or cylindric only in part; poles sometimes unsymmetrical **478**

478a   Cells fusiform or sickle-shaped 479

478b   Cells irregularly cylindrical, curved or bent; poles unsymmetrical. Fig. 344 ...
.... *Bumilleriopsis*

Figure 344

Figure 344   *Bumilleriopsis brevis* Pascher.

Cells in this genus have many yellow-green, disclike chloroplasts, and occur singly or several together in incidental clusters (not colonial). The irregularly curved cells (rarely somewhat fusiform) with two poles different in shape help to identify *Bumilleriopsis*.

479a   Cells broadly fusiform, abruptly narrowed at the poles. Fig. 345 ....
.... *Pleurogaster*

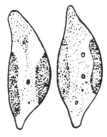

Figure 345

Figure 345   *Pleurogaster lunaris* Pascher.

The chief difference between this genus and *Bumilleriopsis* (Fig. 344) is the definitely fusiform shape, and the two poles of the cells being similar in shape. There are 2 laminate chloroplasts and usually several refractive globules. Two of the 4 known species occur in the United States.

479b   Cells narrowly fusiform, spindle-shaped or sickle-shaped. Fig. 346 ....
.... *Chlorocloster*

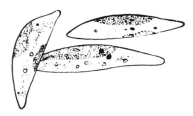

Figure 346

Figure 346   *Chlorocloster pyreniger* Pascher.

In this genus the cells are narrowly spindle-shaped and usually are distinctly curved or even sickle-shaped (according to species). They are found intermingled sparingly among other algae in the tychoplankton. The chloroplasts are similar to those of *Pleurogaster* (Fig. 345). Only 1 of a dozen species is reported from the United States.

480a   (468) Cells elongate-cylindric, with a spine at one or both poles. 481

480b   Cells short-cylindric, spherical or angular 482

481a   Cells nearly straight or only slightly bent, with a spine at each pole. Fig. 347 .... *Centritractus*

Figure 347

Figure 347 *Centritractus belanophorus* Lemm.

Species in this genus are either short or relatively long cylinders with a long spine at either pole. The cells are either straight or slightly curved. Some species resemble *Ophiocytium* (Fig. 331) but are not S-shaped nor coiled. The wall is in 2 sections which overlap in the mid-region, or there may be a 'cap' formed near either end of the cylinder. There are usually 2 laminate chloroplasts but these may fragment and appear as several. There is 1 nucleus whereas *Ophiocytium* is multinucleate and has several parietal chloroplasts. Apparently reproduction in *Centritractus* is still unknown except that zoospores have been reported with some question.

481b   Cells coiled, S-shaped, or hooked at one end. Fig. 331. . . . . *Ophiocytium*

482a   (480) Cells spherical 483

482b   Cells some other shape. 484

483a   Cell wall saccate at the margins, the surface of the cell showing broad depressions (sometimes faintly seen). Fig. 348. . . . . *Arachnochloris*

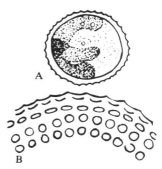

Figure 348

Figure 348 *Arachnochloris minor* Pascher. (A) cell showing chloroplasts (sectional view); (B) sample of wall showing circular thin areas.

Cells in this genus are spherical; the wall regularly scrobiculate so that the margin appears spiny. The projections are the tops of ridges formed by the depressions in the wall. There are 2 tonguelike or bandlike, parietal chloroplasts. Two of the 9 known species occur in the United States. Plants appear rarely, intermingled in tychoplanktonic algae especially in soft water lakes and bogs which favor the development of many other genera of the Xanthophyceae.

483b   Cell wall bearing curved or straight spines. Fig. 349 . . . .
. . . . . *(Acanthochloris) Meringosphaera*

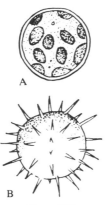

Figure 349

Figure 349 *Meringosphaera spinosa* Presc. (A) optical section of cell showing chloroplasts; (B) exterior of cell to show spiny wall.

Although this is a marine genus, 1 species which seems to belong here has been reported from the United States. The cells are spherical, have spiny walls and many parietal chloroplasts. In many respects it is similar to *Acanthochloris* and P. Bourrelly has appropriately suggested that *M. spinosa* Presc. should be assigned to that genus. *Acanthochloris* with spherical cells and a single, bandlike chloroplast has not been reported from the United States otherwise.

484a (482) Cells broadly fusiform or subtriangular, narrowed abruptly at one or both poles to form spinelike projections. Fig. 345 . . . . *Pleurogaster*

484b Cells other shapes 485

**485a Cells circular with lobes in one view, narrow and crescent-shaped when seen from the side; membrane ornamented with scrobiculations. Fig. 350 . . . .
. . . . *Chlorogibba***

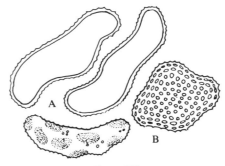

Figure 350

Figure 350 *Chlorogibba ostreata* Pascher. (A) cell shapes in side view; (B) exterior of cell in front view showing wall scrobiculations.

The cells in this genus are flattened; circular in outline in one view, compressed and curved or hemispherical as seen from the side. The wall is scrobiculate. There are about 6 species, differentiated by cell shape, occurring in both Europe and the United States. Like other related genera they are to be sought in the tychoplankton of shallow ponds and swamps.

485b Cells shaped otherwise 486

**486a Cells subcylindrical. Fig. 351 . . . .
. . . . *Chlorallantus***

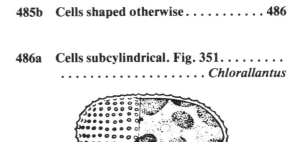

Figure 351

Figure 351 *Chlorallantus oblongus* Pascher, semidiagrammatic to show portion of scrobiculate wall, and chloroplasts.

These cells are oblong to short-cylindric and have a wall that is deeply scrobiculate in a regular pattern of transverse rows. There are numerous disclike chloroplasts.

486b Cells other shapes **487**

487a Cells oval or biconvex-elliptic. Fig. 352 . . . . *Trachychloron*

Figure 352

Figure 352 *Trachychloron biconnicum* Pascher. (Optical section).

Like *Chlorallantus* (Fig. 351) cells in this genus have walls with deep depressions. The cells are broadly elliptic to fusiform in shape and contain a gracefully curved yellow-green parietal plate, or several smaller plates. Two of the 9 known species occur in the United States.

487b Cells triangular, pyramidal or tetragonal **488**

488a Cells pyramidal or tetragonal. Fig. 353 . . . *(Tetragoniella) Tetraedriella*

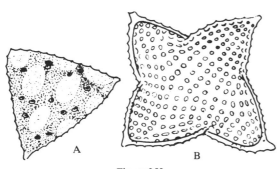

Figure 353

Figure 353 *Tetraedriella gigas* Pascher. (A) optical section view of cell to show chloroplasts; (B) exterior of cell to show scrobiculate wall.

The cells of this genus are beautifully sculptured by regularly arranged rows of depressions. According to the position in which the cells are viewed they exhibit different shapes, triangular, tetragonal or cushion-shaped. This genus should be compared with *Pseudostaurastrum* in which there are similar shapes, but with wall smooth or without orderly decorations. *Tetragoniella* is regarded as synonymous with *Tetraedriella*. The chloroplasts are several to many parietal discs.

488b Cells flattened, appearing triangular in front view, fusiform when seen from the side. Fig. 354. . . . . *Goniochloris*

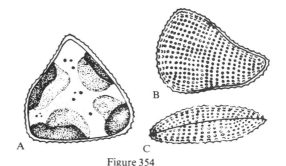

Figure 354

Figure 354 *Goniochloris sculpta* Geitler. (A) optical section to show platelike chloroplasts; (B) exterior of cell to show regularly arranged depressions in the wall; (C) lateral view of cell to show junction of the 2 wall sections (sometimes discerned with difficulty).

Like *Tetraedriella* (Fig. 353) cells in this genus are variously shaped according to their position in view. They are usually elliptic when seen from the side, or flattened cushion-shaped. Whereas the former is tetraedrical this genus is polyedrical. The wall is sculptured with rectilinear rows of circular pits. There are 3 or 4 curved, platelike chloroplasts. There are about 12 known species but apparently only 1 has been reported from the United States.

489a (455) **Cells attached by a stipe** **490**

489b **Cells without a stipe.** **491**

490a **Cells globular on a long slender stalk; plants ordinarily solitary but often so densely clumped as to appear colonial; epiphytic on filamentous algae, especially desmids. Fig. 332** . . . . *Peroniella*

490b **Cells cylindrical on a short stipe; colonies formed by zoospores germinating on the rim of mother-cells, repeatedly to form arbuscular chains. Fig. 331** . . . . *Ophiocytium*

491a (489) **With a mucilaginous sheath.** . . **492**

491b **Without a sheath** **497**

492a **Cells oval, many in a gelatinous matrix** **493**

492b **Cells shaped and arranged otherwise** **494**

493a **Cells in 2's and 4's in a single layer at the periphery of a saclike or globular, gelatinous matrix, with their long axes practically parallel. Fig. 355** . . . *Chlorosaccus*

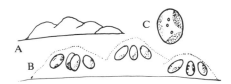

Figure 355

Figure 355 *Chlorosaccus fluidus* Luther. (A) diagram of colony shape epiphytic on aquatic plants; (B) cell arrangement within the colony; (C) single cell.

This rare alga occurs as macroscopic, gelatinous balls or mounds, with oval cells more or less oriented at the periphery, with longitudinal axes at right angles to the surface. The colonies occur on aquatic plants or on stems of *Chara*. Not infrequently the colonies are found adrift. The cells have from 2 to 6 parietal, yellow-green chloroplasts. There is but 1 known species.

493b **Cells irregularly arranged in a globular mucilage, each cell or pair of cells surrounded by a sheath; cells containing 2 contractile vacuoles. Fig. 356** . . . . *(Gloeochloris) Heterogloea*

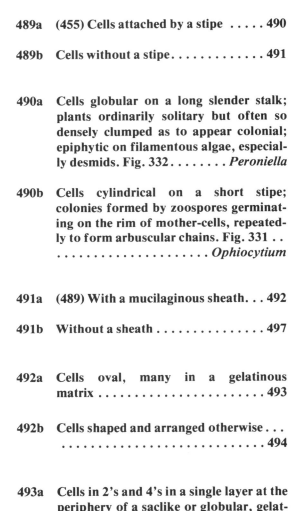

Figure 356

Figure 356 *Heterogloea endochloris* Pascher, cells within the colonial mucilage, and a single cell to show chloroplasts.

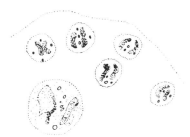

The oval cells (usually colonial), occur in the tychoplankton. There are several yellowish chloroplasts and 2 contractile vacuoles, characteristic of the Heterocapsales in the Xanthophyceae. Reproduction is by cell division or biflagellated zoospores. *Heterogloea endochloris* Pascher has appeared in collections of algae from midwest and western United States, and from one station on the Atlantic coast (as *Gloeochloris*).

494a (492) **Cells solitary or in pairs at the ends of branched, gelatinous tubes. Fig. 317 . . . . *Mischococcus***

494b **Cells arranged otherwise 495**

495a **Cells spherical, 2 within a globular envelope. Fig. 336 . . . . . . *Chlorobotrys***

495b **Cells spherical, many within a gelatinous matrix. 496**

496a **Cells compactly arranged to form a hollow, palmelloid colony; chloroplast parietal and radiate (star-shaped), turned outwardly against the free wall of the cell, each showing a central granule of leucosin. Fig. 357 . . . . . . *Phaeaster* ***

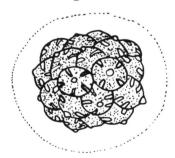

Figure 357

Figure 357 *Phaeaster pascheri* Scherf., cells in colonial arrangement. (Redrawn from Korschikoff.)

In this genus the cells are grouped or definitely colonial, enclosed in a gelatinous sheath. *Phaeaster pascheri* Scherf. is one of the species which has a stellate chloroplast bearing a central pyrenoid and a granule of leucosin. The cells are so arranged that the chloroplast is displayed outwardly. Reproduction is by uniflagellate zoospores bearing a chloroplast similar in shape to that of the vegetative cell. Another species does not have a pyrenoid.

496b **Cells dispersed within a spherical, gelatinous sheath, chloroplasts 2 or 3 parietal plates. Fig. 358 . . . *Gloeobotrys***

Figure 358

Figure 358 *Gloeobotrys limneticus* (G.M. Smith) Pascher.

The chief difference between this genus and *Chlorobotrys* (Fig. 336) is the presence of a definite mucilaginous sheath about the cells so that regular colonies are formed. The species illustrated was assigned to *Chlorobotrys* at one time; described from open-water plankton. Six species have been reported from the United States, differentiated by size of the cells and of the colonies, and the number of chloroplasts.

*See note on p. 174.

497a (491) Cells forming a loose cushion or a stratum of a few cells adherent to other filamentous algae. **498**

497b Cells arranged otherwise **499**

498a Cells hemispherical, in 2's and 4's, forming an expanse on the substrate, the cells in tetrads and relatively compact; wall smooth, chloroplasts several, without pyrenoids. Fig. 359 . . . . *(Chlorellidiopsis) Chlorellidium*

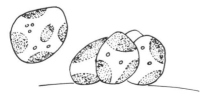

Figure 359

Figure 359 *Chlorellidium (Chlorellidiopsis) separabilis* (Pascher) Bourr.

This genus is apparently synonymous with *Chlorellidiopsis*. The cells (spherical or spheroidal) are usually closely grouped in clusters and adherent to filamentous algae, but may occur singly. Plants have been collected also from subaerial habitats. There are 2 parietal chloroplasts and at least 1 dark red oil spot in mature cells. Reproduction is by autospores and by zoospores.

498b Cells oval or nearly spherical, in groups of 2 or 4; colony sessile on a substrate; chloroplasts 2 parietal plates. Fig. 360 . . . . *Lutherella*

Figure 360

Figure 360 *Lutherella obovoidea* Pascher.

The globular or pyriform cells may be solitary but belong to a family in which members are gregarious or adjoined to form epiphytic layers. *Lutherella* with its 2 parietal chloroplasts occurs in 2's and 4's on the walls of other algae. The cells are sessile and there seems to be no attaching disc. The habit of having the top half of the cell lift away as a cap for the escape of zoospores is suggestive of *Chlorothecium* (Fig. 327). The zoospores are unequally biflagellate. In this country the plants seem to occur more frequently in bog waters than in lakes, and where there is a miscellany of other Chrysophyta.

499a (497) Wall in 2 sections which separate (transversely) to liberate spores and which persist nearby; cells may be incidentally colonial because of gregarious habit. Fig. 337 . . . . *Diachros*

499b Wall in 1 piece, breaking down irregularly to liberate spores. **500**

500a Cells curved, sausage-shaped, arranged in 4's, adherent to other quartets by the remains of old, mother-cell walls in a radiate fashion; enclosed in a mucilage without an evident external sheath. Fig. 361 . . . . *Dichotomococcus*

Figure 361

Figure 361 *Dichotomococcus elongatus* Fott. (Redrawn from Thompson.)

This genus is questionably placed in a family in which the cells are colonial and clumped, and in which reproduction is by autospores or zoospores. The genus reproduces when the elongate-ovoid or wedge-shaped cells produce 2 autospores. Upon liberation these remain adjoined to other pairs of autospores by the gelatinous material from the mother-cell wall and in a somewhat radiate fashion. There is 1 parietal chloroplast lacking a pyrenoid. The entire colony of several generations is enclosed in mucilage, but the limitations do not form a definite sheath. Both of the 2 known species have been reported from the United States.

500b **Cells spherical, clustered to form a colony; no mother-cell wall remains evident. Fig. 340** . . . . *Botrydiopsis*

501a **(431) Plant a branched, feathery but gelatinous thallus, the protoplasts crowded in linear series within tough, tubular envelopes; plants of cold streams. Fig. 362** . . . . *Hydrurus*

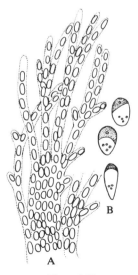

Figure 362

Figure 362 *Hydrurus foetidus* (Vill.) Trev. (A) portion of branched, gelatinous colony showing cell arrangement; (B) 3 cell shapes showing chloroplast in the anterior end.

Usually in high mountain or very cold streams this organism forms brown, stringy, gelatinous masses, attached to stones or wood. The bushy, brown tufts contain oval cells arranged in linear series within tubelike strands. The strands branch and a well-developed thallus is 'feathery'. Sometimes a small stream actually will be choked with dense growths of *Hydrurus*. The disagreeable odor is responsible for the specific name. The cell may metamorphose into curiously shaped, pyramidal zoospores that have 1 flagellum. Two species have been reported from the United States, but the one illustrated is by far the more common.

501b **Plant not a feathery, gelatinous thallus** **502**

502a **Cells amoeboid, many protoplasts interconnected by filliform pseudopodia; chloroplast faint, yellowish. Fig. 363. . .** . . . . *Chrysarachnion*

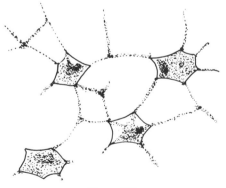

Figure 363

Figure 363 *Chrysarachnion insidians* Pascher. (Diagrammed from Bourrelly.)

The cells in this rhizopodlike genus are interconnected by fine, protoplasmic strands so as to form a network. The chloroplast is a pale, golden-yellow color and there is a contractile vacuole in each unit. The genus belongs to a family in which the rhizopodal form is the predominant phase; flagellation, if present, being only temporary. The organism is to be expected in collections from the surface of ponds and bog water.

502b   Plants not as above; cells arranged otherwise, or solitary 503

503a   Thallus encrusting and adherent, forming a pseudoparenchymatous growth from which compactly arranged, erect branches arise; chloroplasts several brown discs. Fig. 364 . . . *Heribaudiella*

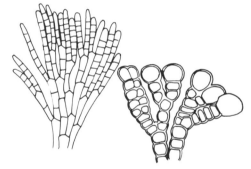

Figure 364

Figure 364   *Heribaudiella fluviatilis* (Aresch.) Sved. Portions of prostrate thallus, and a section of the erect branches bearing sporangia. (Redrawn from Gomont.)

This is one of the few freshwater members of the Phaeophyta. The genus, presently assigned to the Lithodermataceae, characteristically forms an encrusting, prostrate expanse of branches in a pseudoparenchymatous mass. From the prostrate portion, closely arranged upright branches develop. Plants live on rocks in swiftly flowing streams. The disclike chloro-

plasts are brownish. Sporangia are formed at the tips of the branches. Apparently 1 species has been reported from the United States and only from Connecticut.

503b   Thallus otherwise. 504

504a   Plants sparsely branched, erect filaments arising from a few-celled, basal cluster of cells; 2 parietal chloroplasts, ochre-green or brownish. Fig. 365 . . . . *Phaeothamnion*

Figure 365

Figure 365   *Phaeothamnion confervicola* Lag.

This member of the Chrysophyceae is 1 of 2 genera in which there is a branched filamentous thallus. The branches scarcely taper at the apices. The plants are relatively small and grow on the walls of larger filamentous algae. Each cell has 1 or 2 parietal, ochre-green to brownish chloroplasts. Reproduction is by from 4 to 8 heterokont zoospores produced in any cell.

504b   Plants otherwise 505

505a   A colony of vase-shaped cones (loricas), 1 or 2 cones arising from within the mouth of another and forming forked series; protoplasts with 2 unequal flagel-

*Possibly includes *Lithoderma* Aresch. and *Lithodora* Klebahn.

la, but the cells often absent from the loricas. Fig. 366 . . . . *Dinobryon*

Figure 366

Figure 366   *Dinobryon sertularia* Ehr.

There are several species of this genus, all of which are characterized by having the motile protoplasts enclosed within a colorless envelope. The loricas are usually contained 1 or 2 within another so that branching chains result. Some species occur solitarily. The genus is one which inhabits mostly hard-water and eutrophic lakes, occurring in the euplankton and sometimes so abundant as to form a 'bloom.' They produce disagreeable odors and tastes in domestic water supplies.

505b   Cells arranged otherwise 506

506a   A colonial cluster (sometimes solitary) of spherical cells, mostly epiphytic on filamentous algae, with 2 parietal, yellow-green chloroplasts. Fig. 339 . . . . *(Epichrysis) Chrysosphaera*

506b   Cells shaped or arranged otherwise 507

507a   A unicell, consisting of a yellowish protoplast contained within a vaselike

envelope; two heterokont flagella; organism often absent from the lorica. Fig. 366 . . . . *Dinobryon*

507b   Cells shaped otherwise, or located differently; solitary or colonial 508

508a   Cells solitary, colonial or filamentous; wall siliceous and etched with grooves or rows of puncta which form definite patterns; wall in 2 sections, 1 part (epivalve) forming a lid over a smaller (hypovalve); oil drops usually conspicuous as shiny globules; especially solitary cells (if elongate or 'cigar'-shaped) showing a gliding, jerky movement. Diatoms. See Fig. 367 655

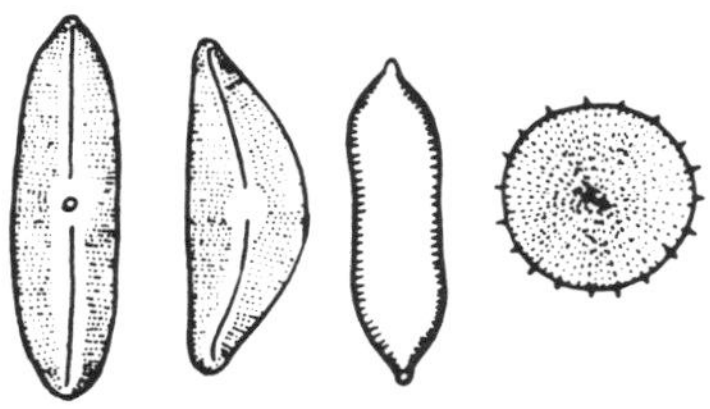

Figure 367

Figure 367   Representative Diatom frustules illustrating markings on siliceous walls.

508b   Cells without decorated siliceous walls; wall not in 2 sections; oil droplets lacking or inconspicuous; not showing gliding movements; if motile, equipped with flagella or moving by pseudopodia (amoeboid fashion) 509

509a   Unicellular. 510

509b   Multicellular or colonial 547

510a   Non-motile, without flagella 511

510b With flagella, but sometimes sedentary rather than swimming 527

511a Cells depressed-globose, attached and sessile; 2 (usually) parietal chloroplasts; often aggregate but frequently found solitary. Fig. 339 . . . . *(Epichrysis) Chrysosphaera*

511b Cells shaped or situated otherwise . . 512

512a Cells tetrahedral or polyhedral; free-floating; with 1 brownish, lobed chloroplast, parietal over most of the wall. Fig. 368 . . . . *Tetragonidium*

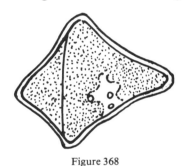

Figure 368

Figure 368 *Tetragonidium verrucatum* Pascher. (Redrawn from Pascher.)

This is a member of the Cryptophyceae which is non-motile in the vegetative state. The cells are tetragonal and have yellow-brown, parietal chloroplasts that fold around most of the wall. Reproduction is by zoospores with unequal flagella.

512b Cells shaped otherwise, with different chloroplasts 513

513a Cells globular or transversely oval, with a raised collar about an apical opening,

the sides of the cell extended into reflexed, long, spinelike processes; chloroplast 1 or 2, parietal. Fig. 369 . . . . *Diceras*

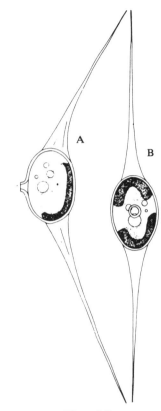

Figure 369

Figure 369 *Diceras phaseolus* Fott. (A) front view; (B) vertical view. (Redrawn from Fott.)

These golden-brown, free-floating cells are enclosed by a close-fitting lorica which bears on either side a long, sharply pointed spine. The body of the cell is transversely oval. The protoplast has 2 fine strands extending through an apical collar. The species illustrated has been found only in Maryland in the United States.

513b Cells shaped otherwise 514

514a Cells globular, with a raised collar about an apical pore, the wall with several, slender, radiating spines, some of which are forked at the tip. (This is thought probably to be the cyst of some other alga. Its classification is uncertain.) Fig. 370 . . . . *Chrysastrella*

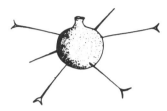

Figure 370

Figure 370   *Chrysastrella paradoxa* Chod.

The species illustrated is 1 of 2 reported from the United States, occurring either in open-water plankton or intermingled with other algae near shore. The test or envelope bears a few long, needlelike setae which are often forked at the tips. It is included in the Chrysostomataceae of the Chrysophyta.

514b Cells shaped otherwise 515

515a Cells attached, sedentary on other algae or on aquatic plants 516

515b Cells free-floating 522

516a Cells with variously shaped loricas, vaselike, with a long or short neck, or hemispherical without a neck but with an apical pore, the protoplast globular, often showing a fine, threadlike protoplasmic extension through the pore. Fig. 371 . . . . *Lagynion*

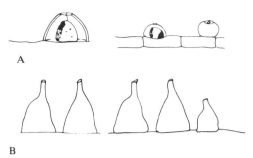

Figure 371

Figure 371 *Lagynion.* (A) *L. reductum* Presc.; (B) *L. triangularis* var. *pyramidatum* Presc.

Organisms in this genus are relatively small and are easily overlooked because they occur as epiphytes on filamentous algae. The shape of the lorica varies from globular to pyramidal and vase-shaped, with or without a neck. In some species the neck is bent or deflexed. Within the lorica is a globular protoplast with a faintly pigmented, yellowish chloroplast. The lorica is flattened against the substrate and is colorless. There are 7 species at least reported from the United States. As far as known division of the protoplast is the only method of reproduction.

516b Cells otherwise. 517

517a Cells with a globular lorica, colorless or yellowish, with a broadly open, apical pore; extended posteriorly into 2 prongs which straddle the host algal cell; protoplast globular with a pseudopodial thread which often shows forkings at the tip; chloroplasts golden-brown. Fig. 372 . . . . *Chrysopyxis*

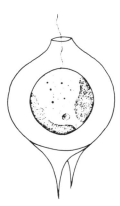

Figure 372

Figure 372   *Chrysopyxis bipes* Stein.

This curious genus seems to be confined to swamps and *Sphagnum* bogs. There is a globular, colorless lorica which has two posterior, hornlike projections that straddle the host cell of some other alga. In the anterior end is a broad opening through which a fine pseudopodium extends. There is a broad, bandlike chloroplast that folds about the wall. In reproduction 2 motile cells are formed. These escape and swim about for a time, then become quiescent and construct a lorica. The protoplast fills only a small part of the envelope. Three species have been reported from the United States.

517b   Cells shaped otherwise. 518

518a   Cells with a vaselike lorica, narrowed to a neck anteriorly in which there is an apical pore, and narrowed posteriorly to a basal attaching portion (sessile). Fig. 373 . . . . *Derepyxis*

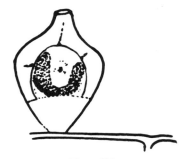

Figure 373

Figure 373   *Derepyxis dispar* Senn, showing membrane across the base of the lorica on which the protoplast rests.

The chief difference between this genus and *Lagynion* (Fig. 372) is the presence of a supporting membrane through the lorica at the base of the protoplast. The lorica is sessile on filamentous algae. Five species are known from the United States, differentiated mostly on the shape of the lorica.

518b   Cells otherwise. 519

519a   Cells inversely triangular or tetrahedral in top view, the angles tipped with 1 or 2 spines, abruptly narrowed below to an attachment. Fig. 374 . . . . . *Tetradinium*

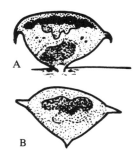

Figure 374

Figure 374 *Tetradinium simplex* Presc. (A) front view of epiphytic cell; (B) vertical view.

The chloroplasts of this sessile member of the dinoflagellates are typically golden-brown. The 3 or 4 corners of the cell are tipped with 2 short, sometimes curved spines. The pyramidal cells have a short stipe attaching them to filamentous algae. This genus should be compared with *Raciborskia* (Fig. 376). The species illustrated and 1 other are the only ones known from the United States.

519b   **Cells other shapes, *or* if inversely triangular then elliptic in vertical view. 520**

520a   **Cells pyriform or oval, epizoic; with rhizoidal attaching organs at the narrowed posterior. Fig. 375. . . . *Oodinium***

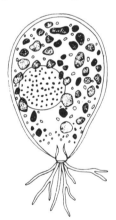

Figure 375

Figure 375 *Oodinium limneticum* Jacobs. (Redrawn from Jacobs.)

The brown, globular or ovate cells have basal rhizoidal extensions which attach the parasite to fish and to other aquatic animals. In reproduction the cells form 2 dinoflagellate-type zoospores. Starch is present in the protoplast. Only 1 species is known for the genus.

520b   **Cells differently shaped, without rhizoidal holdfasts. 521**

521a   **Cells inversely triangular or transversely elliptic in front view, elliptic in top view, the angles spine-tipped; narrowed posteriorly to a short, attaching stipe. Fig. 376 . . . . *Raciborskia***

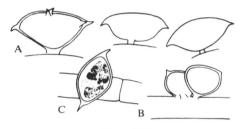

Figure 376

Figure 376 *Raciborskia bicornis* Wolosz. (A) front view showing the attaching stipe; (B), (C) side and vertical views.

These are elliptical cells, sessile on a short stalk and attached to filamentous algae or aquatic mosses. There is a single spine at each pole of the cell as well as a pair of spines (at times) on the dorsal wall. One species only has been reported from the United States. This genus should be compared with *Tetradinium* (Fig. 374).

521b Cells globular with a long, narrow stipe that has an attaching disc at the base, and an inflation immediately below the cell body; protoplast with many brownish chloroplasts and a red spot; often showing a dinoflagellate-type of a partial transverse furrow. Fig. 377 . . . . *Stylodinium*

Figure 377

Figure 377 *Stylodinium globosum* Klebs.

These globular cells have a relatively long, slender stipe and an attaching disc. They occur as epiphytes on filamentous algae. The membrane is relatively thick and encloses a protoplast in which there are numerous disclike chloroplasts and a red oil globule. Frequently the protoplast shows a partial transverse furrow characteristic of the dinoflagellates. In reproduction the cell forms 2 dinoflagellate zoospores which escape by a rupture of the wall. Two species are known for the United States, rare but widely distributed.

522a (515) Cells oval or elliptic, the wall impregnated and covered with variously shaped siliceous scales (appearing like chain armor) which bear long bristles; with an apical opening (usually obscured by wall scales); with a single flagellum, the flagellum often difficult to discern, the cell quiescent in microscope mounts. Fig. 378 . . . . *Mallomonas*

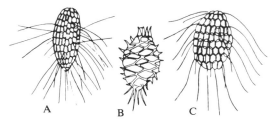

Figure 378

Figure 378 *Mallomonas*. (A) *M. caudata* Iwanoff; (B) *M. pseudocoronata* Presc.; (C) *M. acaroides* Perty.

These species occur in open-water plankton of mostly hard water lakes, frequently in abundance. They are differentiated from one another by the shape and arrangement of the scales in the membrane and by the morphology and arrangement of the bristles. Electron microscopy has demonstrated some highly refined characteristic features of the scales and bristles which are useful in taxonomy. Some species occur regularly in lakes in which there is a high degree of pollution. Although motile, the single flagellum is hardly observable unless the cells are recently collected and viewed under favorable optical conditions. The scales and spines are siliceous. Thirty-three species have been reported from the United States and Canada.

522b Cells otherwise. 523

523a Cells irregularly shaped, more or less isodiametric, amoeboid, with pseudopodia; protoplast very pale yellow-green or yellowish brown 524

523b Cells otherwise. 525

524a Pseudopodia long and needlelike, sometimes cells in a chain by interlocking of pseudopodia. Fig. 379 . . . .
. . . . . *Rhizochrysis*

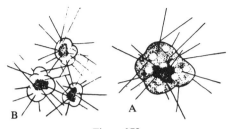

Figure 379

Figure 379 *Rhizochrysis limnetica* G.M. Smith. (A) one cell with needlelike pseudopodia; (B) cells in temporary colonial arrangement.

This amoeboid member of the Chrysophyceae has long, slender, needlelike pseudopodia. There are 1 or 2 golden-yellow chloroplasts. The cells are ordinarily solitary but may occur in loose, temporarily united groups. Two species are known from the United States.

524b Pseudopodia relatively short, spurlike, a single flagellum sometimes present after metamorphosis of the cell; solitary but sometimes appearing in chains through interjoining of pseudopodia. Fig. 380 . .
. . . . *Chrysamoeba*

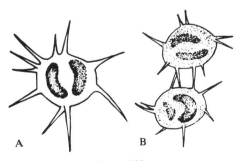

Figure 380

Figure 380 *Chrysamoeba radians* Klebs. (A) single cell with chloroplasts; (B) temporarily adjoined cells to form a 'colony'.

This rhizopodial form occurs more frequently than *Rhizochrysis* (Fig. 379). Like that genus *Chrysamoeba* has pseudopodia but they are short and thornlike rather than needlelike. The cells ordinarily occur in an amoeboid condition but may develop a single flagellum as a locomotory organ. There are two pale yellow chloroplasts. The species illustrated is widely distributed in the United States.

525a (523) Cells variously curved and crescent-shaped with pointed, sometimes reflexed or twisted horns; many brown chloroplasts and sometimes a red spot. Fig. 381 . . . . *Cystodinium*

Figure 381

Figure 381 *Cystodinium cornifax* (Schiller) Klebs.

These are free-floating, lunate, cystlike members of the Dinoflagellata. Species are differentiated on the basis of various shapes and the direction and morphology of the polar horns. Several species of *Tetraedron* (Fig. 209) have been described erroneously from members of the genus *Cystodinium*. The protoplast is distinctly the dinoflagellate type and usually there is a red oil droplet. Five species have been reported from the United States.

525b Cells otherwise. 526

526a Cells globular with numerous, disclike chloroplasts in stellate clusters; often showing a single dinoflagellate protoplast with the external cell wall serving as a cyst membrane. Fig. 382. . . . .
. . . . *Hypnodinium*

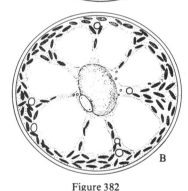

Figure 382

Figure 382 *Hypnodinium sphaericum* Klebs. (A) cyst with dinoflagellate type of daughter cell; (B) cyst. (Redrawn from Thompson.)

The brown, ellipsoid chloroplasts are arranged in numerous rosettes within these globular cells. Species are free-floating and have no flagella but the protoplast shows a typical transverse furrow and a red oil globule. The single species is rare and has been found only in Maryland in this country.

526b Cells oval or globular, solitary or in a few-celled cluster within a lamellate

sheath; cells often showing a conspicuous transverse furrow; golden-brown chloroplasts somewhat radiately arranged but not in clusters. Fig. 383 . . . .
. . . . *Gloeodinium*

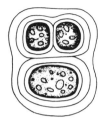

Figure 383

Figure 383 *Gloeodinium* sp., cyst.

Although these plants exist in a palmelloid, nonmotile condition they clearly show their dinoflagellate affinity by the transverse furrow and the radiately arranged, golden-brown chloroplasts. The cells are enclosed in a lamellate, gelatinous sheath, from 2 to 8 in a clump. Two species have been reported from the United States.

527a (510) Cells with flagella but sedentary, either attached or if free, quiescent . . . .
. . . . 528

527b Cells with flagella; freely swimming . . .
. . . . 530

528a Cells with a colorless, pyriform or elliptical lorica, widely open anteriorly, sessile on other algae but without a visible attaching organ. Fig. 384 . . . .
. . . . *Epipyxis**

*One-celled forms of *Dinobryon* have been referred to this genus. According to some studies, *Epipyxis* is synonymous with *Hyalobryon* (Fig. 385).

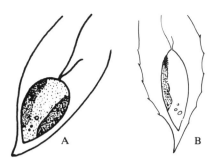

Figure 384

Figure 384 *Epipyxis.* (A) *E. tabellariae* (Lemm.) G.M. Smith, cell in lorica, with heterokont flagella; (B) *E.* sp., cell in lorica which shows marginal projections formed by previous generations of lorica-formations.

The cells in this genus have a cone-shaped lorica of various forms, but one which shows projections along the margins, representing the cones of previous generations. It is sometimes treated as synonymous with *Hyalobryon* (Fig. 385). The cells are sessile and epiphytic, solitary or colonial. The protoplast within the lorica has 2 parietal chloroplasts and sometimes a stigma or eye-spot. Nearly 25 species have been reported from the United States (including species referable to *Hyalobryon*).

528b   Cells with other types of loricas 529

529a   Envelope cone-shaped, with smooth or slightly wavy margins, free-floating, but sometimes attached; protoplast with 2 flagella and 2 yellowish chloroplasts. Fig. 366 . . . . *Dinobryon*

**529b   Lorica cone-shaped, with marginal bristlelike projections caused by transverse "growth rings." Fig. 385 . . . .
. . . . *Hyalobryon***

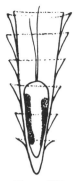

Figure 385

Figure 385 *Hyalobryon mucicola* (Lemm.) Pascher.

In this genus the cells are solitary and the cone-shaped envelopes have margins which appear as is bristled. The markings represent the remains of envelopes of previous generations of cells. Whereas *Dinobryon* (Fig. 366) is freely swimming, this genus is solitary (or colonial) as an epiphyte on filamentous algae. At least 3 species have been reported from the United States.

530a   (527) Cells with a broadly oval, colorless lorica, truncate at the apex with an apical pore; protoplast elliptic, with a single flagellum extending from the narrowly opened apex; 2 brown chloroplasts and an eye-spot. Fig. 386 . . . . *Chrysococcus*

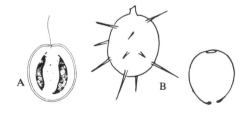

Figure 386

Figure 386 *Chrysococcus rufescens* Klebs. (A) cell showing chloroplasts; (B), (C) other shapes of loricas of 2 species.

The globular loricas of this genus have a small, often inconspicuous pore at the anterior and through which a single flagellum extends. The protoplast does not nearly fill the cavity of the lorica. There are 2 golden-brown chloroplasts. There are at least 17 species reported from the United States, many of them from the Ohio valley only.

530b Cells without a lorica **531**

531a **Cells oval or elliptic, with the wall impregnated or covered by overlapping, siliceous scales which bear long bristles; flagellum 1, often inconspicuous. Fig. 378** . . . . *Mallomonas*

531b **Cells without siliceous scales in the wall** **532**

532a **Cells oval in front view, flattened and narrow when seen from the side, lobed or 2-lipped at the apex; 2 flagella not the same length. Fig. 26** . . . . . *Cryptochrysis*

532b **Cells differently shaped** **533**

533a **Cells broadly circular in front view, flattened and narrow when seen from the side; without a true wall; flagella 3, slightly subapical in attachment; with 2 golden-yellow, lateral chloroplasts. Fig. 387** . . . . *Chrysochromulina*

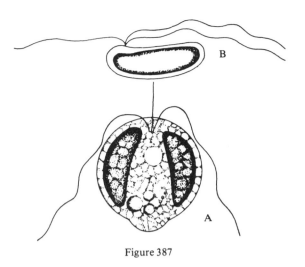

Figure 387

Figure 387 *Chrysochromulina parva* Lackey. (A) front view; (B) vertical view. (Redrawn from Lackey.)

The cells in this genus are without a wall; have 3 somewhat laterally attached flagella arising near the anterior end. The cells are broadly rounded in front view, but narrowly oval and somewhat reniform when seen from the side. One species only has been reported from midwest America.

533b **Cells differently shaped with a cell wall; flagella 2** **534**

534a **Cells oval in front view, with an apical depression; 2 flagella arising apically; cell flattened when seen from the side and showing the 2-valve form of the wall; transverse furrow lacking; 2 brown chloroplasts. Fig. 388** . . . . *Exuviella*

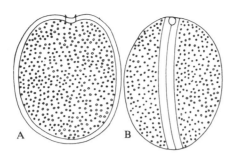

Figure 388

Figure 388 *Exuviella compressa* Ostenf. (A) front view; (B) lateral view showing suture formed by 2 adjoining valves.

This is an unusual member of the Dinoflagellata in that there is no transverse furrow, and the flagella are apical rather than lateral. Although almost entirely marine, at least 1 species has been found in brackish water. When seen from the side the two valves that compose the cell wall can be seen, with a suture where they adjoin.

534b   Cells otherwise. 535

535a   Cells pyriform, with a median depression apically; wall not bivalved; 2 apical flagella arising from a gullet; 2 yellowish, or yellow-green chloroplasts. Fig. 16 . . . . *Cryptomonas*

535b   Cells otherwise. 536

536a   Cells bean-shaped, with 2 flagella that are laterally attached; 1 or 2 lobed, parietal, brownish chloroplasts and an eyespot. Fig. 302. . . . . *Protochrysis*

536b   Cells shaped otherwise 537

537a   Cells oval or distinctly pyriform, with 2 flagella of nearly equal length, arising from the broad anterior end; 1 folded, broad, platelike chloroplast, brownish or olive-green (sometimes reddish). Fig. 15 . . . . *Rhodomonas*

537b   Cells otherwise. 538

538a   Cells pyriform with 2 distinctly unequal flagella from the broad anterior end; 2 small, yellowish chloroplasts; without a gullet. Fig. 335 . . . .
. . . . . *(Chlorochromonas) Ochromonas*

538b   Cells otherwise. 539

539a   Cells with a long, anterior horn and 2 or 3 posterior horns. Fig. 389 . . . *Ceratium*

Figure 389

Figure 389 *Ceratium hirundinella* (O.F.M.) Duj.

This genus possesses such a distinct shape that it can be identified readily. It occurs either in the eu- or tychoplankton, but especially in open-water, sometimes producing a 'bloom.' Water of small lakes may be gray-chocolate in color from the profuse number of individuals. There is a prominent transverse furrow that divides the cell into an epicone and a hypocone. When seen from the ventral surface the longitudinal, rather broad sulcus is discernible. Here the 2 flagella are attached, one of which trails, the other (a band type) is wound about the cell and vibrates in the transverse furrow. There are numerous, brown chloroplasts and a red eyespot (?). The plates which compose the cell wall are marked with close reticulations. *Ceratium* is the most common of all the freshwater dinoflagellates. About ten species have been reported from fresh and brackish water of the United States.

539b Cells without prominent horns as above. 540

540a Cells without a true wall, but with a membrane which may be either delicate or thick and firm; smooth, without evident plates 541

540b Cells with a true wall, with a pattern of definitely arranged plates usually evident (in some the boundary of the plates is seen with difficulty; reduce or modify illumination); a transverse furrow encircling the cell completely or incompletely, and a longitudinal (ventral) sulcus . . 544

541a Cell oval; the transverse furrow spirally descending. Fig. 390. . . . . . . *Gyrodinium*

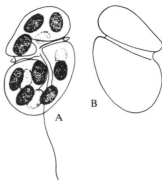

Figure 390

Figure 390 *Gyrodinium pusillum* (Schill.) Kofoid & Swezy. (A) ventral view, showing relatively large chloroplasts; (B) dorsal view showing position of transverse furrow.

The chloroplasts of this genus are relatively large. Identification can be made by the descending, spiral, transverse furrow which produces a hypocone much larger than the epicone. Species occur in marine or brackish water. Five species have been reported for the United States.

541b Cells top-shaped or fusiform; transverse furrow mostly at right angles to the long axis, or occurring as a 'V' near the apex. 542

542a Transverse furrow dividing the cell approximately into 2 equal parts (epicone and hypocone). Fig. 391 . . . . *Gymnodinium*

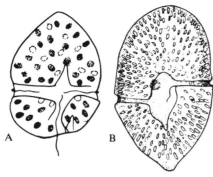

Figure 391

Figure 391 *Gymnodinium.* (A) *G. palustre* Schilling; (B) *G. fuscum* (Ehr.) Stein.

Species of this genus are numerous and are of wide occurrence in both fresh and salt water. They are mostly intermingled with other algae in freshwater, whereas in the sea they produce veritable 'blooms', the most common cause of the infamous "red tide." There is no true cell wall (hence the genus name). The transverse furrow extends around the cell in a slightly downward direction. Over 20 species have been reported from freshwater habitats in the United States.

542b **Transverse furrow dividing the cell into a hypocone and epicone of unequal size** **543**

543a **Cells mostly top-shaded, the epicone definitely larger than the hypocone. Fig. 392** . . . . *Massartia*

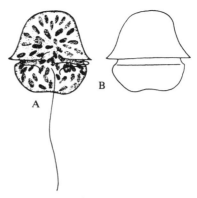

Figure 392

Figure 392 *Massartia musei* (Dan.) Schiller; (A) ventral view; (B) dorsal view.

These dinoflagellate cells, without a wall, are broadly oval but truncate at the poles. The transverse furrow divides the cell into an epicone distinctly larger than the hypocone. The brown chloroplasts are somewhat radiate in arrangement. Compare with *Gymnodinium* (Fig. 391).

543b **Cells subquadrate, the epicone smaller than the hypocone and appearing as an apical lobe of the cell. Fig. 393** . . . . *Amphidinium*

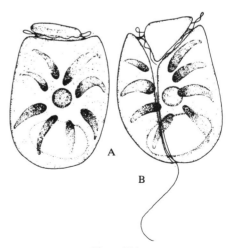

Figure 393

Figure 393 *Amphidinium klebsii* Kof. (A) dorsal view; (B) ventral view.

This genus is mostly marine in distribution but may be found in brackish water, or in freshwater near the sea. The chloroplasts are large and radiately arranged. The cell is unusual in that the transverse furrow is located near the apex so that the epicone is but a small lobe.

544a (540) Wall thick; plates easily discerned, with a suture (usually) between the plates; transverse furrow completely encircling the cell. **545**

544b Wall thin; plates discerned with difficulty (especially in filled and living cells); transverse furrow completely encircling the cell or not. **546**

545a Wall with 2 antapical plates (the plates at the posterior pole, to be seen in posterior end view); cell slightly flattened dorsiventrally in most species; posterior pole sometimes extended into a horn or into short projections. Fig. 394 . . . . ***Peridinium***

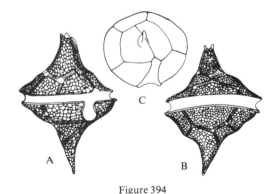

Figure 394

Figure 394 *Peridinium wisonsinense* Eddy. (A) ventral view showing longitudinal sulcus; (B) dorsal view; (C) posterior view showing 2 antapical plates.

This genus is represented by more species in freshwater than any of the other dinoflagellates. They are differentiated by shape and size of the cell, and by the number, shape and arrangement of the wall plates. The 2 posterior or antapical plates can be determined by patiently rolling the cell so that it can be seen from the end. Often the apical view shows a terminal pore. Most species are some variety of top-shaped; others oval or globular; many are slightly flattened when seen from the side. Some species have dorsal flanges or hornlike processes. In most species the plates show a distinct reticulation. Over 40 species have been reported from freshwater in the United States. Cysts often occur (Fig. 395).

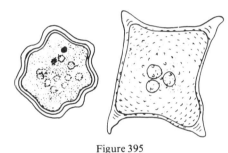

Figure 395

Figure 395 Representative Dinoflagellate cysts.

545b Wall with 1 antapical plate; cell not flattened dorsiventrally, round in vertical view. Fig. 396 . . . . ***Gonyaulax***

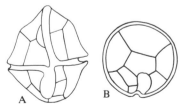

Figure 396

Figure 396 *Gonyaulax apiculata* (Pén.) Entz. (A) ventral view showing longitudinal sulcus; (B) antapical view showing the 1 antapical plate (an).

These almost spherical cells are differentiated from *Peridinium* (Fig. 395) by the single antapical plate and by the slightly spiral direction of the transverse furrow. Some authorities regard the species illustrated as belonging to *Peridinium*. *Gonyaulax* is more prevalent in marine waters; is sometimes the cause of 'red tides' and is related to fish poisoning.

546a (544) Cells strongly flattened dorsiventrally; plates not evident; transverse furrow not encircling the cell completely, usually located in the posterior part of the cell. Fig. 397 . . . . . . *Hemidinium*

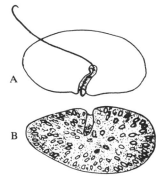

Figure 397

Figure 397 *Hemidinium nasutum* Stein. (A) ventral view; (B) dorsal view.

Cells in this genus are oval or elliptic when seen from the broad side but are much flattened in lateral view. The transverse furrow extends only part way around the cell. Species are differentiated by shape and size of the cell and by the pattern of the plates in the wall which usually are very delicate and difficult of determination. The 2 species reported from the United States are rare; are to be found in the tychoplankton, especially in shallow ponds and swamps.

546b Cells not at all or but very little flattened dorsiventrally (nearly round as seen in

end view); plates evident in the wall, especially clear when the cells are empty; transverse furrow completely encircling the cell. Fig. 398 . . . . *Glenodinium*

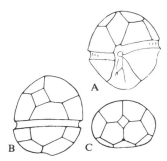

Figure 398

Figure 398 *Glenodinium kulczynski* (Wolsz.) Schiller. (A) ventral view showing longitudinal furrow or sulcus; (B) dorsal view; (C) apical view.

The several species of this genus reported from the United States are differentiated by cell size and shape. It is differentiated from other dinoflagellates by having the plates in the epicone variable in number and arrangement but by having in the hypocone 5 or 6 postcingulars (the plates lying next to the transverse furrow) and 2 antapical plates. Mostly the species are broadly oval to nearly round, with very thin walls. The cells must be rotated to various positions to determine the plate number and arrangement, using empty cells for observation.

547a (509) Colony a disc of horizontally arranged cells, usually showing as radiating filaments. Fig. 318 . . . .
. . . . *Phaeodermatium*

547b Colonies otherwise. 548

548a Colony motile; cells flattened. 549

548b Colony non-motile, or if moving, by rhizoidal processes (pseudopodia) . . 555

549a Colony globose or subglobose (oval); cells ovoid or pyriform, compactly arranged by forming a hollow sphere . . . .
. . . . 550

549b Colony not globular; cells shaped otherwise 554

550a Cells bearing 2 long, rigid, rodlike processes at their anterior ends. Fig. 399 . . . . *Chrysosphaerella*

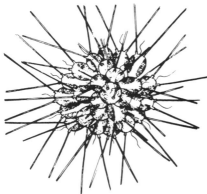

Figure 399

Figure 399 *Chrysosphaerella longispinum* Lauterb., cells bearing siliceous rods.

This distinctive genus is readily identified by the curious long rods borne in pairs on each cell. The rods are siliceous, have a collar at their bases, with a single flagellum between. The chloroplasts are brownish-yellow. Although widely distributed the single species is rare in occurrence.

550b Cells without such rods 551

551a Cells ovoid, at the periphery of an oval, gelatinous sheath, attached on branched, radiating threads. Fig. 400 . .
. . . . *Uroglena*

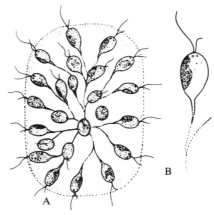

Figure 400

Figure 400 *Uroglena volvox* Ehr. (A) colony of cells on radiating, branched threads; (B) single cell with heterokont flagella and chloroplast.

Although rare, *Uroglena volvox* Ehr., the only species, is widely distributed in the United States and throughout the world. It is to be expected in the euplankton of lakes. The elliptic cells with the posterior portion abruptly narrowed, are arranged at the periphery of a mucilaginous sheath. The cells are attached at the ends of fine, radiating and branched threads which can be seen after the application of a stain. The chloroplast is a parietal, folded plate, golden-yellow in color, but usually faint. The cell bears 2 flagella of unequal length. Identification is aided by the distinctive oblong shape of the colony. Conrad and other authors advocate the inclusion of *Uroglena* with *Uroglenopsis* (Fig. 403).

551b Cells shaped otherwise 552

552a Cells pyriform, arranged rather compactly within a wide, gelatinous sheath in which conspicuous granular particles occur; flagella of unequal length. Fig. 401 . . . . *Syncrypta*

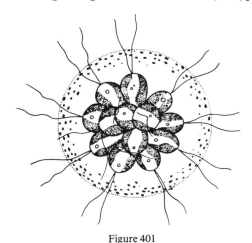

Figure 401

Figure 401 *Syncrypta volvox* Ehr., showing pebbled granulations in the colonial mucilage.

Syncrypta is a motile colony of radiately arranged, pyriform cells enclosed in a gelatinous envelope in which there are numerous, small granular bodies. The chloroplasts and the cell-shape are similar to *Synura* (Fig. 402) but the cell membrane is smooth, without siliceous spicules. The 2 flagella are of unequal length. The species illustrated is the only one known in the United States of the ten that have been described.

552b Cells shaped or arranged otherwise . . . .
. . . . 553

553a Cells elongate-ellipsoid or elongate-pyriform, rather compactly arranged to form a globular colony which is not enclosed in a gelatinous sheath; cell wall with minute siliceous scales in the anterior end; flagella 2, of equal length but dissimilar structurally; chloroplasts golden-brown. Fig. 402 . . . . . .
. . . . *Synura*

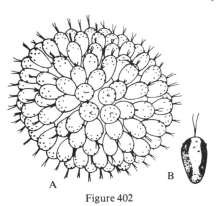

Figure 402

Figure 402 *Synura uvella* Ehr. (A) colony; (B) single cell showing siliceous spicules in anterior region of cell wall and the isokont flagella.

This genus is very common in hard-water lakes and may be so abundant as to produce disagreeable odors and tastes in water supply reservoirs. The chloroplasts are 2, parietal, golden-brown plates and often so dense as to mask the small siliceous spicules in the anterior wall. The spicules can be determined by proper focusing on quiescent colonies. The species illustrated and another, *S. adamsii* G.M. Smith are common and widely distributed in the United States, although the latter is less frequently encountered.

553b Cells ovoid or egg-shaped, separated and evenly spaced within a colonial envelope; without scales in the wall; flagella 2, of unequal length; chloroplast pale yellow. Fig. 403 . . . . . *Uroglenopsis*

Figure 403

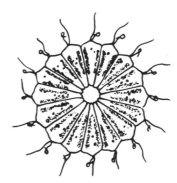

Figure 404

Figure 404   *Cyclonexis annularis* Stokes. (Redrawn from Stokes.)

Figure 403   *Uroglenopsis americana* (Calkins) Lemm.

This genus frequents water that is contaminated with nitrogenous wastes. The colonies are large and contain several hundreds of cells; are sometimes mistaken for *Volvox* (Fig. 68), but are readily differentiated by the yellowish-brown color of the platelike (not cup-shaped) chloroplasts. The cells have 2 flagella of unequal length. It is interesting that this organism is sometimes the dominant member of the plankton in arctic lakes. There is some justification for including this genus with *Uroglena* (Fig. 400).

The flat, disclike colony of compactly arranged and radiate, pyriform cells is very distinctive. The flagella are relatively coarse and can be seen readily when the colony is quiescent. There are 2 elongate, lightly pigmented chloroplasts. Of the 3 species known, 1 is reported from the United States.

554b   Cells not as above; contained in a vase-shaped envelope; *or* 2 such envelopes arising from the mouth of another to form a branched series. Fig. 366 . . . . *Dinobryon*

554a   (549) Cells elongate-ovoid or pyriform, compactly arranged side by side in a radiate fashion in 1 plane to form a plate with a small central opening; motile by 2 flagella. Fig. 404. . . . . *Cyclonexis*

555a   (548) Individuals with pseudopodia, the colony sometimes loosely formed and only temporary 556

555b   Individuals without organs of locomotion; colony non-motile 560

556a   Cells enclosed in a brown, shell-like lorica in which there are pores through which fine pseudopodia extend to meet pseudopodia of neighboring cells, thus forming a meshwork over *Sphagnum*; chloroplast pale, brownish or nearly colorless. Fig. 405 . . . . *Heliapsis*

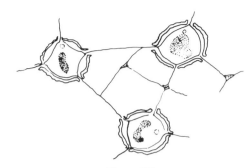

Figure 405

Figure 405   *Heliapsis mutabilis* Pascher. (Diagrammatic.)

Although this genus belongs to the rhizopodial Chrysophyta, the protoplasts are enclosed in a thick-walled, globular lorica. In the walls of this there are pores through which pseudopodial, protoplasmic strands extend, joining others to form a meshwork. There is a brownish chloroplast and a contractile vacuole. The single species of the genus has been reported from the United States.

556b   Cells otherwise. **557**

557a   Cells   amoeboid   with   fine,   firm pseudopodia, the pseudopodia adjoining those of others to form a chain in 1 series. Fig. 406 . . . . *Chrysidiastrum*

Figure 406

Figure   406   *Chrysidiastrum   catenatum* Lauterb. (Redrawn from G.M. Smith.)

Although this pseudopodial organism may occur singly it is seen frequently adjoined in loose colonies by the interconnecting pseudopodia. There is 1 platelike or disc-shaped chloroplast, usually faintly pigmented. Plants are to be expected intermingled with other algae in swamps and bogs. The species illustrated is the only one reported from the United States.

557b   Cells otherwise. **558**

558a   Colony definite, formed of amoeboid cells with knotted protoplasmic strands adjoining so as to form a network; chloroplast pale and small. Fig. 363 . . . .
. . . . *Chrysarachnion*

558b   Colony   incidental   and   transitory, formed by union of other types of pseudopodia. **559**

559a   Pseudopodia   numerous,   radiating needles; colony-formation mostly temporary and incidental. Fig. 379 . . . .
. . . . *Rhizochrysis*

559b   Pseudopodia short protoplasmic extensions, joining other individuals to form temporary colonies. Fig. 380 . . . .
. . . . *Chrysamoeba*

560a   (555) Colony consisting of vase-shaped envelopes, 1 or 2 such envelopes arising from the mouth of another to form forked series (organisms actually motile by 2 flagella, but often appearing quiescent, with the flagella completely invisible in microscopic mounts). Fig. 366 . . .
. . . . *Dinobryon*

560b   Individuals not cone-shaped; colony formed otherwise. **561**

561a Thallus composed of a compact layer of epiphytic cells in rectilinear series; mucilaginous sheath wanting. Fig. 407 . . . . .
. . . . *Phaeoplaca*

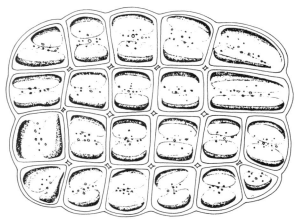

Figure 407

Figure 407 *Phaeoplaca thallosa* Chod. (Redrawn from Thompson.)

This genus has angular cells with relatively thick walls, compactly arranged to form subquadrangular thalli; the cells in 1 layer. The cells contain yellow-brown chloroplasts. Of the 2 known species 1 has been reported from the United States.

561b Thallus composed of cells arranged otherwise; mucilaginous sheath present . . .
. . . . 562

562a Thallus a sparsely branched, gelatinous cylinder or a mucilaginous network, with cells arranged in 1 to several irregular, linear series. Fig. 408 . . . .
. . . . *(Phaeosphaera) Tetrasporopsis*

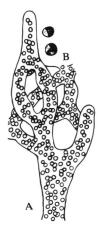

Figure 408

Figure 408 *Tetrasporopsis perforata* (Whitf.) Bourr. (A) small portion of perforate, gelatinous colony; (B) cells with parietal chloroplast.

The golden-brown cells of this genus occur in gelatinous masses of irregular shape and of macroscopic size. The thallus may be a stringy mass of mucilage, occurring in skeins or meshworks. Unlike *Tetraspora* (Fig. 79) with which it should be compared, the cells are not arranged in groups of 4 but occur in irregular fashion, or in series throughout the gelatinous strands; neither are there pseudocilia. The plant is widely distributed in North Carolina and in the Arctic.

562b Thallus not a branched, gelatinous strand or network 563

563a Cells as many as 150 within an epiphytic, gelatinous investment from which a tuft of fine to stout, branched hairs extends. Fig. 409 . . . . *Naegelliella*

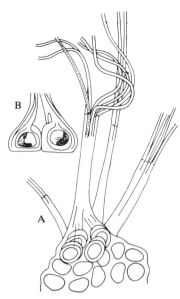

Figure 409

Figure 410

Figure 410 *Chrysocapsa planctonica* (W. & W.) Pascher. (Redrawn from G.M. Smith.)

The colonies are globular or nearly so; the mucilage clear and transparent, the cells with brown chloroplasts. The species illustrated and one other have been reported from the tycho-plankton of ponds and swamps in the United States. But *Chrysocapsa paludosa* (W. & W.) Pascher is sometimes found in the euplankton; has oval rather than round cells.

Figure 409 *Naegeliella*. (A) *N. britannica* Godward, clump of cells in a sheath which bears gelatinous hairs; (B) *N. flagellifera* Correns. (Redrawn from Correns.)

This rare genus occurs as an epiphyte on aquatic plants. The cells are arranged in 1 or 2 layers within a mucilaginous matrix. From the surface of the colony 1 or 2 tufts of fine, gelatinous hairs arise. One species only has been reported from the United States.

563b   Cells arranged otherwise; colony not bearing a tuft of hairs 564

564a   Cells in a cluster of 2-4-8-16 within an irregularly globose colonial sheath. Fig. 410. . . . . *Chrysocapsa*

564b   Cells 16-32-64 within a wide, flat, colonial matrix, the mucilage impregnated with granular substances which appear as shiny granules. Fig. 411 . . . . *Chrysostephanosphaera*

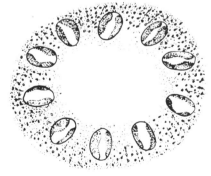

Figure 411

Figure 411 *Chrysostephanosphaera globulifera* Scherff.

The colony of this genus is flat and disclike. The oval cells, arranged in a circle, are enclosed by a wide, gelatinous matrix which contains dark granules. The interior of the colony is clear and homogeneous. There are 2 laminate, yellow-brown chloroplasts. The species illustrated is the only one of 3 known for the genus that occurs in the United States.

565a **(4) Plants filamentous or pseudofilamentous, threadlike (the thread of cells called a trichome; trichome and sheath, if present, together called a filament)...**
..........................**566**

565b **Plants not filamentous; cells globular, rod-shaped, or angular from mutual compression, *or* forming cushionlike masses of isodiametric cells in which a suggestion of a filamentous arrangement may be apparent** **619**

566a **Trichomes coiled or spiralled in a regular fashion** **567**

566b **Trichomes otherwise, straight or irregularly twisted, not forming a regular spiral (occasionally, however, *Oscillatoria* (Fig. 424) may be found twisted about itself in a regular spiral fashion)** **570**

567a **Trichomes unicellular (without cross walls). Fig. 412** ....*Spirulina*

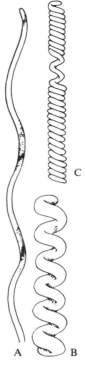

Figure 412

Figure 412 *Spirulina*. (A) *S. laxissima* G.S. West; (B) *S. princeps* (W. & W.) G.S. West; (C) *S. subsalsa* Oersted.

Although essentially unicellular, this genus is threadlike and is included with the filamentous Oscillatoriaceae of the Cyanophyta. Some species are solitary but they are often found in masses, either by themselves or intermingled with *Oscillatoria* (Fig. 424). Species are differentiated by size and shape of the coils or by the type of coiling, close or loose. The plants are activly moving as seen microscopically. The movement in this and other blue-green genera is accomplished by the extrusion of mucilage and by the flow of mucilage along the trichome.

567b **Trichomes multicellular, i.e., with cross walls** **568**

568a **Trichomes composed of beadlike or barrel-shaped cells, with heterocysts present (cells located here and there in the trichomes which are larger and different in shape from the vegetative cells). Fig. 413** .... *Anabaena*

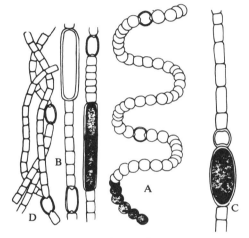

Figure 413

Figure 413 *Anabaena.* (A) *A. spiroides* var. *crassa* Lemm.; (B) *A. subcylindrica* Borge; (C) *A. sp. showing oval akinetes;* (D) *A. subcylindrica* Borge showing heterocysts.

A few species in this genus are planktonic; others are epiphytic, or form gelatinous masses on submersed substrates. Among the planktonic forms several are coiled and more or less regularly; others are straight or merely curved and entangled. When clumped, or colonial in a soft mucilage they may be differentiated from *Nostoc* (Fig. 451) by the fact that there is no definite tegument or pellicle enclosing the mucilage. The colony is soft and formless whereas *Nostoc* colonies are firm and keep a definite shape. Within the trichome there are heterocysts, and when mature akinetes (gonidia) reproductive cells which have a thick wall and a dense concentration of food storage material. The akinetes are different in shape from other cells in the trichome; may be borne adjacent to the heterocyst or remote from it. Some of the planktonic species are responsible for 'blooms' and are capable of producing lethal toxins, causing the death of animals and birds which drink infested water.

568b **Cells not beadlike but rectangular or cylindrical, usually wider than long and disclike** **569**

569a **Trichome with a close sheath. Fig. 414** .... *Lyngbya*

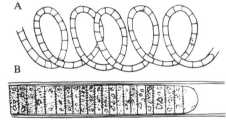

Figure 414

Figure 414 *Lyngbya.* (A) *L. contorta* Lemm.; (B) *L. birgei* G.M. Smith.

Most species are straight and rigid or sometimes curved and entangled, whereas *Lyngbya contorta* Lemm. is regularly spiralled. Trichomes are mostly rounded or conical at the apex, seldom constricted at the joints, do not exhibit the apical differentiation and the tapering that is characteristic of *Oscillatoria* (Fig. 424) which has no definite sheath.* Compare *Lyngbya* with *Phormidium* (Fig. 429) which has a very thin and sticky sheath and a genus in which compact mats are formed by many species.

569b   Trichomes without a sheath, regularly spiralled and actively moving. Fig. 415 . . . . *Arthrospira*

Figure 415

Figure 415   *Arthrospira jenneri* (Kütz.) Stiz.

The plants in this genus are multicellular but at times the cross partitions are difficult to discern and short section of a trichome may be mistaken for *Spirulina* (Fig. 413). Species are differentiated mostly by size and by the form of the spiral (closely coiled or loose). The plants show a spiral motion although perhaps not as active as *Spirulina*. *Arthrospira* occurs intermingled with *Spirulina* but is also common in *purées* of miscellaneous algae in the tychoplankton.

570a   (566) Trichomes with cells all alike in shape and size (although the trichome may taper slightly at one or both ends, or the apical cell may be slightly swollen [capitate]); heterocysts lacking. **571**

570b   Trichomes with differentiated cells (heterocysts present), the heterocysts appearing empty and with polar plugs of mucilage; other cells enlarged, with thick walls serving as akinetes (gonidia or spores). **592**

571a   Trichomes decidedly tapering throughout at one or both ends. **572**

571b   Trichomes not decidedly tapering, the same diameter throughout (but some species narrowed slightly in the extreme apical portion). **574**

572a   Trichomes tapering toward both ends, with U-shaped, false branches. Fig. 416. . . . . *Hammatoidea*

*Recently F. Drouet has combined *Lyngbya* with *Oscillatoria*.

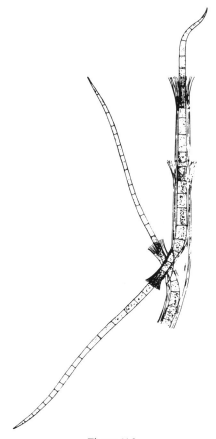

Figure 416

Figure 416 *Hammatoidea yellowstonensis* Copeland. (Redrawn from Copeland.)

This genus has false branches which form U-shaped loops lateral to the main axis. Like other members of the Rivulariaceae the branches taper toward the apices, but there are no heterocysts as in that family. The plants usually occur in the mucilaginous sheaths of other algae. Two species have been reported from the United States, one from the hot springs of Yellowstone Park.

572b **Trichomes decidedly tapering from a base to a long or short hairlike apex. . . .**
. . . . 573

573a **Trichomes aggregated, tapering from a base which is incorporated in a prostrate cushion of cells. Fig. 417 . . . *Amphithrix***

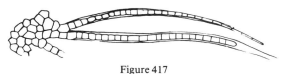

Figure 417

Figure 417 *Amphithrix janthina* (Mont.) Born. & Flah. (Redrawn from Bornet & Flahault.)

These are tapering filaments, arranged in somewhat parallel fashion so as to form clusters, but without conspicuous mucilage, attached to substrates. There is a weakly-developed, prostrate expansion of the thallus from which the upright trichomes arise. The lack of heterocysts makes this a somewhat anomalous member of the Rivulariaceae which characteristically have a heterocyst at the base of a filament. Two species have been reported from the United States and are not widely distributed.

573b **Trichomes solitary or loosely clustered (2 or 3 together, sometimes forming stellate clusters), epiphytic on larger algae or on aquatic plants. Fig. 418 . . . .**
. . . . *Calothrix*

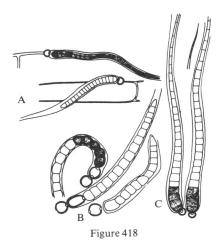

Figure 418

Figure 418   *Calothrix.* (A) *C. epiphytica* W. & W.; (B) *C. atricha* Frémey; (C) *C. braunii* Born. & Flah.

Whereas most species of *Calothrix* possess heterocysts there are a few which do not. The tapering filaments of this genus are often found as solitary plants, or they may be loosely clustered, 2, 3 or 4 together. In some instances after a profuse growth they may be gregarious and form an extensive expanse. There is a basal heterocyst and in most species there is an akinete adjacent to the terminal heterocyst. Species are differentiated on the basis of size, presence or absence of a heterocyst, and presence or absence of an akinete. Also the degree of tapering and the details of the sheath, which usually extends beyond the end of the trichome, are considered. In some species the trichomes are short and the tapering very abrupt. At least 41 species have been reported from the United States.

574a   (571) Filaments branched (examine several specimens; branching may be sparse). 575

574b   Filaments unbranched 578

575a   Branches false (a branch formed by a proliferation of a broken trichome which pushes to one side of the main axis; not branching by the lateral division of a cell in the main axis); branching often sparse. Fig. 419 . . *Plectonema*

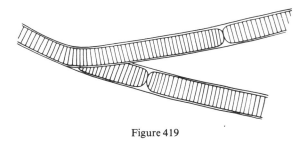

Figure 419

Figure 419   *Plectonema wollei* Farlow.

The false habit of branching places this genus in the Scytonemaceae, but unlike other members of the family there no heterocysts. The species illustrated is a common one, occurring in brownish-green or black, cottony masses at or near the surface of the water; is a relatively large species, being up to 50 μm in diameter. Many of the smaller species form clumps, or are intermingled with other algae. Several specimens should be examined throughout the length of a filament in making determinations because branching may be sparse and the plant might be mistaken for *Lyngbya*. There is a dozen or more species reported from the United States, all rather widely distributed.

575b   Branching true; the branches formed by lateral division of cells in the main axis of the trichome 576

576a   Filaments with prostrate and erect portions; dichotomous; sheaths transversely lamellate. Fig. 420. . . . . . . *Colteronema*

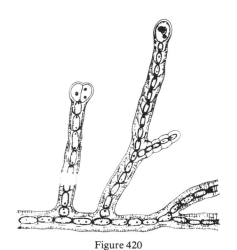

Figure 420

Figure 420 *Colteronema funebre* Copeland. (Redrawn from Copeland.)

This plant is similar to *Albrightia* (Fig. 422) but the branching is dichotomous. The branches arise vertically from a more or less prostrate axial filament. The rather soft, gelatinous sheath has transverse lamellations. There are no heterocysts. The single species is known only from Yellowstone Park.

576b   **Filaments without a prostrate and an erect portion; sheath not transversely lamellate** **577**

577a   **Filaments with disjunct cells which are transversely oval, arranged in a homogeneous, gelatinous sheath which may branch or show anastomosing. Fig. 421** . . . .*Heterohormogonium*

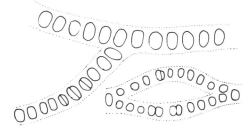

Figure 421

Figure 421 *Heterohormogonium schizodichotomum* Copeland. (Diagrammed from Copeland.)

This genus has a wide, gelatinous sheath in which oval cells are arranged in a linear series, but not in contact. The filament may be variously branched and may anastomose. The genus is known only from hot springs.

577b   **Cells elongate-oval or sausage-shaped, arranged in continuous series within a gelatinous sheath. Fig. 422** . .*Albrightia*

Figure 422

Figure 422 *Albrightia tortuosa* Copeland. (Redrawn from Copeland.)

The branched filaments of this plant resemble strings of sausages enclosed in a relatively firm, colorless sheath. There are no heterocysts and reproduction occurs only by cell division as far as is known, this occurring in the apical region of the trichome. *Albrightia* is known only from the hot springs of Yellowstone National Park.

578a   **(574) Trichomes without a sheath. . . 579**

578b   **Trichomes with a sheath** **581**

579a Trichome short, from 3 to 10 (20) cells long. Fig. 423 . . . . *Borzia*

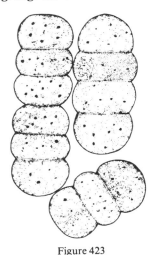

Figure 423

Figure 423 *Borzia trilocularis* Cohn. (Redrawn from Daily.)

This rare plant occurs as very short, hormogonialike trichomes of about 8 cells. The terminal cells are hemispherical. *Borzia* has been reported from several stations across the United States. There are only 2 species known thus far.

579b Trichomes longer, composed of many cells 580

580a Trichomes solitary or intermingled with other algae, often in aggregated clusters but not lying parallel in bundles, the trichomes sometimes tapered slightly toward the anterior end, or with the apical cell swollen (capitate). Fig. 424 . . . . .
. . . . *Oscillatoria*

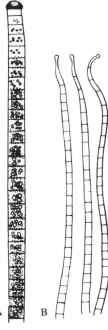

Figure 424

Figure 424 *Oscillatoria*. (A) *O. rubescens* DeCand.; (B) *O. splendida* Grev.

One of the chief characteristics of species in this genus is their oscillating movement, and another is their lack of a definite, firm sheath (as in *Lyngbya,* Fig. 414). One interpretation of the genus disregards the sheath characteristic and unites *Lyngbya* and *Oscillatoria*. A mass of *Oscillatoria* left in a shallow dish in the laboratory will creep up the sides or spread itself over the bottom. There are numerous species (subject to various interpretations), differentiated on the basis of size, cell proportions and the morphology of the apical region. Some species taper slightly toward the anterior end, there being a slight basal-distal differentiation, and the apical cell may be swollen or capitate, sometimes with a calyptra (a capping membrane). *Oscillatoria* occurs both in water and on moist subaerial substrates such as soil or dripping rocks. A few species, such as *O. rubescens* (D.C.) Gom. are planktonic and at certain times of the year this species may cause the water to become blood-red. The color is related to the fact that the cells have pseudovacuoles that refract the light rather than to pigment coloration.

580b Trichomes lying in parallel bundles, not tapering toward the anterior end; apical cell never capitate; planktonic. Fig. 425 . . . . *Trichodesmium*

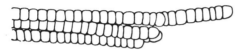

Figure 425

Figure 425 *Trichodesmium lacustre* Klebahn.

This is a genus of uncertain position because it has an *Anabaena*like filament but is without heterocysts. The trichomes are arranged in parallel bundles that occur as free-floating, dark green flakes. *Trichodesmium erythraceum* Ehr., a marine species, because of the refractibility of its pseudovacuoles, gives the characteristic color to the Red Sea.

581a (578) Sheath containing 1 trichome **582**

581b Sheath containing more than 1 trichome **589**

582a Sheath firm and definite, not adhering to sheaths of adjacent filaments **583**

582b Sheath soft and sticky, often adhering to sheaths of adjacent plants and intermingling, confluent with them **585**

583a Sheaths purplish or reddish, conspicuously stratified. Fig. 426 . . . . *Porphyrosiphon*

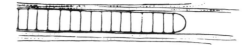

Figure 426

Figure 426 *Porphyrosiphon notarisii* (Menegh.) Kütz.

The purple color of the lamellated sheath in this genus (especially *P. notarisii* (Menegh.) Kütz. accounts for the brightly colored patches on damp soil in subtropical United States. Denuded soil in the South frequently shows extensive purple-red mats when colonized by this species. Occasionally the plants become involved with fungal threads to approximate a lichen.

583b Sheaths colorless or yellowish **584**

584a Trichomes short, from 2 to 20 cells long; sheath colorless; homoegeneous. Fig. 427 . . . . *Romeria*

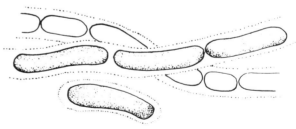

Figure 427

Figure 427 *Romeria elegans* var. *nivicola* Kol, from Olympic Mountain snowfields.

This genus consists of sausage-shaped cells arranged in short trichomes, usually within a thin sheath. There are neither heterocysts nor akinetes formed as far as known. In the United States the plant has been found only in snow fields.

584b **Trichomes long, composed of many cells; sheaths colorless or yellowish, sometimes stratified. Fig. 414** . . . .
. . . . *Lyngbya*

585a **(582) Filaments forming an expanded plant mass, sometimes developing erect tufts.** **586**

585b **Filaments not forming expanded plant masses on a stratum** **587**

586a **Plant mass having erect tufts. Fig. 428.** .
. . . . *Symploca*

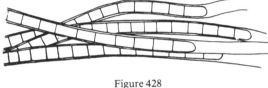

Figure 428

Figure 428 *Symploca muscorum* (Ag.) Gom.

Filaments of the species illustrated occur in erect tufts on moist, subaerial substrates. Species which have thin, sticky sheaths would be compared with *Phormidium* (Fig. 430) with which they may be confused if seen individually and not in colonial mass. At least 15 species have been reported from the United States.

586b **Plant mass without erect tufts. Fig. 429**
. . . . *Phormidium*

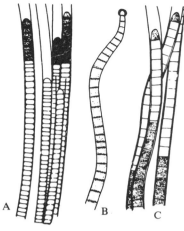

Figure 429

Figure 429 *Phormidium*. (A) *P. ambiguum* Gom.; (B) *P. favosum* (Bory) Gom.; (C) *P. inundatum* Kütz.

The sheaths of the trichomes in this genus are very thin and sticky and the plants often adhere together, forming a compact mat, or a skein over submersed surfaces or on dripping rocks. The plant masses are blue- or black-green and feel slimy or slippery to the touch. There are numerous species (assigned to *Oscillatoria* by F. Drouet). They are differentiated by size, sheath characteristics, and by the morphology of the apical cell. Plants should be compared with *Lyngbya* (Fig. 414). Some of the many species which occur in the United States are found in hot springs.

587a **(585) Filaments lying parallel in free-floating bundles. Fig. 425** . . . .
. . . . *Trichodesmium*

587b **Filaments not in parallel bundles** . . . **588**

588a Filaments composed of rectangular or cylindrical cells in continuous series within a close sheath, irregularly intermingled with one another. Fig. 429 . . . .
. . . . *Phormidium*

588b Filaments composed of a wide, gelatinous sheath (or tube) in which transversely oval cells are arranged in disjunct fashion. Fig. 421 . . . .
. . . . *Heterohormogonium*

589a (581) Sheaths soft and sticky, without an even or smooth outer boundary 590

589b Sheath firm and definite, not softly mucilaginous 591

590a Sheath containing 2 or 3 trichomes. Fig. 430 . . . . *Hydrocoleum*

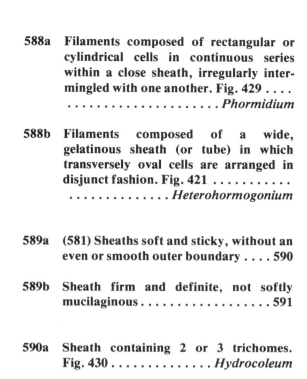

Figure 430

Figure 430 *Hydrocoleum oligotrichum* A. Braun.

In this genus there are only 3 (sometimes 4) trichomes within a wide, lamellate, gelatinous sheath. The filaments may be solitary or spread in a thin layer on damp soil. *Hydrocoleum oligotrichum* A. Braun is lime-encrusted whereas *H. homeotrichum* Kütz. is not. About 15 species have been reported from the United States, differentiated on the basis of size and characteristics of the sheath and habit of growth.

590b Sheaths containing many trichomes. Fig. 431 . . . . *Microcoleus*

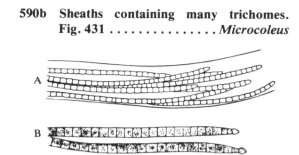

Figure 431

Figure 431 *Microcoleus*. (A) *M. vaginatus* (Vauch.) Gom.; (B) *M. lacustris* (Rab.) Farlow.

Unlike *Hydrocoleum* (Fig. 430) there are many intertwined trichomes within a rather definite sheath. Usually the trichomes show an active slithering motion over one another, may emerge from the sheath and then retract. The thallus is often macroscopic as it grows on damp soil. Some species, however, are more often found on submersed substrates. Species are differentiated by size and by the morphology of the apical region of the trichome. There are 14 species reported from the United States.

591a (589) Sheaths wide, containing 2 or 3 loosely arranged trichomes (often relatively short). Fig. 432 . . . . . . *Dasygloea*

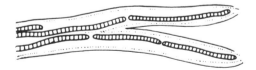

Figure 432

Figure 432 *Dasygloea amorpha* Berk.

The sheaths of this plant are rather firm and definite in outline; contain but 1 to 3 trichomes. The sheaths are usually forked at the ends (as they are also in *Schizothrix* (Fig. 433), with which *Dasygloea* should be compared. Only 1 species has been reported from the United States but is widely distributed.

591b **Sheaths close, usually containing several, crowded trichomes. Fig. 433 . . . .** *Schizothrix******

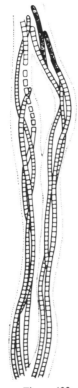

Figure 433

Figure 433 *Schizothrix tinctoria* Gom.

In this genus there are but few trichomes within a definite and rather firm sheath. The plant masses are of macroscopic size and often form

extended films and wefts over submersed vegetation. Several of the species quickly disintegrate when stored in a covered container for a short time without a preservative. Under this situation the plants liberate a copious amount of the pigment phycocyanin. In an aqueous solution the pigment shows a distinct fluorescence. There are at least a dozen species reported from the United States, differentiated by size, by cell proportions and by the characteristics of the sheath which often is forked.

592a **(570)Trichomes definitely tapering at one or both ends 593**

592b **Trichomes not tapering, or rarely tapering slightly near the apex 599**

593a **Trichomes tapering at both ends, short, 20 cells or less long; heterocysts wanting; akinetes present. Fig. 434 . . . .** *Raphidiopsis*

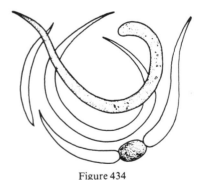

Figure 434

Figure 434 *Raphidiopsis curvata* Fritsch & Rich.

The short trichomes taper at one or both ends. The plants are solitary, curved and twisted, or sigmoid and sickle-shaped. There are no het-

*See F. Drouet 1968 for his revision of this genus.

erocysts but akinetes occur. Thus far 1 species has been reported from a few localities in the United States (Ohio and Florida), although another species (possibly *R. mediterranea* Skuja) has been found in Minnesota. Plants are either eu- or tychoplanktonic.

593b   **Trichomes tapering from base to apex with basal-distal differentiation; with a heterocyst and often with an akinete at the base** **594**

594a   **Filaments enclosed within abundant mucilage; forming a globular or hemispherical body; attached or free-floating** **595**

594b   **Filaments not enclosed by abundant mucilage to form a thallus of definite shape** **597**

595a   **Sheath containing 2 or more trichomes. Fig. 435** . . . . *Sacconema*

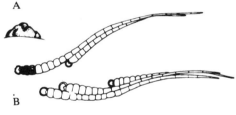

Figure 435

Figure 435 *Sacconema rupestre* Borzi. (A) habit of colony; (B) filaments with basal heterocysts.

Trichomes in this genus are tapering from a basal heterocyst as in *Gloeotrichia* (Fig. 436), but there is more than 1 trichomes within the filamentous sheath, and the gelatinous colony is very irregular in shape as it occurs on stones (sometimes in very deep water). The sheaths are

wide, lamellate, and are flaring at the outer end. The species illustrated seems to be the only one reported from the United States, and possibly the only one known for the genus. Some students include *Sacconema* with *Calothrix* (Fig. 418).

595b   **Sheaths containing 1 trichome** **596**

596a   **With cylindrical spores adjoining a basal heterocyst; colonial mucilage soft in floating species, firm in attached species which form hemispherical or globular thalli 1-3 mm in diameter. Fig. 436** . . . . *Gloeotrichia*

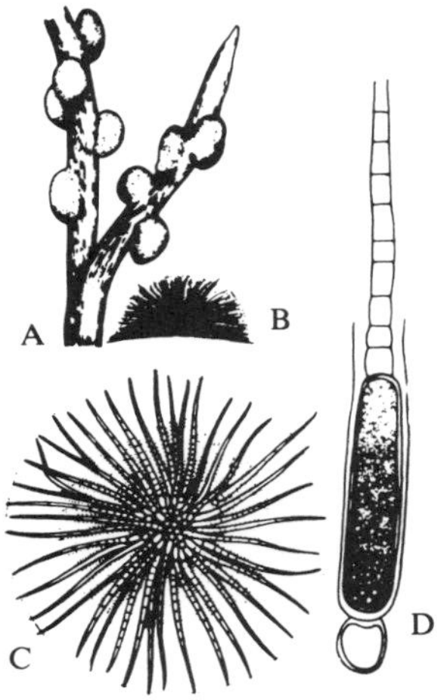

Figure 436

Figure 436   *Gloeotrichia*. (A) *G. pisum* (Ag.) Thur., habit on *Ceratophyllum;* (B) diagram of filament arrangement; (C) *G. echinulata* (J.E. Smith) Richter, diagram of filament arrangement in colony; (D) single filament showing terminal heterocyst and akinete.

In this genus the tapering trichomes are encased in mucilage which is usually relatively soft in the planktonic species, but firm and hard in the attached forms. The trichomes are radiately arranged in the mucilage, but are not so closely compacted as in *Rivularia* (Fig. 437). *Gloeotrichia* has filaments with large, cylindrical akinetes adjoined to the basal heterocyst. When immature, species may be mistaken for *Rivularia,* which never produces akinetes. Doubtless many of the records of *Rivularia* are *Gloeotrichia* in which the akinetes have not yet developed. One of the more common species is *G. echinulata* (J.E. Smith) P. Richter which occurs in abundance in the plankton of hardwater lakes. The colonies are globular and appear as 'tapioca' grains, making the water buff-colored. When abundant along bathing beaches this plant causes a severe skin irritation among some persons which has been mistaken for 'swimmer's itch'. *Gloeotrichia natans* (Hedw.) Rab. is also fairly common. It begins development as an attached thallus but later appears at the surface in brown, gelatinous and amorphous masses, either expanded and flat or somewhat globular. *Gloeotrichia pisum* Lag. forms hard, green or black balls, 1 or 2 mm in diameter on submersed vegetation, sometimes completely covering the host plant. Nine species have been reported from the United States.

596b   Spores absent; trichomes embedded in hard mucilage to form globular thalli which may coalesce, thus producing a continuous, lumpy stratum; trichomes radiate, or more often subparallel and densely packed within the colonial mucilage. Fig. 437 . . . . *Rivularia*

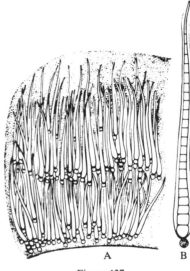

Figure 437

Figure 437   *Rivularia* sp. (A) diagram of portion of attached colony to show arrangement of filaments; (B) one filament showing basal heterocyst.

This genus may be differentiated from *Gloeotrichia* (Fig. 436) by its lack of akinetes at the base of the trichome, by the compact, almost parallel arrangement of the trichomes, and by the extreme firmness of the colonial mucilage. All species are attached, mostly to stones and logs, sometimes forming extensive, pebbled patches, blackish-green in color. Some larger colonies show a 'zonation' resulting from successive generations of false branches. Twenty-four species have been reported from the United States, but many of the names seem to be confused with *Gloeotrichia*.

597a   (594) Filaments freely, falsely branched, the trichomes lying 2 or more within the sheath of the main filament for some distance, then diverging. Fig. 438 . . . . .
. . . . . *Dichothrix*

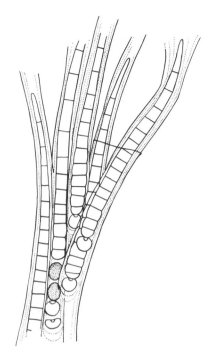

Figure 438

Figure 438 *Dichothrix gypsophila* (Kütz.) Born. et Flah., false branches, with branches lying partly within the sheath of the primary trichome.

In this genus the tapering trichomes are enclosed 2 or 3 together within the branching sheaths. Bushlike tufts are produced by their habit of growth and these sometimes attain macroscopic proportions. The species are differentiated by size, by sheath characteristics and type of arrangement of the false branches. Plants are customarily found intermingled with miscellaneous algae in the tychoplankton, sometimes attached and form tufted mats over submersed plant stems. Eleven species have been reported from the United States, most of which are widely distributed.

597b Filaments not freely branched, if branched at all the branches not lying parallel within the sheath of the main filament. 598

598a Branching at regular intervals, solitary or in pairs, the branches tapering; heterocysts mostly terminal, but also intercalary. Fig. 439 . . . . . . *Scytonematopsis*

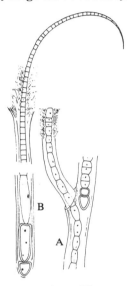

Figure 439

Figure 439 *Scytonematopsis hydnoides* Copeland. (A) habit of branching; (B) heterocyst and tapering trichome. (Redrawn from Copeland.)

This is the only member of the Rivulariaceae in which the tapering, false branches arise at regular intervals. (Some species of *Rivularia* are a possible exception.) Heterocysts are both basal and intercalary, and usually branches, solitary or in pairs, arise immediately below them. Plants are very similar to *Tolypothrix* (Fig. 454) except for the decided tapering of the trichomes. The only species reported from the United States has been found in the hot springs of Yellowstone National Park.

598b Branching absent or scarce and irregular, the heterocysts basal. Fig. 418
.... *Calothrix*

599a (592) Trichome branches formed by the lateral division of cells in the main axis (true branching). 600

599b Trichomes unbranched or with false branches (sections of trichomes developing in a series of cells to one side of a break in the main axial row of cells). ....
.... 604

600a Individual trichome sheath not apparent; colony of trichomes invested by a mucilage; heterocysts usually on the ends of short branches. Fig. 440 ......
.... *Nostochopsis*

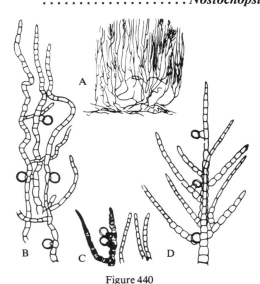

Figure 440

Figure 440 *Nostochopsis lobatus* Wood. (A) diagram of filament arrangement within colony; (B)-(D) filaments showing lateral heterocysts.

The branched trichomes in this genus are enclosed in a rather firm mucilage which gives an erect, strandlike or tubelike form to the thallus. The trichomes are composed of *Anabaena*like cells and bear true branches. The heterocysts are borne laterally along the trichome or on the ends of short branches (rarely intercalary also). The trichomes are more or less parallel and erect in the gelatinous matrix. In flowing water the thalli may lie prostrate, whereas in quiet water they stand erect, reach the surface and then flatten out. Apparently only 1 species has been reported from the United States, is widely distributed.

600b Individual trichome sheath evident; heterocysts in the same series with the main axis, or cut off laterally from them but not on the ends of branches 601

601a Filaments closely aggregated, forming an attached, gelatinous thallus, 1-2 mm in diameter. Fig. 441 .... *Capsosira*

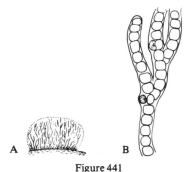

Figure 441

Figure 441 *Capsosira brebissonii* Kütz. (A) habit of attached colony; (B) portion of filament showing lateral heterocysts. The sheath is thin and soft, without a definite limiting tegument.

In this genus there are small, bulbous, mucilaginous colonies attached to submersed substrates. The individual trichomes which bear true branches are surrounded by individual sheaths that are yellowish. Heterocysts are usually lateral, but may be intercalary in the trichomes which do not taper toward the apices. There is but 1 species reported from the United States and from only 1 station, in Connecticut.

601b **Filaments not forming a definitely shaped, gelatinous thallus but spreading irregularly** **602**

602a **Filaments with more than 1 series of cells within a wide, gelatinous sheath; heterocysts small, cut off laterally from the vegetative cells and scarce. The branches often have cells in 1 series when young. Fig. 442** ...... *Stigonema*

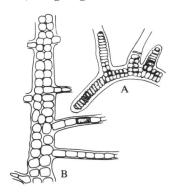

Figure 442

Figure 442 *Stigonema.* (A) *S. muscicola* Borzi (*Fischerella muscicola* [Thur.] Gom.); (B) *S. turfaceum* (Berk.) Cook.

Although there are several species reported from the United States, *Stigonema turfaceum* (Berk.) Cook and *S. ocellatum* (Dillw.) Thur. are by far the most common. The latter is one which frequently does not show the multiseriate arrangement of cells. The sheath is wide and distinctly lamellate, the cells showing individual sheaths. The heterocysts typically are cut off laterally from vegetative cells; often are scarce and difficult of discernment. Usually they are olive-brown. The cells in some species are con-

nected by narrow processes, similar to cell morphology in the Rhodophyta. *Stigonema* forms brownish, olive-green or blue-green growths on submersed reed stems, on exposed roots and other aquatic vegetation. On moist soil or wet rocks and concrete some species form velvety expanses. *Stigonema ocellatum* invariably is found in soft-water lakes and desmid habitats.

602b **Filaments with 1 series of cells; heterocysts intercalary, not lateral** .. **603**

603a **Branches extending parallel with the main axial trichome. Fig. 443** .... *Thalpophila*

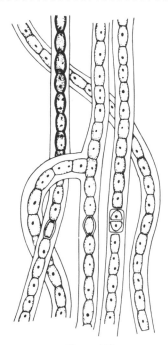

Figure 443

Figure 443 *Thalpophila imperialis* Copeland. (Redrawn from Copeland.)

As shown in the illustration, *Thalpophila* trichomes have true branching as in *Hapalosiphon* (Fig. 444), but the branches are predominantly unilateral and lie parallel with the main axis. The branching habit produces a cordlike thallus with each trichome within its own sheath. *Thalpophila* has been found only in the geyser waters of Yellowstone National Park.

603b   Branches arising and extending at right angles to the main filament. Fig. 444 . . .
. . . . *Hapalosiphon*

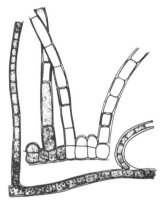

Figure 444

Figure 444   *Hapalosiphon hibernicus* W. & W.

This genus is differentiated readily from *Stigonema* (Fig. 442) by the cells being arranged in a single series, by the intercalary heterocysts, and by the close, relatively firm and sometimes lamellate sheath. The cells are mostly rectangular or short cylindric rather than oval or globose. The habit of branching separates it from *Thalpophila* (Fig. 443). The plants mostly sprawl over a substrate, branching and rebranching. The younger branches especially when growth is vigorous, possess cell characteristics and sheath features different from the main axis and hence students have been led to identify species from examining outer portions of the thallus, leading to erroneous conclusions. *Hapalosiphon* occurs more frequently in soft-water habitats. At least 1 common species is terrestrial. Twenty species have been reported

from the United States, some of which are widely distributed; others decidedly localized.

604a   (599) Trichomes unbranched 605

604b   Trichomes with false branches 615

605a   Individual trichome sheath firm and definite, heterocysts basal (rarely intercalary also). Fig. 445 . . . . . *Microchaete*

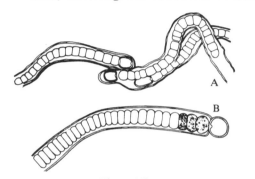

Figure 445

Figure 445   *Microchaete*. (A) *M. diplosiphon* Gom.; (B) *M. robusta* Setch. & Gard.

Plants of this genus are mostly epiphytic or loosely-adjoined to filamentous algae, with part of the filament lying parallel with the substrate, then curving away. The trichomes do not taper, although there may be a slight reduction in diameter toward the apices. Heterocysts are mostly basal, 1 to 3 in a series. Nine species are reported from the United States, differentiated by size and by habit of growth.

605b   Individual sheath soft, often indistinct and confluent with colonial mucilage; heterocysts either all terminal or all intercalary 606

606a   Heterocysts terminal 607

606b Heterocysts intercalary 608

607a Spores adjacent to the heterocysts which are at one end of the trichome, rarely at both ends. Fig. 446 . . *Cylindrospermum*

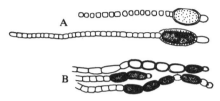

Figure 446

Figure 446 *Cylindrospermum.* (A) *C. majus* Kütz.; (B) *C. marchicum* Lemm.

The chief characteristic of this genus is the location of the heterocyst and akinetes—always terminal and usually only at one end. There may be a series of several akinetes. The trichomes do not taper, lie in a somewhat parallel fashion within a soft, amorphous mucilage, forming films or skeins over submersed vegetation; usually are bright blue-green. Frequently a dense 'nest' of spores will be found where there has been a colony of *Cylindrospermum*. Of the 21 species reported from the United States a few are terrestrial.

607b Spores not adjacent to the heterocysts; heterocysts regularly at both ends of the trichome. Fig. 447 . . . . . . *Anabaenopsis*

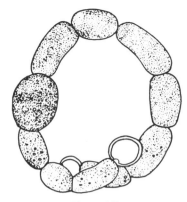

Figure 447

Figure 447 *Anabaenopsis elenkinni* Miller. (Redrawn from G.M. Smith.)

The trichomes of this genus are usually relatively short (8 to 20 cells), and mostly coiled or S-curved. Heterocysts are terminal and the akinetes are formed remote from them. Five species have been reported from the plankton in the United States. Species seem to be confined to waters rich in nitrogen.

608a **(606)** Thallus composed of many trichomes, usually parallel within the colonial mucilage 609

608b Plant a solitary trichome, or if aggregated, not parallel but entangled within the colonial mucilage. 611

609a Trichomes enclosed in abundant mucilage, arranged to form a hollow, attached, tubular thallus. Fig. 448 . . . . . .
. . . . *Wollea*

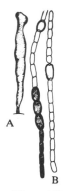

Figure 448

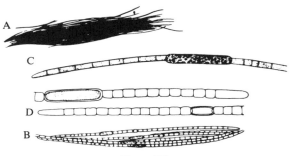

Figure 449

Figure 448 *Wollea saccata* (Wolle) Born. & Flah. (A) habit of colony; (B) trichomes to show heterocysts and akinetes in a series.

Trichomes of this species lie more or less parallel in long, gelatinous, tubelike strands which grow vertically from the bottom of standing water. The cells are barrel-shaped or *Anabaena*like, and the intercalary heterocysts are oval or subcylindric. The species illustrated is the only one known.

609b Thallus not a gelatinous, saclike tube ..
. 610

610a Trichomes parallel, forming a free-floating flakelike bundle, each trichome containing near the midregion a single heterocyst and an akinete (rarely 2 but not adjacent to the heterocyst), the akinetes appearing at maturity; trichomes sometimes tapering slightly at the extremities. Fig. 449 . . . .
. . . . *Aphanizomenon*

Figure 449 *Aphanizomenon flos-aquae* (L.) Ralfs. (A) diagram of colony; (B) a few trichomes from the colony; (C), (D) trichomes to show medial akinete and trichome.

Species of this genus have trichomes that lie parallel in bundles of macroscopic proportions and usually occur so profusely that the water seems to be filled with bits of chopped grass. The cells are short cylindric or barrel-shaped and are the same diameter throughout except that there is a slight tapering at the apices. Each trichome contains 1 heterocyst and 1 akinete (rarely 2 each) produced near the midregion. The cells contain pseudovacuoles which give the plants great buoyancy and this accounts for the fact that profuse growths become concentrated at the surface where floating scums result. This leads to deterioration of the plants and a release of cell-contents, including endotoxins (apparently). This condition leads to increased bacterial action and either directly or indirectly *Aphanizomenon* becomes a spoiler of water for both domestic and recreational purposes and in some instances leads to a catastrophic fish kill. Three species (or varieties) have been reported from the United States. The one illustrated is the most common but is known to occur in a number of 'strains' as evidenced by laboratory culture.

610b Trichomes not parallel, or if so, forming indefinitely shaped flakes or clumps; mostly not macroscopic. Fig. 413 . . . . .
. . . . . *Anabaena*

611a (608) Trichomes planktonic, solitary . . .
. . . . 612

611b Trichomes colonial, in a gelatinous mass 613

612a Vegetative cells and heterocysts compressed, wider than long; disc-shaped. Fig. 450 . . . . *Nodularia*

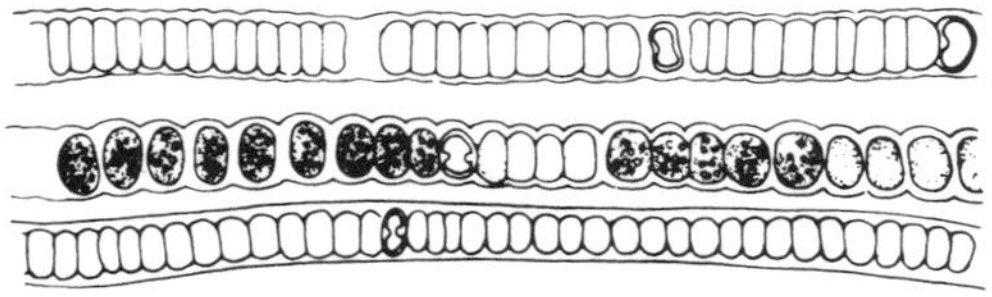

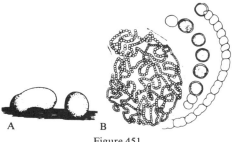

Figure 451 *Nostoc*. (A) *N. pruniforme* Ag., habit and shape of colonies; (B) *N. linckia* (Roth) Born. & Flah., with 2 trichomes in detail showing heterocysts and akinetes.

Figure 450

Figure 450 *Nodularia spumigena* Mert.

Filaments of this genus are at once recognizable by the very short, compressed (almost disclike) cells and heterocysts. The sheath is rather thin and close; is mucilaginous and sometimes not discerned inmediately. Plants are either eu- or tychoplanktonic. The species illustrated is the most common of 6 that have been reported from the United States.

612b Cells globose to cylindric or barrel-shaped, not compressed as above. Fig. 413 . . . . *Anabaena*

613a (611) Plant mass definite in shape, usually globular, bounded by a firm, gelatinous tegument (sometimes forming an expanded, gelatinous rubbery sheet); colonies microscopic or macroscopic. Fig. 451 . . . . *Nostoc*

This genus is characterized by the firm outer tegument of the mucilage which encloses numerous, coiled threads of beadlike cells. The colonies retain a definite shape, mostly globular although a few species form rubbery or leathery expansions, especially *Nostoc commune* Vauch. which is widely distributed but is especially abundant in alpine meadows and on the tundra of the Arctic. The colonies vary in size from microscopic to thalli 10 cm in diameter (*N. amplissimum* Gard., known as Mare's Eggs); *N. caeruleum* Lyngb. is planktonic; occurs as minute, blue balls sometimes very abundant especially in soft-water lakes. *N. parmeloides* Kütz. forms small, olive-green, shelving thalli on the down stream side of rocks, especially in mountain brooks. The thalli, up to 1.5 cm across, invariably contain the eggs or larva of a midge fly. Colonies of *Nostoc* growing on damp soil are used as food by orientals.

613b Plant mass not definite in shape; mucilage soft, not bounded by a firm tegument 614

614a Trichomes forming small bundles, often occurring solitary, within a gelatinous sheath, either entangled or parallel. Fig. 452 . . . . *Aulosira*

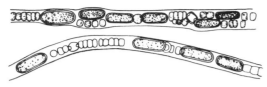

Figure 452

Figure 452   *Aulosira laxa* Kirch.

This genus is much like *Microchaete* (Fig. 445) and is sometimes classified with it. Some authorities differentiate it on the basis of the soft sheath, the intercalary heterocysts and the akinetes which also are intercalary and about the same diameter as the heterocysts. Two species have been reported from the United States.

614b Trichomes   not   forming   bundles. Fig. 413. . . . . *Anabaena*

615a (604) Branches arising in pairs about midway between 2 heterocysts (branches also rarely solitary). Fig. 453 . . . .
. . . . *Scytonema*

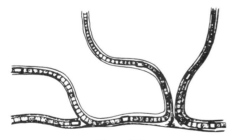

Figure 453

Figure 453   *Scytonema archangelii* Born. & Flah.

This genus exhibits false branching, the branches arising either singly or in pairs where there has been a break in the trichome of the main axis. Sheaths are either close and thin, or wide and lamellate; often are yellow or brown. Heterocysts intercalary. Plants occur intermin-

gled with other algae in the tychoplankton or form clumps attached to the stems of aquatic plants. Some species are terrestrial. *Scytonema* should   be   compared   with   *Tolypothrix* (Fig. 454) in which branches arise only at a heterocyst or at the base of a series of them. Twenty-five or more species have been reported from the United States, most of them widely distributed.

615b Branches always arising singly immediately below a heterocyst or a series of them; branching sometimes rare and not regular, requiring a search through a number of plants to determine this character 616

616a Sheath close and firm; 1 trichome in a sheath 617

616b Sheath usually wide and soft; at least more than 1 trichome in a sheath . . . 618

617a Branches frequent, arising immediately below a heterocyst which is always intercalary. Fig. 454. . . . . *Tolypothrix*

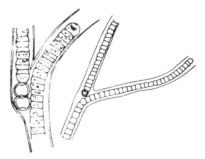

Figure 454

Figure 454   *Tolypothrix distorta* Kütz.

617b Branches   rare;   heterocysts   terminal (rarely intercalary also). Fig. 445. . . . . . .
. . . . *Microchaete*

618a (616) Trichomes parallel within a fairly wide sheath; plant mass developing bushy tufts; heterocysts basal in the trichome. Fig. 455 . . . . *Desmonema*

Figure 455

Figure 455   *Desmonema wrangelii* (Ag.) Born. & Flah.

The falsely branched filaments of this genus differ from others in the Scytonemataceae by having several trichomes within a single sheath. The filaments are gregarious and form masses of macroscopic size on moist subaerial substrates. They usually show erect tufts. The species illustrated is the only one reported from the United States.

618b Trichomes twisted and entangled in a wide sheath; heterocysts intercalary. Fig. 456 . . . . *Diplocolon*

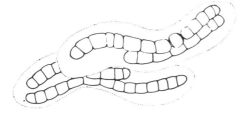

Figure 456

Figure 456   *Diplocolon heppii* Näg. (Redrawn from G.M. Smith.)

This plant forms an expanse on moist subaerial substrates such as dripping rocks and banks. The trichomes are falsely branched, have intercalary heterocysts and are enclosed several together in a wide, gelatinous sheath; often form rather short hormogonia. Only 1 species is known for the United States.

619a (565) Plants attached; cells club-shaped or globular-angular isodiametric; gregarious, forming cushionlike masses or horizontal expanses, *or* solitary (especially when cylindric and club-shaped); epiphytic or growing on shells; cells usually showing endospores (segments of the protoplast rounded up and forming reproductive bodies which are sporelike). 620

619b Plant not attached (or only incidentally adherent; cells mostly spherical, hemispherical or rod-shaped; not forming cushionlike masses or horizontal expanses; plants free-floating in the tychoplankton or euplankton *629*

620a Cells erect, club-shaped or subcylindric, straight or slightly curved. 621

620b Cells some other shape; gregarious, forming horizontal expanses which may show a slight tendency to elongate as short filaments, *or* as cushions 622

621a Protoplast dividing by cleavage planes in the apex to form endospores which are cut off successively. Fig. 457 . . . . *Chamaesiphon*

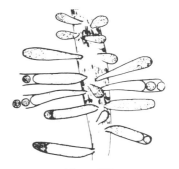

Figure 457

Figure 457 *Chamaesiphon incrustans* Grun.

These club-shaped or cylindrical plants grow as epiphytes on filamentous algae and whereas they may be solitary they usually occur in gregarious patches. When mature, the ends of the protoplasts become cut off, the segments forming endospores which drift away as regenerative elements. A patch of the plants will show many different stages of development from these spores.

621b Protoplast divided throughout its length to form endospores simultaneously. Fig. 458. . . . . *Stichosiphon*

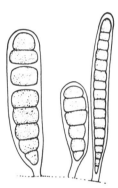

Figure 458

Figure 458 *Stichosiphon regularis* Geitler, showing simultaneous cleavage of entire protoplast to form endospores.

Plants in this genus are similar to those of *Chamaesiphon* (Fig. 457) but differ in that the entire protoplast become segmented by simultaneous cleavages to form endospores. The club-shaped plants are epiphytic on filamentous algae especially members of the Cladophoraceae.

622a (620) Plant mass composed of cells in 1 layer. (See *Chlorogloea*, Fig. 462, however) 623

622b Plant mass in the form of a cushion with the cells arranged in vertical rows or as false filaments 626

623a Colony not attached; cells closely arranged in packets. Fig. 459. . . . . *Myxosarcina*

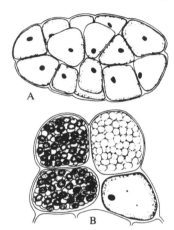

Figure 459

Figure 459 *Myxosarcina amethystina* Copeland. (A) colony; (B) cells with endospores. (Redrawn from Copeland.)

In this genus the plants are essentially unicellular but, by cell division, packets of angular cells are formed which are enclosed in a common mucilage. Some of the cells are usually seen filled with endospores. The plants are free-floating mostly. The only species reported from the United States occurs in the hot springs of Yellowstone National Park.

623b Cells not arranged as above; colony sessile 624

624a Plant mass composed of a few closely arranged, pyriform cells; endospores formed by a cleavage in 3 planes. Fig. 460 . . . . *Dermocarpa*

Figure 460

Figure 460 *Dermocarpa.* (A) *D. rostrata* Copeland; (B) *D. prasina* (Reinsch) Born. & Flah. (A, redrawn from Copeland; B, redrawn from Reinsch.)

Dermocarpa occurs as a solitary cell but often individuals are closely aggregated, forming compact clumps on aquatic plants or other submersed substrates. The cell frequently shows the contents divided into numerous spherical endospores. The plants are represented by several species in the United States, differentiated by cell size and shape.

624b **Plant mass a definite colony of many cells** **625**

625a **Cells globular or slightly angular from mutual compression, forming prostrate, pseudofilamentous growths on substrates. Fig. 461** . . . . *Xenococcus*

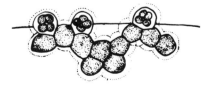

Figure 461

Figure 461 *Xenococcus schousbei* Thur.

This genus is largely marine but there are at least 4 freshwater species in the United States. The plants occur as patches of blue-green cells, compactly arranged as epiphytes on filamentous algae. Cells form endospores although they may multiply rapidly by fission.

625b **Cells compactly arranged in a globular mass forming tubercular growths epiphytic or endophytic, enclosed by a common sheath and involving a large number of small round or oval cells. Fig. 462** . . . . *Chlorogloea*

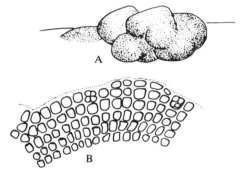

Figure 462

Figure 462 *Chlorogloea microcystoides* Geitler. (A) colony; (B) optical section of colony to show cell arrangement.

Like *Xenococcus* (Fig. 461) this genus is both marine and freshwater. The cells are globular or oval and are loosely arranged within a wide, gelatinous sheath. The thallus may have short upright extensions in some species. The plant appears much like an attached colony of *Chroococcus* (Fig. 473) but the cells produce endospores in reproduction. Only 1 species has been reported from the United States and seems to be rare.

626a **(622) Cells surrounded by a sheath; plant mass thick, cartilaginous, usually macroscopic. Fig. 463** . . *Chrondrocystis*

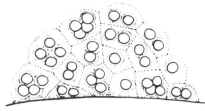

Figure 463

Figure 463 *Chondrocystis schauinslandii* Lemm.

The species illustrated forms extensive, cushion-like masses on exposed, subaerial surfaces and often are heavily encrusted with lime. The colonial mass is enclosed by a tough mucilage in which families of cells are surrounded by individual sheaths. This is the only species of the genus reported from the United States, occurring in the western section.

626b **Cells not enclosed by a thick sheath; plant mass usually macroscopic in well-developed thalli** **627**

627a **Plant mass forming a flat layer or cushion** **628**

627b **Plant mass cushionlike; cells forming erect series which are similar to branched filaments. Fig. 464** . . . . **(Oncobyrsa) Hydrococcus**

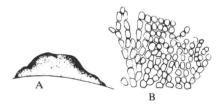

Figure 464

Figure 464 *Hydrococcus (Oncobyrsa)* sp. (A) habit of colony; (B) diagram of cell arrangement.

Three species of this genus have been described (synonymous with *Oncobyrsa*), from the United States, including arctic Alaska. The species illustrated occurs on mosses and other aquatic plants at high altitudes and in the Arctic. The thallus is a mound of cells encased in a tough mucilage. Although the general habit is that of members of the Chamaesiphonaceae, there have been no endospores observed.

628a **Thallus a pseudofilamentous growth in the shells of molluscs. Fig. 465** . . **Hyella**

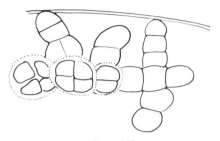

Figure 465

Figure 465 *Hyella fontana* Huber & Jadin., some cells with endospores.

This plant presents itself as a prostrate cushion of indefinite and entangled short filaments, growing within the shells of molluscs. There are both relatively long, horizontally growing filaments and short, upright ones. Almost any cell in the thallus may develop endospores. About half of the known species are marine, but 2 freshwater forms have been reported from the United States.

628b **Thallus an encrusting growth with rhizoidal extensions on the ventral side, the upper or outer portion consisting of uniseriate or multiseriate branches, all very compact with sheaths fused together. Fig. 466** . . . . *Pleurocapsa*

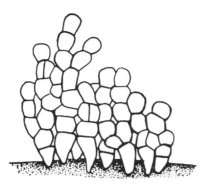

Figure 466

Figure 466   *Pleurocapsa minor* Hansg.

In this genus the attached plant mass is essentially filamentous but the cells are so closely appressed that the filamentous branching habit cannot be determined readily without dissecting the colony. Encrusting thalli often show some differentiation between the lower or inner cells and those near the surface in which endospores are produced. Nine species have been reported from the United States.

629a   (619) Cells globose or hemispherical because of recent cell division 630

629b   Cells some other shape. 644

630a   Cells enclosed in mucilage and bearing long, gelatinous hairs. Figs. 313, 467. . . . . *Gloeochaete*

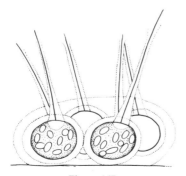

Figure 467

Figure 467   *Gloeochaete wittrockiana* Lag., cells in clumps of 4 adherent to filamentous algae.

630b   Cells not in a sheath which bears a hair 631

631a   Protoplasts occurring as bright blue-green, vermiform bodies, radiately arranged or scattered within *Oocystis*like cells (See *Oocystis,* Fig. 146.); enclosed by mother-cell wall. Fig. 315. . . . . *Glaucocystis*

631b   Cells not as above 632

632a   Cells solitary or grouped in small families of 2-4-8 (rarely as many as 16) individuals, if more than 16, then in a flat plate 633

632b   Cells aggregated in large numbers; enclosed by a mucilaginous sheath . . 638

633a   Cells solitary or in pairs, without a gelatinous sheath. Fig. 468 . . . . *Synechocystis*

Figure 468

Figure 468   *Synechocystis aquatilis* Sauv.

This is a rather rare although simple plant which probably is more common in occurrence than is evidenced by records of it from the

United States. The cells are globular or oval, solitary or in pairs, without a mucilaginous sheath being apparent. The densely granular central body of the cells seems to be more complex than for other genera of the Chroococcaceae. *Synechocystis* is found in the tychoplanktin. Five species are reported from the United States, 1 from hot springs.

633b Cells enclosed by a mucilaginous sheath which is sometimes indistinct **634**

634a Cells arranged in rectilinear series . . **635**

634b Cells not arranged in rectilinear series **636**

635a Cells arranged to form a flat plate; cell division in 2 directions in 1 plane. Fig. 469. . . . . *Merismopedia*

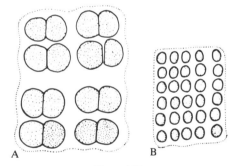

Figure 469

Figure 469 *Merismopedia.* (A) *M. elegans* var. *major* G.M. Smith; (B) *M. glauca* (Ehr.) Näg.

This genus is readily distinguished by the definite, rectilinear arrangement of spherical or oval cells in rectangular plates. Colonies increase in size by fission of cells in 2 directions. *Merismopedia convoluta* Bréb. is an uncommon species in which the colonies become very large (visible to the unaided eye) sheets with enrolled margins. There are several species in the United States, differentiated by cell size, presence or absence of pseudovacuoles, and by size of the colony.

635b Cells arranged to form cubical colonies of many individuals; cell division in 3 planes; cubes of cells enclosed by sheaths. Fig. 470 . . . . *Eucapsis*

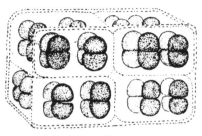

Figure 470

Figure 470 *Eucapsis alpina* Clements & Schantz.

This colonial blue-green alga is regarded by some as being a *Chroococcus*. But the cubical, sarcinalike arrangement of the cells with separate sheaths around groups of 8 or 16 cells distinguishes *Eucapsis*. Cell division occurs regularly in 3 planes. Only 1 species and its varieties are known for the United States although 2 or 3 are reported from Europe.

636a (634) Cells heart-shaped or round, occurring at the ends of short, radiating, gelatinous strands (focus down into the colony and reduce illumination to detect presence of radiating strands). Fig. 471 . . . . *Gomphosphaeria*

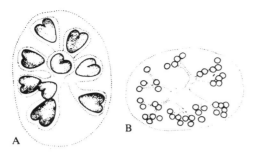

Figure 471

Figure 471 *Gomphosphaeria.* (A) *G. aponina* Kütz.; (B) *G. lacustris* Chod.

These plants are characterized by having cells in globular or oval colonies that are free-floating, the cells closely or distantly arranged at the ends of mucilaginous strands that radiate and branch from the center of the colony. *Gomphosphaeria lacustris* Chod. is frequently found in the euplankton, whereas *G. aponina* Kütz. occurs mostly in the tychoplankton. In the latter species the cells often appear heart-shaped because fission is initiated at the outer wall and the cells are lobed at the apex.

636b Cells not at the ends of radiating strands. 637

637a Groups of many cells enclosed in concentric layers of mucilage; colonial investment intermingling (confluent) with the sheaths of other groups and so forming gelatinous masses, mostly on moist, subaerial substrates; sheaths showing definite rings, often colored brown or red. Fig. 472. . . . . *Gloeocapsa*

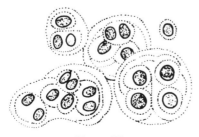

Figure 472

Figure 472 *Gloeocapsa punctata* Näg.

This is a genus in which globular cells are enclosed, many 'families' together, within gelatinous masses of considerable size. Cells, pair of cells, or clusters are surrounded by lamellate sheaths. Many species, especially when few cells are involved, can scarcely be differentiated from some species of *Chroococcus* (Fig. 473) and there is a disposition among some specialists to place the 2 genera together. Plants are either aquatic, or more commonly on moist subaerial substrates. Some species have highly pigmented sheaths and appear as red or orange-colored masses on damp soil and rocks. Forty-four species are indefinitely reported from the United States.

637b Colonial mucilage not intermingling with that of other colonies; families of few (2 to 8) cells separated from one another; usually free-floating but commonly inhabitating soil and moist substrates of all kinds, sometimes adherent to aquatic plants; colonial sheath usually not showing concentric layers, colorless. Fig. 473. . . . . *Chroococcus*

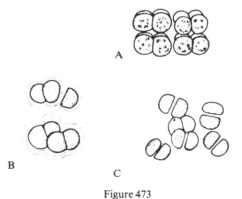

Figure 473

Figure 473 *Chroococcus.* (A) *C. prescottii* Drouet & Daily; (B) *C. limneticus* var. *distans* G.M. Smith; (C) *C. limneticus* Lemm.

There are numerous species in the genus, many of them inadequately described and differentiated. The genus is separated from *Gloeocapsa* (Fig. 472) mostly on the basis of the fewness of cells in a colony and by the fact that 'families' of cells are not enclosed in a much-lamellated sheath. The colonies are composed of 2, 4, or 8 (sometimes 16) cells. A few species are adherent and apparently epiphytic, but most are eu- or tychoplanktonic. Some are commonly found on moist subaerial substrates. *C. turgidus* (Kütz.) Näg. is a large species in which cells occur in 2's and 4's within a slightly stratified sheath and one that is invariably found in desmid habitats in soft water.

638a (632) **Cells arranged to form a flat plate** 639

638b **Cells forming irregularly globular or oval colonies, sometimes clathrate** . . 640

639a **Cells arranged in rectilinear series. Fig. 469.** . . . . *Merismopedia*

639b **Cells irregularly arranged, not in rectilinear series. Fig. 474** . . . . . *Holopedium*

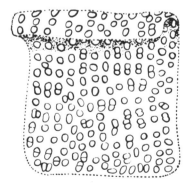

Figure 474

Figure 474 *Holopedium irregulare* Lag.

This genus differs from *Merismopedium* (Fig. 469) by having cells irregularly arranged in a flat or curled plate. Five species have been reported from the United States, differentiated by cell-size and nature of the colonial sheath.

640a (638) **Colony globular, rather definite in shape** 641

640b **Colony irregular in outline** 643

641a **Cells very numerous and crowded within the colonial mucilage (in some species showing false [pseudo-] vacuoles which refract the light so that the cells appear brownish, black or purplish). Fig. 475.** . . . . *Microcystis*

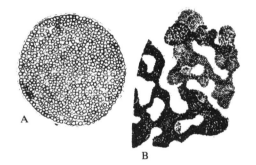

Figure 475

Figure 475 *Microcystis*. (A) *M. flos-aquae* (Wittr.) Kirch.; (B) *M. aeruginosa* Kütz. emend. Elenkin.

The marblelike cells of this genus are closely compacted and irregularly arranged in definitely shaped, but mostly irregular colonies, enclosed in mucilage. The mucilage is often not clearly seen in preserved material. *Microcystis flos-aquae* (Wittr.) Kütz., however, occurs in nearly globular colonies, whereas *M. aeruginosa* Kütz. colonies are highly irregular and clathrate when mature. Some species contain pseudovacuoles (gas pockets) and float high in the water. Hence they produce surface scums and, like *Aphanizomenon* (Fig. 449) cause a great deal of disturbance in lakes and reservoirs. Dense growths may lead directly or indirectly to the death of fish through suffocation or by poisoning, and the toxin produced by some species causes the death of cattle and birds. It is interesting that where some species occur *(M. aeruginosa, e.g.)* the habitat is completely dominated by the plant to the exclusion of almost all other forms of Cyanophyta. It has been noted that a lake may be densely overgrown with either *Microcystis* or with *Aphanizomenon,* but seldom, if ever, the two together. There are several species reported from the United States, differentiated by cell-size, presence of pseudovacuoles, by the nature of the sheath, colony shape, *etc.*

641b **Cells densely crowded but evenly spaced or regularly arranged 642**

642a **Cells in 1 layer at the periphery of the mucilage of a globular or lobed mucilaginous sheath. Fig. 476 . . . .**
. . . . *Coelosphaerium*

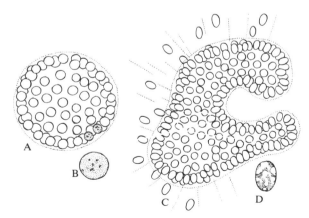

Figure 476

Figure 476 *Coelosphaerium.* (A) *C. kuetzingianum* Näg. (B) *C. kuetzingianum* cell without pseudovacuoles; (C) *C. naegelianum* Unger, an expanding colony; (D) *C. naegelianum* cell showing pseudovacuoles.

In this genus the cells are arranged (as the name suggests) in a hollow colony, the cells forming a peripheral layer immediately within the colonial sheath. Cells are either round, oval or elliptic and somewhat radiately arranged. Two species are commonly found in the euplankton of hardwater lakes. *Coelosphaerium naegelianum* Unger is readily identified by the presence of pseudovacuoles which cause the colony to appear dark brownish, purplish or even black.

642b **Cells distributed throughout the colonial mucilage; cells spherical, evenly spaced, often in pairs. Fig. 477 . . . *Aphanocapsa***

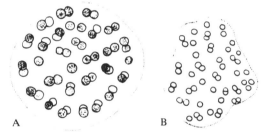

Figure 477

Figure 477 *Aphanocapsa.* (A) *A. grevillei* (Hass.) Rab.; (B) *A. elachista* W. & W.

Whereas *Microcystis* (Fig. 475) has cells compactly arranged or crowded within a gelatinous sheath, in this genus the cells are evenly spaced, often occurring in pairs. The cells do not have pseudovacuoles. Some are minute and intergrade with colonial bacteria, are sometimes mistaken for them. *Aphanocapsa* is mostly euplanktonic, several species being reported from the United States. The genus is included under *Anacystis* by some authors.

643a (640) Cells crowded, usually with refractive pseudovacuoles. Fig. 475 . . . . .
. . . . *Microcystis*

643b Cells evenly spaced within the mucilage; false vacuoles lacking. Fig. 477 . . . .
. . . . *Aphanocapsa*

644a (629) Cells quadrangular, arranged in flat plates. Fig. 478. . . . . *Tetrapedia*

Figure 478

Figure 478   *Tetrapedia reinschiana* Archer.

Tetrapedia reinschiana Arch. is a rare and dubious plant that has quadrangular cells arranged in multiples of 4 to form a flat, rectangular plate. Other species may be solitary and either 3- or 4-angled, some with minute spines. Five of the 11 known species occur in the United States.

644b Cells some other shape. 645

645a Cells solitary or in colonies of few cells 646

645b Cells numerous within a globular, amorphous, gelatinous matrix 652

646a Without a gelatinous sheath. Fig. 479 . .
. . . . *Synechococcus*

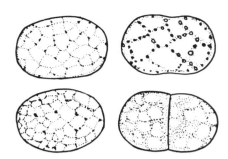

Figure 479

Figure 479   *Synechococcus aeruginosus* Näg.

This is a solitary oval unicell which does not possess a gelatinous sheath. Cells may be in pairs as a result of recent fission. They are relatively large for Cyanophyta of this type (may be up to 35 μm in length) and are often conspicuous in the microscope mount because of their bright blue color. Five species and their varieties have been reported from the United States.

646b With a gelatinous sheath (sometimes discerned with difficulty), or enclosed by a gelatinous matrix 647

647a Cells elongate, pointed at the ends, fusiform. Fig. 480 . . . *Dactylococcopsis*

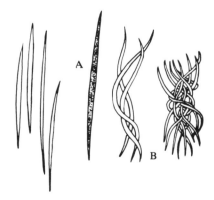

Figure 480

Figure 480 *Dactylococcopsis.* (A) *D. acicularis* Lemm.; (B) *D. fascicularis* Lemm.

These are fusiform-shaped cells arranged with their long axes mostly parallel with that of the fusiform colonial sheath. Some species, however, do not exhibit a definite sheath. Mostly colonial, some forms occur as solitary cells, either straight, slightly bowed or spirally twisted. They should be compared with *Ankistrodesmus* (Fig. 125) in which some species have similarly shaped cells. At least 1 species has been described from snow fields; 8 have been reported from the United States.

647b Cells not pointed at the ends 648

648a Cells heart-shaped, or oval; at the ends of radiating strands of mucilage; colonies globular. Fig. 471. . . . . *Gomphosphaeria*

648b Cells not at the ends of radiating strands. 649

649a Cells radiately disposed. Fig. 481. . . . . *Marssoniella*

Figure 481

Figure 481 *Marssoniella elegans* Lemm.

In this genus there are pyriform cells which are more or less definitely arranged in a radiate fashion about a common center, enclosed by mucilage that does not exhibit a definite sheath. The narrow end of the cells is directed outward. The genus is to be expected in the euplankton. Only 1 species is reported from the United States; is widely distributed.

649b Cells not radiately disposed 650

650a Individual cell sheath distinct; cells elongate to rod shaped, many in a common sheath. Fig. 482 . . . . . . *Gloeothece*

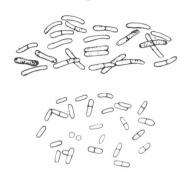

Figure 482

Figure 482 *Gloeothece linearis* Näg.

In this genus the cells are elongate cylinders, bacilliform, or may be merely elongate ovals. The cells are enclosed many together within a common mucilaginous sheath and each cell or pair of cells has an individual sheath. The cells are often crowded and irregularly arranged, much like *Microcystis* (Fig. 475) with rod-shaped cells rather than spheres. Species should be compared with *Aphanothece* (Fig. 485), a genus with which some authors combine *Gloeothece*.

650b Cells bacilliform, enclosed in a mucilage but without individual sheaths 651

651a Cells elongate-cylindric and curved; few (4 to 10) within a wide gelatinous sheath. Fig. 483. . . . . *Rhabdoderma*

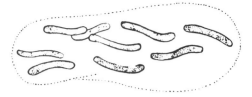

Figure 483

Figure 483 *Rhabdoderma lineare* Schm. & Lauterb.

In this genus the cells are elongate and cylindrical, even vermiform and are much like *Gloeothece* (Fig. 482) except that there is no individual sheath, and the colonies are composed of but a few cells. The plants occur in either eu- or tychoplankton, intermingled with miscellaneous algae. About 5 species have been reported from the United States, differentiated by size of cell and cell proportions.

651b Cells elongate-cylindric, solitary or a few attached end to end in a wide, gelatinous sheath; individual sheaths of cells often indistinct. Fig. 482 . . . *Gloeothece*

652a (645) Cells arranged at the periphery of a globular, gelatinous matrix; colony hollow. Fig. 476 . . . . *Coelosphaerium*

652b Colony formed otherwise. 653

653a Cells numerous within a tubular sheath which is pointed at the ends. Fig. 484. . . . . *Bacillosiphon*

Figure 484

Figure 484 *Bacillosiphon induratus* Copeland. (Redrawn from Copeland.)

The cells in this genus are elongate, bacilliform (similar to *Aphanothece*, Fig. 485), but are arranged with their long axes parallel in gelatinous strands which may be lime-impregnated. The single species has been reported from hot springs in Yellowstone National Park.

653b Colony otherwise. 654

654a Cells irregularly scattered throughout a shapeless mucilage, or in a somewhat globular mucilage which often attains macroscopic size; colonial mucilage firm, the colonies retaining their shape when removed from the water; individual cell sheaths lacking. Fig. 485 . . . . *Aphanothece*

Figure 485

Figure 485 *Aphanothece castagnei* (Bréb.) Rab.

Cells are short-cylindric or elongate oval, occurring in colonies which may be either microscopic or macroscopic. Some species form firm, olive-green, gelatinous balls remindful of *Nostoc* (Fig. 451) and are sometimes mistaken for them when not examined microscopically. The cells (similar to some species of *Gloeothece*, Fig. 482) do not have individual sheaths; occur in much larger colonies. The colonies often develop on lake bottoms, become loosened and float to the surface and then are washed ashore, sometimes forming a 'soupy' mass at the shore line. Sixteen species are recognized from the United States, but some are the reports confused with *Gloeothece*.

654b Individual cell sheaths distinct. Fig. 482 . . . . *Gloeothece*

655a (508) Frustule (diatom shell)* elongate, rod-shaped, boat-shaped, rectangular or wedge-shaped, 2 or more times longer than wide....................668

655b Frustule isodiametric, or nearly so; round, triangular, or oval, but less than twice the diameter in length 656

656a Frustules rectangular in side (girdle)* view, joined in chains by interlocking of long, slender, spinelike horns which arise from the corners of the valves; frustules without a raphe; horns hollow, or solid. Fig. 486 .... *Chaetoceros*

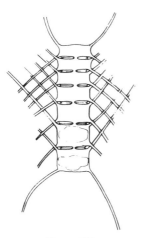

Figure 486

Figure 486 *Chaetoceros elmorei* Boyer, girdle view showing only portions of the very long polar horns. (Redrawn from Boyer.)

This genus is well-named because of the long, hornlike processes, one at either pole of the

*Diatoms exhibit two general expressions of form and according to a standard classification system there are two Orders. One is the elongate, 'cigar'-shaped, 'boat'-shaped or wedge-shaped cell with bi-

lateral symmetry in the arrangement of wall markings, constituting the Pennales. The other includes round, 'pill'-box-shaped, or nearly isodiametric cells with radiate ornamentations, the Centrales.

The wall or shell of the diatom is called a frustule. It is not a plain, simple envelope but is composed of two sections, one slightly larger and overlapping the smaller much as a box lid. The larger or upper section is called the epitheca, the smaller, lower or inner section the hypotheca. The 'lid' and the 'box' are adjoined by overlapping side pieces, the cingula (singular cingulum). The flattened or broad surface is called a valve and has marginal flanges to which the side pieces, the cingula fit. In some genera there may be supplementary or additional connecting bands making up the sides of the frustule. These are call intercalary bands. When the shell is seen from the top or bottom it is said to be in valve view; when seen from the side so that the line shows that is formed by the overlapping of the connecting bands, the frustule is said to be in girdle view. The cell may be quite different in appearance in these two views.

The siliceous shells have decorations and markings in an almost endless variety; linear etchings, rows of puncta (dots), costae (ribs), vertical canals (pores). Externally there may be ridges or flanges on the valves; internally, partitions or septations (septa). In certain Pennales genera the valves (or at least one of the two) may contain a longitudinal groove or furrow, the raphe (seen in valve view) as a distinct line, usually straight and in the midregion, but it may be sigmoid or to one side of center. Sometimes the raphe lies within a marginal rib or keel. The raphe is not continuous throughout the full length of the valve but is interrupted at the midpoint by an internal swelling on the inside of the wall called the central nodule. There may be nodules at either pole of the cell (polar nodules). The central nodule may be large and lobed and referred to as a stauros. A false or pseudoraphe results when the wall decorations form lines (striae) in from the margin but leave a narrow, linear smooth central region. This line or smooth field may appear on one or both valves. Care and patience must be used in diatom identification to determine the presence or absence of a raphe or pseudoraphe on one or both valves.

The Centrales which are circular or triangular in valve view do not possess a raphe; may have spines or horns. The radiate markings may be evenly disposed or interrupted by smooth zones, or form patterns according to the size and shape of the smooth and marked areas.

Observations by electronmicroscopists in recent years have disclosed refinements in the decoration and markings of the frustule. Lines under the light microscope actually may be rows of minute puncta, *etc*. These disclosures have necessitated a change in terminology relating to wall markings. The terms employed in the accompanying key are based on observations made by the customarily used light microscope.

oval cells as seen in valve view.* In girdle view the cells are quadrate, with a horn at each angle. The horns of adjacent cells interlock so that filaments are formed. Most species are marine but some occur in brackish water such as Devils Lake, North Dakota.

656b   Frustules without spinelike horns. . . 657

657a   Frustules triangular in valve view; seldom seen lying in girdle view; raphe and pseudoraphe lacking. Fig. 487 . . . .
. . . . . _Hydrosera_

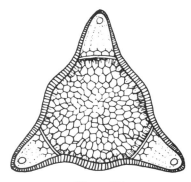

Figure 487

Figure 487   _Hydrosera triquetra_ Wall., valve view.

These beautifully ornamented frustules are triangular in valve view, quadrate in girdle view and in cross section are somewhat orbicular. The genus is mostly marine but species may be found in estuaries and in coastal ponds. _Hydrosera_ is regarded as synonymous with _Triceratium_ by some diatomists.

657b   Frustules some other shapes. 658

658a   Frustules circular in valve view. 659

658b   Frustules oval, broadly elliptic, slipper-shaped or rhomboid. 662

659a   Valve with a single, undecorated, mammillate protrusion or thickening immediately within the margin; (frustules sometimes broadly elliptic or rhomboid). Fig. 488 . . . . . _Actinocyclus_

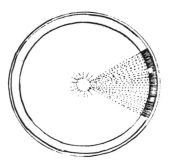

Figure 488

Figure 488   _Actinocyclus_ sp., valve view showing portion of wall ornamentation.

The pill-box frustule (member of the Centrales) has a large, pustulelike swelling immediately within the margin of the epivalve. The puncta or areolae are radiately arranged as in _Coscinodiscus_ (Fig. 490), but which are usually larger and more conspicuous in the center of the valve. Some species are broadly elliptic in valve view rather than circular. Most members of the genus are marine, but the one illustrated occurs in Lake Erie and presumably throughout the Great Lakes region.

659b   Valve without intramarginal protrusions 660

660a   Valve with an intramarginal zone of costae encircling the cell; smooth or finely punctate within the marginal circle of costae; (markings in the center of the valve various, but mostly smooth). Fig. 489 . . . . _Cyclotella_

*See note on page 241 for the preparation of diatom mounts.

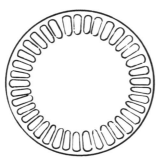

Figure 489

Figure 489 *Cyclotella meneghiniana* Kütz., valve view of a species in which the central region is smooth and the marginal zone costate and punctate.

These circular cells are narrowly rectangular in girdle view and often lie parallel with one another to form chains or short filaments. In valve view there is a zone of radiate costae within the valve margin and a central smooth or finely punctate area. Species are planktonic, often occurring with *Stephanodiscus* (Fig. 491).

660b Valve marked by rows of puncta radiating from a central area to the margin; frustules drum-shaped or rectangular in girdle view 661

661a Valves evenly ornamented by rows of puncta forking as they extend to the margins from a central region where they are irregularly disposed; with an intramarginal circle of fine teeth; plants euplanktonic, or tychoplanktonic. Fig. 490. . . . . *Coscinodiscus*

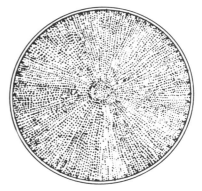

Figure 490

Figure 490 *Coscinodiscus lacustris* Grun., valve view showing radial striae of coarse puncta and minute marginal spines.

These are pillbox-shaped frustules, circular in valve view, narrowly rectangular in girdle view. Radiating from the center where there is an area or irregularly arranged puncta are decussating rows of areolae or puncta. Immediately within the margin in many species is a circle of short, sharp spines. The radiating puncta do not quite converge in the midregion. Compare with *Stephanodiscus* (Fig. 491). There are about 450 species in the plankton of both marine and freshwaters. This genus and *Stephanodiscus* are common within blue-green algal blooms and seem to be more prevalent in hard-water lakes than in soft waters.

661b Valves unevenly ornamented with radiating rows of puncta, converging at the center but interrupted by smooth, radiating zones between the rows of puncta; intramarginal circle of coarse spines which extend beyond the edge of the valve. Fig. 491 . . . . . *Stephanodiscus*

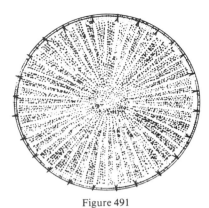

Figure 491

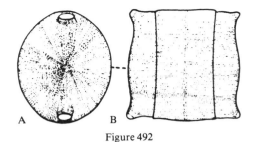

Figure 492

Figure 491 *Stephanodiscus niagarae* Fhr., valve view; alternating smooth and punctate radial zones; marginal teeth prominent.

These are relatively large, drum-shaped cells, usually showing a circle of prominent spines immediately within but extending beyond the margin of the valve. In valve view there are radiating rows of puncta (sometimes multi-seriate rows) alternating with clear, smooth zones. The central area of irregularly arranged puncta is usually less extensive than in *Coscinodiscus* (Fig. 490) and the rows of puncta nearly converge. The girdle view (rectangular) is smooth and there are no intercalary bands. The species illustrated is common in hard or basic lakes; occurs mostly in the euplankton. About 5 species have been reported from the United States; widely distributed.

662a (658) Frustules broadly elliptic or oval in valve view but commonly short-rectangular, the corners protruding and outturned; no raphe or pseudoraphe. Fig. 492 . . . . *Biddulphia*

Figure 492 *Biddulphia laevis* Ehr. (A) valve view; (B) girdle view.

This genus contains variably shaped cells, mostly rectangular in girdle view, broadly oval in valve view. The chief characteristic is a blunt process extending from the ends of the valves. There is a broad intercalary band. The areolae are in radiating, linear series in valve view and in parallel series in girdle view. Most species are marine, usually solitary but sometimes adjoined in chains.

662b Frustules with a raphe or a pseudoraphe, oval or rhomboidal in valve view . . . 663

663a Frustules rhomboidal to circular in valve view, arched and saddle-shaped in girdle view; pseudoraphe in 1 valve at right angles to that of the other valve. Fig. 493 . . . . *Campylodiscus*

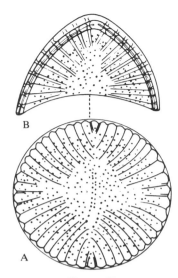

Figure 493

Figure 493 *Campylodiscus hibernicus* Ehr. (A) valve view; (B) girdle view, showing marginal keel in which the raphe lies. (Redrawn from G.M. Smith.)

The clear areas are regarded by some diatomists as not being justifiably called pseudoraphes. When seen vertically these cells are circular in outline but are folded or bent to form a 'saddle' when seen from the side. Costae extend inward from the edge of the valve. In girdle view the cell takes various shapes according to the angle of observation. The cells occur solitarily; mostly marine but at least 3 species have been reported from freshwater in the United States.

663b **Frustules oval or broadly elliptic; not bent nor saddle-shaped** **664**

664a **Frustules broadly elliptic or slipper-shaped, with prominent marginal costae; raphe lateral in a marginal keel** **665**

664b **Frustules oval or narrowly elliptic, with a pseudoraphe, or with a raphe not in a marginal keel** **666**

665a **Frustules slipper-shaped, surface of valve transversely undulate, seen when the cell is viewed from the side; transverse striae often faint, occurring in zones or bands across the valve; some species much longer than wide. Fig. 494.** . . . . *Cymatopleura*

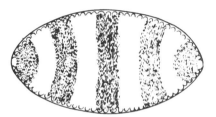

Figure 494

Figure 494 *Cymatopleura elliptica* (Bréb.) W. Smith, valve view; the surface undulations represented by clear and shaded areas. (Redrawn from G.M. Smith.)

Although sometimes linear, most species are broadly elliptic or football-shaped in valve view. The cells are rectangular in girdle view but have wavy margins because the valves are transversely undulate. In this view the marginal, often short costae are very prominent. The valve appears zoned or banded when the frustules are seen in valve view. There is a keel along the valve margins in which the raphe lies. A pseudoraphe can be determined in some species. The cells are solitary; are both marine and freshwater.

665b **Frustules broadly oval, egg-shaped, or slipper-shaped, the surface of the valve not undulate; costae extending inward**

from the margin showing prominently; some species twisted in girdle view. Fig. 495 . . . . *Surirella*

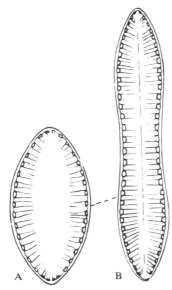

Figure 495

Figure 495 *Surirella*. (A) *S. ovalis* Bréb.; (B) *S. oblonga* Ehr., both in valve view.

These boat-shaped or oval cells are usually identified readily by the very prominent costae which extend from the margin as seen in valve view, with a clear, linear region along the axis. Some species are spirally twisted. The raphe is marginal in both valves. Cells are solitary in both eu- and tychoplankton; are relatively large diatoms.

666a   **(664) Raphe in hypovalve; pseudoraphe in epivalve; with central and polar nodules in the hypovalve; cells epiphytic. Fig. 496 . . . . *Cocconeis***

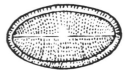

Figure 496

Figure 496   *Cocconeis pediculus* Ehr., valve view showing enlarged clear area in the mid-region of the hypovalve.

The frustules are broadly ovoid-elliptic in valve view. The epivalve shows an axial pseudoraphe and is convex whereas the hypovalve is concave, or flat and shows a raphe; with a central and polar nodules. The valves have prominent transverse striae but the pattern differs on the 2 valves. In many there is a clear marginal band formed by an interruption of the striae on the valve which has the raphe. The frustules are epiphytic on filamentous algae or on aquatic plants, sometimes occurring so abundantly as to form a coating over the host surface.

666b   **Raphe in both epi- and hypovalve; frustules oval or variously shaped . . 667**

667a   **Frustules with transverse septa which show as bands across the cell in valve view; raphe in a canal, the canal with pores. Fig. 497 . . . . *Denticula***

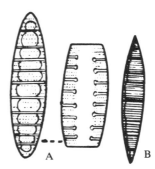

Figure 497

Figure 497 *Denticula*. (A) *D. elegans* Kütz.; (B) *D. tenuis* Kütz.

These cells are subrectangular in girdle view but the margins are slightly convex. The girdle is smooth and there are intercalary bands. The transverse septa are seen extending from the margin to the junction of the valves and the girdle. These show as costae or ribs. In valve view the frustules are narrowly elliptic or lanceolate with a keel next to one margin. There are transverse septa and the wall has fine, transverse striae. The cells may be solitary or in short ribbons.

667b   Frustules without transverse septa; raphe in a central axis of the valve; a prominent central nodule which extends both directions on either side of the raphe; the valve with costae which may be marked by longitudinal ribs. Fig. 498 . . . . *Diploneis*

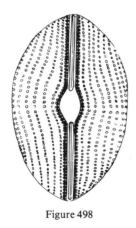

Figure 498

Figure 498 *Diploneis elliptica* (Kütz.) Cl., valve view.

The frustules in this genus are broadly elliptic in valve view; rectangular in girdle view. In some the cells are elongate in valve view and with a median constriction. There are no septa. The central nodule where the valve wall is distinctly thickened is quadrangular and has elongated projections which extend as ribs on either side of the raphe, a feature not clearly evident in freshwater species. The axial field is enlarged in the midregion. Across the valve are somewhat converging costae between which are prominent pores so arranged as to form longitudinal lines. Free-floating or sedentary species occur; about 12 being reported from the United States.

668a   (655) Frustules bilaterally undulate in valve view; the poles capitate; some species in girdle view somewhat triangular, the septa showing as inward projecting processes extending to the intercalary bands. Fig. 499 . . . . *Terpsinoe*

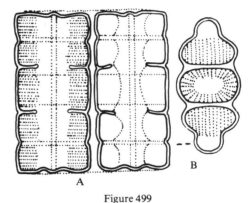

Figure 499

Figure 499 *Terpsinoe americana* (Bailey) Ralfs. (A) girdle view; (B) valve view. (Redrawn from Schmidle.)

The frustules, although quadrangular in girdle view are conspicuously undulate-elliptic or even triangular in valve view. The septa are evident in valve view where they show as perpendicular extensions from the wall. The cells may be solitary or in chains. Apparently only 2 species have been reported from the United States. *T. musica* Ehr. is common in southeastern states where it has been used selectively in the building of aquatic insect larval cases.

668b Frustules shaped otherwise, without septa, or if present, not so arranged. . . .
. . . . **669**

669a Frustules capsule-shaped, cylindrical or quadrangular in girdle view, often with 1 or 2 spinelike extensions at the poles; often filamentous **670**

669b Frustules without spines at the poles . . .
. . . . **672**

670a Frustules with intercalary bands; cells solitary **671**

670b Frustules without intercalary bands; cells arranged in filaments of cylindrical cells; the walls often coarsely punctate and in some species the poles with spines or teeth. Fig. 500 . . . . *Melosira*

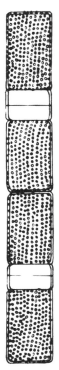

Figure 500

Figure 500 *Melosira granulata* (Ehr.) Ralfs, girdle view of 2 cells in a filament.

These are capsulelike cells, cylindrical as they lie end to end in relatively long filaments in girdle view. The valves are either flat or convex in which instance there are teeth at the poles which aid in disjoining cells. Cells are round in valve view typical of the Centrales. In some there is a sulcus or ringlike incision around the midregion, the girdle being smooth. The wall is punctate, coarsely or faintly so. The species illustrated shows rows of very prominent puncta, whereas others seem to be smooth until examined with oil immersion. The girdle is also ornamented when there is no sulcus. The genus frequently occurs in abundance in the euplankton.

671a Frustules rectangular in girdle view showing many intercalary bands forming curved, imbricated lines; with 2 spines at each pole. Fig. 501 . . . *Attheya*

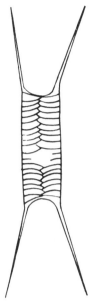

Figure 501

Figure 501 *Attheya zachariasi* Brun., girdle view showing intercalary plates. (Redrawn from Hustedt.)

This curiously shaped frustule is remindful of a miniature squid egg-case. The numerous intercalary plates separate the valves to the extent that a rectangular shape is produced in girdle view—with the 4 corners extended into long, diverging spines. The 2 valves are at either end of the rectangle. In some species there is a horn borne midway between the other two on the valve poles. Compare with *Rhizosolenia* (Fig. 502).

671b Frustules extended at the poles to form a single spine; intercalary bands forming

straight imbricated lines; wall markings usually lacking. Fig. 502 . . *Rhizosolenia*

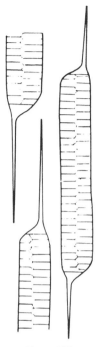

Figure 502

Figure 502 *Rhizosolenia eriensis* H.L. Smith, girdle view. (Redrawn from G.M. Smith.)

In this genus the cylindrical cells lie in girdle view with the 2 valves widely separated by intercalary bands. These may be faint and difficult of discernment. The valves are mostly caplike in this view and terminate in a long, slender spine. The walls are not decorated but the imbricate intercalary bands form a pattern. The walls are delicate and do not withstand the usual (especially acid) cleaning treatment by which diatom frustules are clarified.

672a **(669) Frustules cylindrical in girdle view (quinine capsule-shaped), attached end**

to end in filaments; polar margins often with denticulations. Fig. 500 . . *Melosira*

672b   Frustules not cylindrical, not attached in filaments 673

673a   Frustules triangularly divided (3-parted) with a pseudoraphe in each valve; frustules non-septate. Fig. 503. . *Centronella*

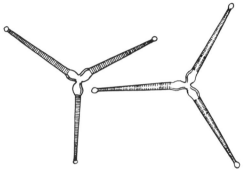

Figure 503

Figure 503   *Centronella reichelti* Voigt, valve view. (Redrawn from Schoenfeldt.)

These uniquely shaped cells have tri-radiate, branched valves. The apices of the valve lobes are slightly capitate. The midregion is smooth but on either side of the pseudoraphe are fine, transverse striae. The frustules are non-septate. Apparently this little understood and dubious diatom has been reported only from Europe.

673b   Frustules not triangularly divided . . 674

674a   Frustules without a raphe; pseudoraphe showing in both valves 675

674b   Frustules with a raphe in at least 1 valve 684

675a   Frustules in girdle view elongate-rectangular, forming a circular colony in which the cells radiate from a common center like spokes of a wheel, the frustules slightly enlarged at the poles. Fig. 504 . . . . *Asterionella*

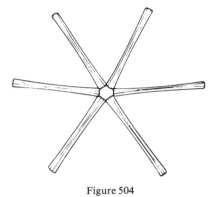

Figure 504

Figure 504   *Asterionella formosa* Hass., colonial arrangement of cells in girdle view.

Species of this genus are planktonic, often very abundant, and are readily identified by the spokelike arrangement of the rectangular frustules about a common center. The poles are enlarged, more so at the adjoined than at the free (outer) ends. The pseudoraphe is narrow and often discerned with difficulty. Septa are lacking. Some species may form a bloom in favorable habitats and often are involved in water spoilage. The common species are usually found in hard-water lakes.

675b   Frustules shaped and arranged otherwise 676

676a   Frustules wedge-shaped in girdle view, adjoined side by side to form flat, circular or semicircular, or fan-shaped colonies, sometimes forming spiral bands. Fig. 505 . . . . *Meridion*

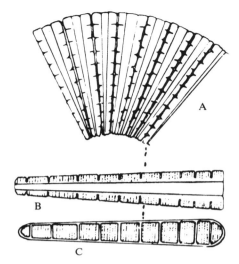

Figure 505

Figure 505 *Meridion circulare* (Grev.) Ag.
(A) portion of circular colony; (B) girdle
view; (C) valve view.

The wedge-shaped cells in this genus have 1 or 2
intercalary bands between the girdles. The frustules lie in girdle view, side by side in a complete
or incomplete, circular plate. In valve view they
are cuneate and show (usually) transverse costae interspersed by striae. The species illustrated is common in waters which support a
luxuriant blue-green algal flora, but is found
also in temporary pools, sometimes coating the
bottom of ditches and trickles of water with a
brown scum.

676b **Frustules other shapes, or without
fanlike arrangement in circular platelike
colonies** **677**

677a **Frustules slightly arcuate or bent in the
long axis** **678**

677b **Frustules not arcuate** **679**

678a **Central smooth area present, extending
to ventral margin in the midregion where
there is a slight swelling on the concave
margin. Fig. 506** . . . . *Ceratoneis*

Figure 506

Figure 506 *Ceratoneis arcus* (Ehr.) Kütz.,
pseudoraphe.

The cells are arched or somewhat boomerang-
shaped, with attenuated or slightly capitate
poles in valve view. In the midregion there is an
enlargement on the ventral (concave) margin,
and a conspicuous clear area in the midregion.
There is a conspicuous pseudoraphe. In girdle
view the frustules are narrowly rectangular with
truncate ends. Transverse striae on the valves
are prominent. Most records are from cold,
mountain streams.

678b **Central smooth area lacking; pseudoraphe narrow throughout the length
of the valve; margins showing pointed**

undulations in valve view. Fig. 507 . . . .
. . . . *Amphicampa*

Figure 507

Figure 507 *Amphicampa eruca* Ehr., valve view. (Redrawn from Ehrenberg.)

Frustules in this genus are mostly slightly arched or bowed as seen in valve view and have both margins undulate, or with toothlike projections. In girdle view cells are elongate-rectangular. The valves are marked by transverse striae. The pseudoraphe lies near the inner (concave) margin. Apparently only 1 species is known from the United States.

679a (677) **Frustules attached in zig-zag chains, sometimes semistellate, rotate or radiate colonies); longitudinal septa present, *straight;* rows of transverse puncta visible in valve view; frustules not showing transverse costae. Fig. 508. . . . . *Tabellaria***

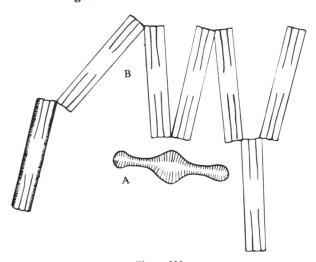

Figure 508

Figure 508 *Tabellaria.* (A) *T.* sp., valve view; (B) *T. fenestrata* (Lyngb.) Kütz., girdle view in zig-zag chains.

Frustules in this genus are short or elongate rectangles as seen in girdle view. They are attached at the corners to one another, forming zig-zag chains or filamentous arrangements. There are narrowly elongate septa which show in girdle view. In valve view the cells, showing longitudinal septa, are narrowly elongate, subcylindrical with capitate poles and a median swelling. There is a pseudoraphe bordered by transverse striae. Species are common in both eu- and tychoplankton.

679b **Frustules not arranged in zig-zag chains, or if so, with *curved* septa 680**

680a **Frustules with curved septa; costae present, appearing as septa; frustules arranged in bands (rarely in zig-zag chains). Fig. 509 . . . . *Tetracyclus***

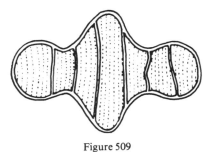

Figure 509

Figure 509 *Tetracyclus lacustris* Ralfs, valve view.

The cells in this genus are rectangular in girdle view, with truncate poles that have rounded corners. The 2 valves are widely separated by intercalary bands, and appear as caps at the ends of the cells. In valve view the cells are oval, elliptic or cruciform and show transverse, false septa. The wall is not ornamented. Three species are reported from the United States, occurring in the tychoplankton, or benthic, forming part of the film on submersed substrates.

680b **Frustules without septa** **681**

681a **Frustules with prominent costae in the valve** **682**

681b **Frustules without prominent costae.** ....
.... **683**

682a **Valve view symmetrical, usually elliptic or subcylindric, often with subcapitate poles; in valve view with a faint pseudoraphe; girdle view rectangular. Fig. 510** .... *Diatoma*

Figure 510

Figure 510 *Diatoma vulgare* Bory var., valve view.

Species in this genus are variously shaped in valve view; oval or elongate, but somewhat rectangular in girdle view. There usually are intercalary bands. There are costae across the valves interspersed by striae. The valve and girdles are finely punctate. In the valves is a narrow pseudoraphe where the wall is not punctate. Frustules usually form zig-zag chains; adjoined by pads at the corners of the cells. Compare with *Tabellaria* (Fig. 509).

682b **Valve view symmetrical, egg-shaped; asymmetrical and wedge-shaped in girdle view; transverse costae conspicuous. Fig. 511** .... *Opephora*

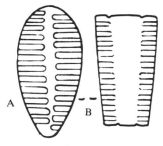

Figure 511

Figure 511 *Opephora martyi* Herib. (A) valve view; (B) girdle view.

These elliptic or egg-shaped cells in valve view are cuneate and symmetrically wedge-shaped in girdle view. There are prominent striae in the valves but no costae. Elsewhere the wall is smooth. In valve view the pseudoraphe shows. Three species have been reported from the United States.

683a **(681) Frustules quadrate or rectangular in girdle view, attached side by side to form ribbons (rarely in chains); valve view fusiform, the poles narrowed from enlarged central region. Fig. 512** ......
.... *Fragilaria*

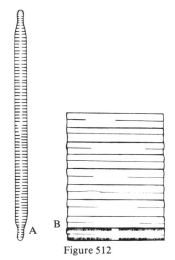

Figure 512

Figure 512 *Fragilaria*. (A) *F*. sp., valve view; (B) *F. capucina* Desm., diagram to show ribbonlike arrangement in girdle view.

The frustules are narrowly elongate, fusiform in valve view (sometimes cross-shaped or triangular); rectangular in girdle view and usually showing intercalary bands. The pseudoraphe is broad and distinct and occurs in both valves. The prominent striae actually are composed of rows of puncta. The most common species forms ribbons with the valves attached side by side, sometimes forming short, zig-zag chains. A dozen or more species occur in both eu- and tychoplankton.

683b   **Frustules elongate and straight (rarely slightly curved), needle-shaped in both views, or with slightly capitate poles; pseudoraphe between transverse striae; frustules solitary or in radiating colonies attached to substrate, singly or in clumps, at one end by short, gelatinous stalks. Fig. 513** . . . . *Synedra*

Figure 513

Figure 513 *Synedra* sp., valve view showing absence of striae in midregion.

These mostly needlelike cells are commonly solitary but often form erect, radiating clusters when attached, or when they are free-floating. Some are fusiform in valve view or are bilaterally undulate. In valve view is a pseudoraphe, narrow or wide, bordered by transverse striae. These are absent in the midregion of the cell in some species. *Synedra* is common in the plankton and in scums on substrates. They often occur in clusters with one pole attached to filamentous algae, forming radiate tufts. There are about 25 species in the United States.

684a   **(674) Frustules lunate or slightly curved in valve view; rectangular or boat-shaped in girdle view** 685

684b   **Frustules some other shape in valve view** 688

685a   **Curvature slight (frustules often nearly straight); frustules bearing a keel near the margin of the valve in which the raphe is enclosed, the location of the keel marked by a row of dots; frustule quadrangular in cross section. Fig. 514** . . . . *Hantzschia*

Figure 514

Figure 514 *Hantzschia amphioxys* Grun.

Like *Nitzschia* (Fig. 529) frustules of *Hantzschia* are beanpod-shaped in valve view. In this view the girdle shows a broad, clear area; rectangular in girdle view. There are fine, transverse striae across the valve. There is a keel along one margin of each valve. Unlike *Nitz-schia* the keeled margin of one valve lies opposite that of the other valve. The raphe lies within the keel and from the fissure a row of pores open into the cell cavity, forming "carinal dots." The frustules are rectangular in optical section. There are only 2 or 3 species reported from the freshwaters of the United States.

685b Curvature decidedly pronounced; frustules not bearing a keel on the valve; asymmetrical in longitudinal axis; the raphe usually lying much closer to the ventral (concave) margin 686

686a Arcuate valve view showing prominent transverse lines of the septa of the frustule (appearing as costae); raphe along the ventral margin and in the midregion bent inwardly to form a 'V' as seen in valve view; frustules epiphytic on filamentous algae and aquatic plants. Fig. 515. . . . . *Epithemia*

Figure 515

Figure 515   *Epithemia* sp., valve view showing V-shaped pattern formed by inbending of smooth axial field.

The frustules are slightly bowed, or decidedly so, with the dorsal (convex) side more strongly curved as seen in valve view. The axial field lies along the concave margin but forms a V-shaped bending in the midregion. The raphe lies within a canal in the axial field. The girdle view of the frustule is rectangular and shows broad, clear (undecorated), smooth girdles. In valve view there are prominent transverse septa (which appear as costae), and rows of puncta. There may be a longitudinal septum also. The frustules are

commonly found as epiphytes on filamentous algae with the concave side down.

686b Frustules without transverse septa (costae) showing in the valve view 687

687a Axial view expanded in the midregion, forming a clear area in the valve ornamentation which extends to the ventral margin of the curved frustule; cells usually with concave margin against a substrate. Fig. 516. . . . . *Amphora*

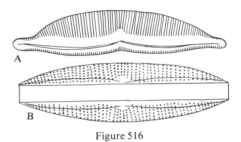

Figure 516

Figure 516   *Amphora*. (A) *A. clevei* Grun. in valve view; (B) *A. ovalis* Kütz. in girdle view.

Frustules in this genus are crescent-shaped in valve view but broadly elliptic with truncate poles in girdle view. The raphe presents two curved lines near the ventral margin of the valve, the two curves meeting over the central nodule which lies next to the ventral margin of the cell. The cells usually are found lying with the concave surface of the hypovalve uppermost when viewed under the microscope, but in nature occurs with the concave face against the substrate (often filamentous algae).

687b Axial field central and small, not expanded as above; frustules forming linear colonies in gelatinous tubes or attached at the ends of branching, gelatinous stalks (often found floating free); lunate, with a slight swelling in the midregion of the ventral (concave) margin. Fig. 517 . . . . *Cymbella*

Figure 517

Figure 517 *Cymbella cistula* (Hempr.) Kirch., valve view showing eccentric raphe, bent inward at the poles toward the concave margin.

These are mostly gracefully curved, crescent-shaped cells which are slightly tumid along the concave margin in the midregion as seen in valve view. The axial field in which the raphe lies is eccentric and enlarged in the midregion. In valve view rows of prominent puncta extend from the margin in a converging pattern. Although usually free-floating, the frustules of some species occur at the ends of branching, gelatinous strands; others enclosed in linear series within long, gelatinous tubes.

688a   (684) **Frustules S-shaped or sigmoid; wall ornamented with transverse and longitudinal striae which make a pattern of intersections. Fig. 518 . . .** *Gyrosigma*

Figure 518

Figure 518 *Gyrosigma acuminatum* (Kütz) Cleve, valve view.

The frustules of *Gyrosigma* are sigmoid in valve view, as is the narrow axial field which is enlarged in the midregion. The valve is marked by intersecting longitudinal and transverse striae. In girdle view the frustules are lanceolate. The ends of the raphe at the central area bend or are hooked in opposite directions. The cells occur singly. *Gyrosigma* is a relatively common genus; has about 20 species reported from the United States.

688b   **Frustules not sigmoid. 689**

689a   **Frustules broadly elliptic, slipper-shaped or boat-shaped in valve view, the margins showing prominent, often short costae; surface of valve undulate; in girdle view elongate but with the sides undulate; pseudoraphe often indistinct. Fig. 494. . . . .** *Cymatopleura*

689b   **Frustule not undulate, not showing marginal undulations in girdle view . . . .**
. 690

690a   **Raphe along both margins of the valve; located within a keel 691**

690b   **Raphe not marginal; keel present or absent 692**

691a   **Valve sharply bent to form a saddle; pseudoraphe in each valve but at right angles to one another; raphe in a marginal keel. Fig. 493 . . . . . .** *Campylodiscus*

691b   **Valve usually twisted, sometimes flat; prominent costae extending from the valve margin toward the smooth pseudoraphe area. Fig. 495 . . . .** *Surirella*

692a (690) Frustule in valve view curved and 'bone'-shaped, 1 pole distinctly more inflated than the other; transverse rows of puncta in valve view. Fig. 519 . . . .
. . . . *Actinella*

Figure 519

Figure 519 *Actinella punctata* Lewis, valve view.

Species are either solitary or colonial, forming stellate clusters. Cells in this genus are all elongate and subcylindric, but the poles are enlarged and unsymmetrically so. In valve view there are numerous transverse rows of puncta. In some species there are fine, intramarginal spines. The raphe lies along the concave margin of the valve, extending only part way diagonally from the polar nodules. The species illustrated is the only one reported from the United States and is rare.

692b Frustules some other shape 693

693a Valve with 'wings', furnished with a sigmoid keel vertical to the face of the valve; boat-shaped in valve view; 8-shaped or hourglass-shaped in girdle view. Fig. 520 . . . . *Amphiprora*

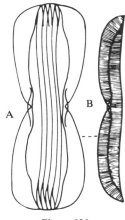

Figure 520

Figure 520 *Amphiprora paludosa* W. Smith. (A) girdle view; (B) valve view.

The frustules are elongate boat-shaped in valve view but 8-shaped or hourglass-shaped in girdle view. The valve has a curious vertical flange or keel which extends in a sigmoid fashion. The raphe lies in the outer margin of the keel. The wall is decorated with longitudinal rows of puncta which form parallel striae. Species may be free-floating or adherent in mucilage on moist substrates. Most species are marine.

693b Valves without a sigmoid keel, some other shape than above in girdle view . .
. . . . 694

694a Pseudoraphe in 1 valve; true raphe in the other 695

694b Raphe in both valves 697

695a Girdle view bent, wedge-shaped with poles usually truncate, attached by stalks or mucilage plugs to substrates; valve view cuneate. Fig. 521 . . . . *Rhoicosphenia*

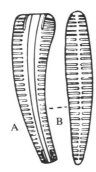

Figure 521

Figure 521 *Rhoicosphenia curvata* (Kütz.) Grun. (A) girdle view; (B) valve view. (Redrawn from G.M. Smith.)

In valve view these cells are elongate-oblanceolate, narrower at one end than the other and somewhat wedge-shaped in both views. The hypovalve shows a raphe and there is a pseudoraphe in the convex epivalve. In girdle view the cells are curved cuneate. The intercalary bands are not decorated. Coarse transverse striae (rows of puncta) show in both views. There are 2 longitudinal septa which are parallel to the face of the valve. The species illustrated is the only one reported from the United States; occurs attached by a gelatinous stalk to algae or to aquatic plants.

695b Girdle view not wedge-shaped 696

696a Frustules broadly elliptic or oval, incompletely septate; valve appressed against substrate; pseudoraphe in epivalve; raphe in hypovalve; valve with raphe having a rim immediately within the margin of the valve. Fig. 496 . . . . *Cocconeis*

696b Frustules narrowly elliptic, with smooth or undulate margins in valve view; bent, rectangular or naviculate in girdle view; valve without rim immediately within the margin as seen in valve view; without septa; attached by stalks to filamentous algae and aquatic plants. Fig. 522 . . . . *Achnanthes*

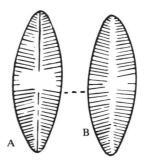

Figure 522

Figure 522 *Achnanthes.* (A) *A. lanceolata* (Bréb.) Grun. hypovalve view showing raphe; (B) epivalve view showing pseudoraphe. *A. coarctata* (Bréb.) Grun. is a common species which occurs on stalks, epiphytizing filamentous algae.

These frustules are symmetrical in valve view but not when seen from the side. They are, in general, elliptic or fusiform in valve view, undulate-rectangular and bent in girdle view. The epivalve shows a pseudoraphe, the hypovalve a raphe. There may be a distinctive lateral, horseshoe-shaped clear area in the midregion of the valve which has the pseudoraphe. The cells may be free, or more commonly attached by a gelatinous stalk to various substrates, sometimes forming packets or filaments. There are over 30 species which have been reported from the United States.

697a (694) Raphe confined to polar regions. 698

697b Raphe evident throughout the length of the valve 699

698a Frustules straight and relatively narrow, sometimes slightly capitate at one pole; margin of frustule smooth in valve view; costae on both lateral margins; raphe extending only 2/3 the distance from the polar nodules. Fig. 523 .... *Peronia*

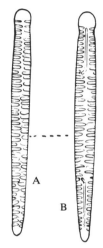

Figure 523

Figure 523 *Peronia erinacea* Bréb. & Arn. (A) valve view without raphe; (B) valve view with short raphe near poles.

These cells are elongate or cigar-shaped with rounded poles. One pole is often larger than the other as seen in valve view. The raphe in one valve extends from the poles only about 2/3 of the distance to the midregion and is slightly curved near the inner end. The other valve has a short pseudoraphe and a short raphe. The valves have prominent marginal costae. Cells are free-floating or benthic; only 2 or 3 species have been reported from the United States.

698b Frustules bent or curved in the apical regions; wavy or undulate on one mar-

gin as seen in valve view; transversely striate; raphe not extending the full length of the valve. Fig. 524 ... *Eunotia*

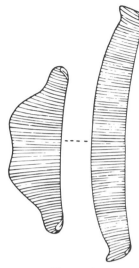

Figure 524

Figure 524 *Eunotia* sp., 2 cell shapes in valve view.

Species of this genus are solitary, or in filaments. The frustules are slightly bowed and are undulate on one or both margins as seen in valve view. In some species the raphe shows only near the polar nodules, and in some a pseudoraphe lies near the ventral margin. The face of the valves is transversely striate. The polar nodules are conspicuous and from them to the ventral (concave) margin a short raphe extends. There are many species common in especially soft-water or desmid habitats. This is a large genus with over 45 species reported from the United States.

699a (697) Raphe located in a canal 700

699b Raphe not located in a canal 705

700a Frustules spirally twisted; raphe accordingly spiral; cells with a keel. Fig. 525 . . . . *Cylindrotheca*

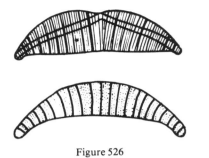

Figure 525

Figure 525   *Cylindrotheca* sp., valve view.

The cells of this rare genus are elongate-fusiform, the apical regions extended into horn-like processes, the main portion of the cell with spiral lines, two of which are coarsely punctate (carinal dots) and enclose the raphe. This genus has been reported from Maryland and from two midwestern states.

700b Frustules not spirally twisted 701

701a Raphe not in a keel 702

701b Raphe in a keel or wing 703

702a Raphe in a canal without pores; frustule broadest on the girdle side and usually seen lying in this position; a clear area in the girdle zone bordered by prominent costae, margins usually with a swelling in the midregion as seen in girdle view. Fig. 526. . . . . *Rhopalodia*

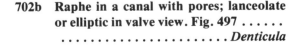

Figure 526

Figure 526   *Rhopalodia* sp., 2 different cell shapes in valve view.

In this genus the cells are much different in shape as seen in girdle and valve views. In valve view the frustule is narrow with the poles bowed, giving the appearance of a 'half-cell' with an enlargement on the midregion of the convex margin. The girdle view is much wider than the valve view. The ventral margin is almost straight, except for the narrow, curved polar region. The valve bears a keel to one side and in this lies the raphe. The valves are coarsely costate, the costae alternating with the striae. In girdle view the cells are elongate, somewhat quadrate in general outline but with a swelling on each side in the midregion. The girdle is smooth, but extending in from the margin are coarse costae which, in the polar regions, converge toward the center of the cell. The genus is entirely freshwater in distribution; but apparently only 2 species have been found in the United States.

702b Raphe in a canal with pores; lanceolate or elliptic in valve view. Fig. 497 . . . . *Denticula*

703a **(701) Keels on margins of valve opposite one another; quadrangular in optical (cross) section. Fig. 514 . . . .** *Hantzschia*

703b **Keels on alternate margins of the valve 704**

704a **Frustules occurring in colonies (ribbons); keel central. Fig. 527 . . . .**
. . . . . *Bacillaria*

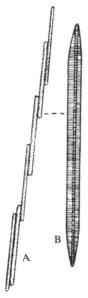

Figure 527

Figure 527 *Bacillaria paradoxa* Gmel. (A) diagram of cells sliding out to form an elongate, pseudofilamentous colony; (B) single cell in valve view.

The species illustrated is placed by some diatomists in the genus *Nitzschia* (Fig. 528). The narrowly elongate cells, with produced apices are united side by side to form ribbons. The cells slide back and forth along the apposed valves and draw the ribbon out into an extended, pseudofilamentous arrangement—then slide back and extend in the opposite direction; hence the name "carpenter's rule" diatom. The keel is central or nearly so and is conspicuously punctate. The raphe is in the keel and is diagonally opposite the raphe in the other valve. Species are both marine and brackish.

704b **Frustules solitary; keel eccentric, diagonally opposite one another; rhomboid in optical (cross) section. Fig. 528 . . . . .**
. . . . . *Nitzschia*

Figure 528

Figure 528 *Nitzschia bilobata* W. Smith, valve view. (Redrawn from Boyer.)

The frustules of this genus are narrowly linear and tapered at the poles in both valve and girdle views. The cells may be straight or slightly sigmoid. Along one margin of a valve is a keel which is opposite the unkeeled margin of the other valve (one raphe diagonally opposite the other). The raphe lies within the keel. The fissure of the raphe has a row of pores that open into the cell cavity. These are called carinal dots. There are prominent transverse rows of punctations on the valve faces. Although commonly solitary, *Nitzschia* often occurs in gelatinous, tubelike strands.

705a **(699) Frustules asymmetrical in either the transverse or the longitudinal axis; key- or wedge-shaped in valve view, slightly larger at one end than the other 706**

705b **Frustules symmetrical in both axes. . 707**

706a **Striae composed of puncta in a double series, interrupted near the margin of the valve so that a longitudinal line is formed; attached. Fig. 529 . . . .**
. *Gomphoneis*

Figure 529

Figure 529 *Gomphoneis herculeana* (Ehr.) Cleve, valve view showing raphe with curvature in the midregion.

These cells are wedge-shaped or cuneate, unsymmetrical in both girdle and valve views (transversely). There is an elongate, axial field which is enlarged in the midregion. The frustules are very similar to *Gomphonema* (Fig. 531) but there is a fine but definite line running parallel with the margin of the valve. The striae are composed of a double row of puncta. Frustules are either attracted (at the smaller pole) or free-floating.

706b **Striae composed of puncta in a single series; attached on branched stalks. Fig. 530. . . . . *Gomphonema***

Figure 530

Figure 530 *Gomphonema lanceolatum* var. *insignis* (Greg.) Cleve, valve view showing eccentric punctum.

Species in this genus have frustules that are wedge-shaped or clavate, larger at one end than the other. In both girdle and valve views the frustules are transversely unsymmetrical. The elongate axial field is enlarged in the midregion where there is often a eccentric, coarse punctum. There are coarse striae extending inward from the valve margins. The puncta form a single row. The cells are usually found attached on branched, gelatinous stalks at their narrow end. Compare with *Gomphoneis* (Fig. 529).

707a **(705) Frustules with septa. 708**

707b **Frustules without septa 709**

708a **Frustules quadrangular in girdle view; usually in zig-zag chains; longitudinally septate, the septa with openings in the center and at the poles. Fig. 531. . . . . *Diatomella***

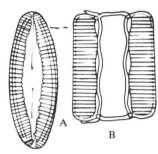

Figure 531

Figure 531 *Diatomella* sp. (A) valve view; (B) girdle view showing intercalary bands.

In valve view frustules are narrowly elongate, broadly rounded at the poles and enlarged in the midregion. In girdle view the cells are rectangular. There are 2 intercalary bands and 2 longitudinal septa in which there are 3 holes, one at each pole and one in the midregion. The valves have marginal transverse rows of striae leaving a wide central, clear longitudinal field. The septa show in both valves. Frustules may be solitary, filamentous, or in zig-zag chains.

708b **Frustules narrowly rectangular in girdle view; naviculoid in valve view; septa with large central opening and parallel linear openings which form 2 lateral series of minute transverse canals. Fig. 532 . . . . *Mastogloia***

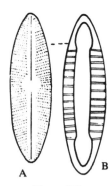

Figure 532

Figure 532 *Mastogloia danseii* Thw. (A) valve view; (B) diagram showing internal septa. (Redrawn from G.M. Smith.)

Frustules in this genus are symmetrical in both longitudinal and transverse axes. They are elongate-elliptic in valve view (sometimes capitate); rectangular in girdle view. The girdle is smooth but the intercalary bands have areolae. There is a raphe in a straight, clear axial field which is conspicuously enlarged in the midregion. There are 2 longitudinal septa. In valve view there are transverse striae on either side of the axial field. Cells are solitary and mostly marine; 4 or 5 species have been reported from the United States.

709a (707) **Raphe extending within a siliceous, riblike thickening.** **710**

709b **Raphe not bordered by siliceous ribs** . . .
. . . . **712**

710a **Valve with an enlarged, undecorated central area in the region of the central nodule; especially evident in large species; frustule broadly elliptic; valve costate. Fig. 498** . . . . *Diploneis*

710b **Valve without an enlarged, clear area in the central region; valve not costate; frustule linear-lanceolate** **711**

711a **Central nodule greatly elongated, the raphe appearing in 2 relatively short sections in the apical region. Fig. 533** . . . . .
. . . . *Amphipleura*

Figure 533

Figure 533 *Amphipleura pellucida* Kütz., valve view showing greatly enlarged central nodule. (Redrawn from Boyer.)

These are elongate and narrow, boat-shaped cells in valve view. The conspicuous differentiating character is the greatly enlarged central nodule that extends 2/3 of the length of the valve and is divided to form 2 parallel extensions toward the poles. The raphe lies between the divisions of the siliceous ribs. The valve has minute transverse striae of punctations which sometimes can be seen only under special optical conditions, hence it appears smooth under ordinary magnification. The puncta may show also forming longitudinal lines. Frustules are solitary and usually benthic in calcareous waters especially.

711b **Central nodule shorter, with 2 siliceous ribs extending toward the apices; raphe lying within the ribs. Fig. 534** . . . .
. . . . *Frustulia*

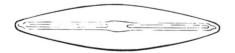

Figure 534

Figure 534 *Frustulia rhomboides* (Ehr.) DeToni, valve view showing raphe between 2 ribs. (Redrawn from G.M. Smith.)

Like *Amphipleura* (Fig. 533) *Frustulia* has an elongate central nodule but it is relatively shorter (less than 1/2 the length of the cell). The cells are boat-shaped in valve view; rectangular in girdle view and without intercalary bands and septa. Two ribs from the central nodule run parallel to the raphe and unite with the polar nodules. There are both transverse and longitudinal striae leaving a small, clear central area. Frustules are either solitary or lie side by side in a gelatinous, tubular colony. At least 5 species are reported from the United States.

712a (709) **Valves with 'wings' bearing a sigmoid keel, 8-shaped in girdle view. Fig. 520** . . . . *Amphiprora*

712b **Valves without a keel** 713

713a **Transverse valve markings interrupted, thus the frustule shows longitudinal lines paralleling the margins of the valve** . . . .
. . . . 714

713b **Transverse markings not so interrupted** 716

714a **Interruption of transverse lines forming a zig-zag pattern or line. Fig. 535** . . . . . .
. . . . *Anomoeoneis*

Figure 535

Figure 535 *Anomoeoneis sphaerophora* var. *sculpta* O.F. Müll. (Redrawn from Reimer.)

The frustules are elongate boat-shaped in valve view with narrowly rounded (sometimes capitate) apices, whereas they are rectangular in girdle view. In the valve view a narrow axial field (clear of wall markings) is enlarged in the midregion. The raphe is narrow and either straight or hooked in the same direction in the central region. Lateral to the central field are fine, transverse striae which are interrupted by clear spaces in such a way that a zig-zag pattern is formed over the face of the valve. This is a freshwater genus; benthic or in the tychoplankton; 8 species reported from the United States.

714b **Interruption not forming zig-zag pattern** 715

715a **Transverse markings formed of puncta. Fig. 536** . . . . *Neidium*

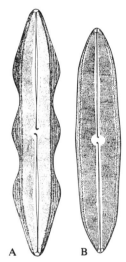

A   B

Figure 536

Figure 536 *Neidium*. (A) *N. magellanicum* Cleve, valve view showing curvature of raphe in midregion; (B) *N. iridis* var. *subundulata* (A.C.-Eul.) Reimer. (A, redrawn from Frenguelli; B, redrawn from Reimer.)

These boat-shaped cells in valve view show prominent, but discontinuous transverse, punctate striae. Although mostly with convex lateral margins, some are straight or even concave. The striae are conspicuously interrupted by blank spaces in two lateral lines immediately within the margins of the valves. There is a clear axial field which is enlarged in the midregion and takes on various shapes. There are no septa. At the central nodule the raphe forms 2 hooks, the hooks of the 2 raphes turning in opposite directions (usually). At the poles the raphe may be divided. The frustules are rectangular in girdle view. Compare with *Caloneis* (Fig. 537), another strictly freshwater genus: There are over 20 species reported from the United States.

715b Transverse markings of continuous lines. Fig. 537 . . . . *Caloneis*

Figure 537

Figure 537 *Caloneis bacillum* (Grun.) Cleve, valve view showing clear area in midregion and curvature of raphe. (Redrawn from Foged.)

Cells are elongate cigar-shaped, but slightly widened in the midregion where the axial field is also enlarged. There are both central and polar nodules. There are parallel transverse striae extending from the lateral margins of the valve. There are no septa nor intercalary bands. The frustules are rectangular in girdle view. Immediately within the margins of the valve is a parallel line which crosses the transverse striae which often appear as costae. These parallel lines are caused by pores in the striae which

open into the cell cavity. There are many species in both fresh and salt water habitats.

716a (713) Clear area in longitudinal axis sigmoid. Fig. 538 . . . . *Scoliopleura*

Figure 538

Figure 538 *Scoliopleura* sp., valve view with sigmoid raphe.

Frustules are elongate-oval to naviculoid or fusiform in valve view; lanceolate in girdle view. The outstanding characteristic of this genus is the sigmoid axial field. Within this lies a sigmoid raphe. There may be a line on either side of the axial field bordering the transverse striae (rows of puncta). There are 4 longitudinal chloroplasts in each cell. Species are both marine and freshwater.

716b Clear area of axial field not sigmoid 717

717a Valves with costae forming the transverse markings; costae punctate. Fig. 539 . . . . *Brebissonia*

Figure 539

Figure 539 *Brebissonia boeckii* (Ehr.) Grun., valve view showing raphe between parallel ridges. (Redrawn from Boyer.)

These cells are attached by branched, gelatinous strands to a substrate, sometimes to aquatic plants. The cells are lanceolate-boat-shaped in valve view; quadrangular in girdle view. The girdles are distinctly ornamented between which are several smooth intercalary bands. The 2 sections of the raphe lie in a clear, longitudinal axial field and between 2 parallel ridges (which are inconspicuous). There are 2 or several intercalary bands between the girdles. Very prominent striae extend from the margins of the valve toward the axial field. The genus is mostly marine but at least 2 species have been reported from freshwaters in the United States.

717b Valves with ornamentation formed by puncta, or without evident markings . . .
. 718

718a Lateral valve markings strongly oblique, interrupted in the midregion by an undecorated area over the central nodule which extends to the margins of the valve (forming a stauros). Fig. 540 . . . .
. . . . . *Stauroneis*

Figure 540

Figure 540 *Stauroneis parvula* var. *prominula* Grun., valve view showing transverse stauros. (Redrawn from Foged.)

These are mostly naviculoid cells in valve view, with either capitate or acutely rounded poles. The conspicuous feature (the stauros) is a broad, central nodule which extends to both lateral margins of the valve. In this region alone there are no transverse striae which throughout the remainder of the valve are prominent and which converge from the lateral margins to the axial field. There is a straight raphe lying in this field. Septa are lacking. The genus is common and there are many (depending on interpreta-

tions of diatomists), widely distributed species in the United States. There also are many marine species.

718b Central undecorated, clear area not extending to the margins of the valve. . 719

719a Transverse ornamentation composed of costae, the axial field usually broad. Fig. 541 . . . . *Pinnularia*

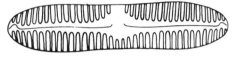

Figure 541

Figure 541 *Pinnularia borealis* Ehr., valve view showing costae. (Redrawn from Foged.)

These mostly solitary and free-floating frustules vary considerably in shape but mostly they are naviculoid with broadly rounded poles in valve view. Some species are enlarged in the midregion. The conspicuous feature is the prominent transverse costae or costaelike striae on either side of a relatively wide, straight axial field. The striae open into the cavity of the cell. The raphe in axial field is sigmoid. There are no septa. In girdle view the frustules are rectangular with truncate poles. Some species have cells which are perhaps the largest among freshwater diatoms. A few species, especially *Pinnularia viridis* (Nitzsch) Ehr. have the characteristic habit of lying side by side (in valve view) in groups of 2 or 4. There are many marine species.

719b Transverse ornamentation composed of puncta, the axial field narrow and linear, the raphe straight. Fig. 542 . . . .
. . . . . *Navicula*

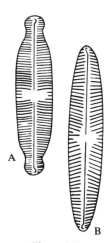

Figure 542

Figure 542 *Navicula.* (A) *N. petersenii* Hustedt; (B) *N. digitoradiata* fa. *minor* Foged. (B, redrawn from Foged.)

This is a large genus in respect to the number of species, occurring in both fresh and marine water; benthic or in the plankton. The cells vary considerably in shape, especially in valve view but in the main they are naviculoid or cigar-shaped, with narrowly rounded or capitate poles; sometimes distinctly rostrate at the apices. There is a raphe in both valves. There may be internal plates but no complete septa. Striae are composed of puncta, sometimes very coarse and form what appear to be costae. The frustules are eu- or tychoplanktonic or may occur in films on submersed substrates. Compare with *Pinnularia* (Fig. 541). The striae or rows of puncta are convergent toward the central, clear area of the axial field which is distinctly enlarged in the midregion. The field in general is straight and narrow and the raphe within it is straight. In girdle view the cells are rectangular.

Appendix

A Brief Synopsis of Unicellular Green Algae not Included in the Pictured Key to Genera Because Differentiating Characters Must be Determined by Culturing and by Observations of Reproductive Details, Both vegetative and Asexual. (Genera Mostly Isolated from Soil.)

I. Tetrasporales: Cells often with pseudocilia, containing contractile vacuoles; solitary or colonial, in copious mucilage; vegetative reproduction by ordinary cell division. See Figs. 79, 80, 86.

Spherical Unicells With Contractile Vacuoles.
Actinochloris Korschikoff. (Hypnomonadaceae)
Cells globular, up to 150 μm in diameter; chloroplast stellate, with the arms or rays flattened against the wall; pyrenoid central; contractile vacuoles several, occurring between lobes of the chloroplast; cells becoming multinucleate in age.

II. Chlorococcales (Including Chlorosphaerales):* solitary, colonial or forming coenobia; reproducing largely by autospores, commonly also by zoospores; contractile vacuoles lacking, although zoospores may show the tetrasporine-volvocine type of contractile vacuole and red eye-spot; sexual reproduction iso-, aniso- or oogamous.

A. Pyrenoid Present.
   1. Chloroplast axial.
      a. *Axilosphaera* Cox & Deason (Chlorococcaceae)
         Cells spherical, up to 18 μm in diameter; chloroplast axial, lobed, with an eccentric pyrenoid; surface view of cell showing ends of lobes, giving the appearance of several parietal plates. From Soil.

      b. *Borodinella* Miller (Chlorococcaceae)
         Cells spherical, solitary or grouped after cell division, floating at the surface of the

*The Order Chlorosphaerales has been suggested by Dr. Walter Herndon for all non-motile, unicellular and colonial forms which carry on vegetative cell division (cell bipartition); reproducing by motile and non-motile spores but which do not produce sarcinalike packets; without contractile vacuoles; includes Chlorosphaeraceae and Coccomyxaceae.

water, forming somewhat tetrahedral colonies after cell division; chloroplast stellate, with a central pyrenoid; reproduction by biflagellate zoospores containing vacuoles.

c. *Borodinellopsis* Dykstra (Chlorococcaceae)
Spherical cells up to 30 μm in diameter; chloroplast stellate, with a pyrenoid, the arms flattened against the cell wall; cell division internal, forming tetrahedrally arranged, nonmotile daughter cells; asexual reproduction by biflagellate zoospores which retain an ovoid shape for some time (unlike *Borodinella*) after becoming stationary. From soil.

2. Chloroplast parietal (continuous or perforate).
a. *Chlorosphaeropsis* (Vischer) Herndon (Chlorosphaeraceae)
Cells spherical, about 17 μm in diameter, solitary at first, becoming clumped; the chloroplast at first a parietal plate but soon becoming reticulate and covering the wall; pyrenoid median in the chloroplast; reproduction by elongate-ellipsoid, biflagellate zoospores. From soil.

b. *Fasciculochloris* McLean & Trainor (Chlorosphaeraceae)
Cells at first oblong-ovoid, becoming spherical, up to 14 μm in diameter; chloroplast a parietal cup with 1 or more pyrenoids; cells dividing to form quartets of daughter cells; asexual reproduction by biflagellate, walled zoospores that have an eye-spot.

c. *Heterotetracystis* Cox & Deason (Chlorosarcinaceae)

Cells spherical, up to 18 μm in diameter, solitary or in 2's and 4's; chloroplast parietal but massive, with a pyrenoid, the chloroplast becoming net-like or reticulate; asexual reproduction by oval-elliptic zoospores which retain their shape for a time after becoming sedentary; in culture often forming orange-colored cysts. From soil.

d. *Neochloris* Starr (Chlorococcaceae)
Cells spherical, up to 65 μm in diameter, with a wall which may be thin, or laminate; chloroplast cup-shaped (somewhat similar to *Gloeocystis*, Fig. 86), with 1 to several pyrenoids; asexual reproduction by biflagellate, ovoid zoospores. From soil.

e. *Spongiochloris* Starr (Chlorococcaceae)
Cells spherical or subsphaerical, up to 100 μm in diameter; chloroplast a parietal reticulum with inward projecting, spongy portions; pyrenoids 1 to several; asexual reproduction by equally-flagellated, naked zoospores which become spherical upon attachment. From soil.

f. *Spongiococcum* Deason (Chlorococcaceae)
Cells spherical, about 20 μm in diameter; chloroplast a fine reticulum, both parietal and spongy (as in *Spongiochloris*) with at least 1 eccentric pyrenoid; reproduction by oval or elongate, walled, *Chlamydomonas*-like zoospores; vegetative cell division by internal cleavage to form several (16 ?) individuals. From soil.

g. *Tetracystis* Brown & Bold (Chlorosphaeraceae)
Cells spherical or subspherical, solitary or aggregated, up to 20 μm in diameter; chloroplast a parietal but massive body which may be perforate; vegetative cell division forming tetrads of daughter cells; asexual reproduction by biflagellate zoospores which become spherical upon attachment. From soil.

B. Pyrenoid Absent.
1. Chloroplasts many, irregular platelets, parietal.
Dictyochloropsis Geitler (Chlorococcaceae)
Cells spherical, subaerial, up to 75 μm in diameter; chloroplasts numerous, irregular, lobed platelets and vermiform bodies; pyrenoids lacking; reproduction by 16-64 autospores.

2. Chloroplast a bilobed, dumbbell-shaped body.
Pulchrasphaera Deason (Chlorococcaceae)
Cells spherical, up to 25 μm in diameter; chloroplast parietal and lobed or dumbbell-shaped, the lobes often perforate, without pyrenoids; cells gregarious but not forming packets following cell division; reproduction by oval, biflagellated zoospores which become spherical immediately upon attaching; eye-spot present in zoospores. Plants from sand.

3. Chloroplast parietal plates, or reticulate, the fenestrations varying in size.
a. *Dictyochloris* Vischer (Chlorococcaceae)
Cells spherical, up to 75 μm in diameter; chloroplast reticulate and covering most of the wall, without a pyrenoid; reproduction by naked, somewhat unequally-flagellated zoospores; cells becoming coenocytic in age. From soil.

b. *Dictyococcus* Gerneck (Chlorococcaceae)
Cells spherical, up to 60 μm in diameter; chloroplasts numerous, angular, parietal plates as seen in surface view, with inward projecting lobes; pyrenoid lacking; cells becoming coenocytic in age; reproduction by biflagellate zoospores. Genus similar to *Bracteacoccus,* Fig. 201.

c. *Friedmannia* Chantanchat & Bold (Chlorococcaceae or Chlorosarcinaceae)
Cells spherical with a parietal, irregularly lobed or segmented chloroplast; pyrenoid lacking; cells dividing to form tetrahedral clumps; asexual reproduction by biflagellate, naked, oval zoospores with an eyespot.

Taxonomic List of Freshwater Algal Genera in Their Families and Orders*

I. PHYLUM (DIVISION) Chlorophyta**

A. SUB-PHYLLUM CHLOROPHYCEAE

1. Order Volvocales

Family Pyramimonaceae

Pyramimonas
p. 44, fig. 57

Family Tetraselmiaceae

Tetraselmis (Platymonas)
p. 41, fig. 49

Family Polyblepharidaceae

Dunaliella
p. 43, fig. 53

Hyaliella ★ (colorless)

Mesostigma
p. 43, fig. 55

Polytomella ★ (colorless)

Quadrichloris
(Tetrachloris)
p. 45, fig. 59

Raciborskiella
p. 46, fig. 62

Spermatozopsis
p. 45, fig. 58

Stephanoptera
p. 42, fig. 52

Family Chlamydomonadaceae

Brachiomonas
p. 39, fig. 42

Carteria
p. 41, fig. 50

Chlamydomonas
p. 26, fig. 17

Chlorogonium
p. 38, fig. 41

Chloromonas
p. 40, figs. 46, 204

Diplostauron ★

Furcilia ★

Gloeomonas ★
Haematococcus
p. 25, fig. 14

Lobomonas
p. 39, fig. 43

Platychloris ★

Polytoma
p. 34, fig. 32

Scherffelia
p. 41, fig. 48

Smithsonimonas
p. 40, fig. 45

Sphaerellopsis
p. 39, fig. 44

Tetrablepharis *

Family Phacotaceae

Cephalomonas
p. 36, fig. 35

Coccomonas
p. 38, fig. 40

Dysmorphococcus
p. 35, fig. 34

Granulochloris
p. 37, fig. 39

Pedinopera
p. 40, fig. 47

Phacotus
p. 37, fig. 38

Pteromonas
p. 35, fig. 33

Thoracomonas
p. 37, fig. 37

Wislouchiella
p. 36, fig. 36

Family Volvocaceae

Astrephomene
p. 50, fig. 73

Eudorina
p. 49, fig. 71

Gonium
p. 46, fig. 61

Pandorina
p. 48, fig. 67

Platydorina
p. 45, fig. 60

Pleodorina
p. 50, fig. 72

Stephanoon
p. 49, fig. 70

Stephanosphaera
p. 49, fig. 69

Volvox
p. 48, fig. 68

Volvulina ★

Family Spondylomoraceae

Corone
p. 47, fig. 65

Pascherina
(Pascheriella)
p. 46, fig. 63

Spondylomorum
p. 47, fig. 64

Uva
(Chlamydobotrys)
(Pyrobotrys)
p. 47, fig. 66

*Genera indicated (★) are not treated in the key.
**For the most part the arrangement follows a plan used by Bourrelly (1966-1970).

*See footnote 30, p. 268.
**Sometimes placed in the
Chlorococcaceae.

*Sometimes placed in its own family, Schizomeridaceae

*Sometimes placed in the Siphonales.

*Possibly should be retained in the Mesotaeniaceae.

III. PHYLUM (DIVISION) PYRRHOPHYTA

A. **CLASS DESMOKONTAE**
   1. **Order Desmomonadales**
      Family Prorocentraceae
      Exuviella
      p. 196, fig. 388

B. **CLASS DINOPHYCEAE**
   1. **Order Dinokontae**
      Family Gymnodiniaceae
      Amphidinium
      p. 199, fig. 393
      Gymnodinium
      p. 198, fig. 391
      Gyodinium, 198
      Massartia
      p. 199, fig. 392

      Family Gonyaulacaceae
      Gonyaulax
      p. 200, fig. 396

      Family Glenodiniaceae
      Glenodinium
      p. 201, fig. 398
      Hemidinium
      p. 201, fig. 397

      Family Peridiniaceae
      Peridinium
      p. 200, fig. 394

      Family Ceratiaceae
      Ceratium
      p. 197, fig. 389

   2. **Order Dinocapsales**
      Family Gloeodiniaceae
      Gloeodinium
      p. 194, fig. 383
      Urococcus
      p. 27, fig. 18

   3. **Order Dinococcales**
      Family Dinococcaceae
      Cystodinium
      p. 193, fig. 381
      Hypnodinium
      p. 194, fig. 382
      Raciborskia
      p. 191, fig. 376

Stylodinium
p. 192, fig. 377
Tetradinium
p. 190, fig. 374

Family Blastodiniaceae
Oodinium
p. 191, fig. 375

IV. PHYLUM (DIVISION) CRYPTOPHYTA (Cryptophyceae)

Family Cryptochrysidaceae
Chroomonas
p. 165, fig. 312
Cryptochrysis
p. 31, fig. 26
Monomastrix
p. 31, fig. 25
Rhodomonas
p. 25, fig. 15

Family Cryptomonadaceae
Chilomonas ★
Cryptomonas
p. 26, fig. 16
Cyathomonas ★

Family Nephroselmidiaceae
Nephroselmis
(Heteromastix?)
p. 43, fig. 54
Protochrysis
p. 158, fig. 302

Family Cryptococcaceae
Tetragonidium
p. 188, fig. 368

V. PHYLUM (DIVISION) CHLOROMONADO-PHYTA

Family Chloromonadaceae
Gonyostomum
p. 30, fig. 24

Merotrichia
p. 30, fig. 23
Trentonia ★
Vacuolaria
p. 29, fig. 22

VI. PHYLUM (DIVISION) CHRYSOPHYTA

A. **SUB-PHYLUM XANTHOPHYCEAE (HETEROKONTAE)**
   1. **Order Chloramoebales**
      Family Chloramoebaceae
      Chloramoeba
      p. 174, fig. 334

   2. **Order Rhizochloridales**
      Family Rhizochloridaceae
      Stipitococcus
      p. 172, fig. 330

      Family Myxochloridaceae
      Chlamydomyxa
      p. 89, fig. 157

   3. **Order Heterogloeales**
      Family Heterogloeaceae
      Heterogloea
      (Gloeochloris)
      p. 182, fig. 356

      Family Pleurochloridellaceae
      Pleurochloridella ★

      Family Malleodendraceae
      Malleodendron
      p. 167, fig. 316

      Family Characidiopsidaceae
      Characidiopsis ★

   4. **Order Mischococcales**
      Family Pleurochloridaceae
      Acanthochloris ★
      Arachnochloris
      p. 179, fig. 348

Family
Chrysocapsaceae
Chrysocapsa
p. 207, fig. 410
Phaeaster
p. 183, fig. 357
Tetrasporopsis
p. 206, fig. 408

Family
Chrysochaetaceae
Chrysochaete ★

Family Hydruraceae
Hydrurus
p. 185, fig. 362

Family
Chrysamoebaceae
Chrysamoeba
p. 193, fig. 380
Chrysostephanosphaera
p. 207, fig. 44

Family
Chrysostomataceae
Chrysastrella
p. 189, fig. 370

Family
Kybotionaceae
Kybotion
p. 171, fig. 326

Family
Myxochrysidaceae
Myxochrysis ★

6. **Order**
Ochromomadales
Family
Ochromonadaceae
Cyclonexis
p. 204, fig. 404
Ochromonas
(Chlorochromonas)
p. 174, fig. 335
Stipitochrysis ★
Syncrypta
(Synuropsis)
p. 203, fig. 401
Uroglena
p. 202, fig. 400
Uroglenopsis
p. 203, fig. 403

Family Dinobryaceae
Dinobryon
p. 187, fig. 366
Epipyxis
p. 194, fig. 384
Hyalobryon
p. 195, fig. 385
Pseudokephyrion ★
(Kephyriopsis)

Family Synuraceae
Chrysosphaerella
p. 202, fig. 399
Mallomonas
p. 192, fig. 378
Synura
(Skadovskiella)
p. 203, fig. 402

Family Ruttneriaceae
Ruttnera ★
Entodesmis ★

Family Naegelliaceae
Naegeliella
p. 206, fig. 409

Family
Phaeothamniaceae
Phaeothamnion
p. 186, fig. 365

Family
Chrysapionaceae
Chrysapion ★
Phaeogloea ★

7. **Order Isochrysidales**
Family
Isochrysidaceae
Chrysidalis ★

Family
Derepyxidaceae
Derepyxis
p. 190, fig. 373

Family
Diacronemataceae
Diacronema ★

8. **Order Prymnesiales**
Family
Prymnesiaceae
Chrysochromulina
p. 196, fig. 387

Family
Coccolithophorida-
ceae
Hymenomonas ★

9. **Order Monosigales**
Family
Monosigaceae
Monosiga ★

Family
Salpingoecaceae
Salpingoeca ★

Family
Phalansteriaceae
Phalansterium ★

C. **SUB-PHYLUM**
BACILLARIOPHYCEAE
(DIATOMACEAE)
1. **Order Centrales**
Family
Coscinodiscaceae
Actinocyclus
p. 242, fig. 488
Coscinodiscus
p. 243, fig. 490
Melosira
p. 248, fig. 500
Cyclotella
p. 242, fig. 489
Stephanodiscus
p. 243, fig. 491

Family
Biddulphiaceae
Biddulphia
p. 244, fig. 492
Hydrosera
(Triceratium)
p. 242, fig. 487
Terpsinoe
p. 247, fig. 499

Family
Chaetoceraceae
Chaetoceros
p. 241, fig. 486

Family
Rhizosoleniaceae
Attheya
p. 249, fig. 501
Rhizosolenia
p. 249, fig. 502

2. **Order Pennales**
Family Fragilariaceae
Amphicampa
p. 252, fig. 507
Asterionella
p. 250, fig. 504
Centronella
p. 250, fig. 503
Ceratoneis
p. 251, fig. 506
Diatoma
p. 253, fig. 510
Fragilaria
p. 253, fig. 512
Meridion
p. 250, fig. 505
Opehora
p. 253, fig. 511
Synedra
p. 254, fig. 513

VII. PHYLUM (DIVISION) PHAEOPHYTA

VIII. PHYLUM (DIVISION) RHODOPHYTA

IX. PHYLUM (DIVISION) CYANOPHYTA

Index and Pictured Glossary

Figures in boldface refer to pages where the genus name appears in the key. Figures in italics refer to pages where species are illustrated. Figures in lightface refer to pages where generic names are mentioned for comparison, or as synonyms. Pages 271 to 280 refer to the taxonomic list of genera in respective families.

A

Acanthochloris, 179, 276, 277
Acanthosphaera, **107**, 273
 zachariasi, 107
Achnanthaceae, 279
Achnanthes, **258**, 279
 coarctata, 258
 lanceolata, 258
ACICULAR: needlelike in shape. Fig. 543

Figure 543

ACID BOG: having soft water, low or lacking in dissolved minerals; pH below neutral (7.0)
Acrochaetium, 163, 279
Actidesmium, **80**, 273
 hookeri, 80
Actinastrum, **78**, 273
 gracillimum, 79
 hantzschii, 79
Actinella, **257**, 279
 punctata, 257
Actinochloris, **268**, 272
Actinocyclus, **242**, 278
Actinotaenium, 55, 275
Aegagropila, **153**, 274
 profunda, 153

AERIAL; SUBAERIAL: habitat on moist soil, rocks, trees, etc.; within a thin film of water; exposed to air.
AKARYONTA, PROKARYONTA, 9, 12
AKINETE: a type of spore formed by the transformation of a vegetative cell into a thick-walled, resting spore containing a concentration of food.
Albrightia, **213**, 280
 tortuosa, 213
ALCOHOL:
 food reserve, 11
 preservative, viii
Algal Paper, 131
ALKALINE WATER: containing a predominating amount of hydroxyl-ions as compared with hydrogen-ions; abundant in electrolytes; hard water lakes ordinarily are alkaline.
Alligator, 7
Allorgeia, 103, 275
 valiae, 103
ALPINE: altitudes above tree line; SUB-ALPINE: upper limits of forest zone.
ALVEOLAE: small cavities; minute chambers.
Ambrosia, 28
AMOEBOID: like an Amoeba; creeping by extensions of highly plastic protoplasm (pseudopodia).
AMORPHOUS: without definite shape; without regular form.
Amphicampa, **252**, 278
 eruca, 252
Amphichrysis, 277
Amphidinium, **199**, 276
 klebsii, 200
Amphipleura, **263**, 279
 pellucida, 263

Amphiprora, **257**, 264, 279
 paludosa, 257
Amphithrix, **211**, 280
 janthina, 211
Amphora, **255**, 279
 clevei, 255
 ovalis, 255
Amscottia, **103**, 275
 mira, 103
Anabaena, 9, **209**, 225-228, 280
 spiroides var. crassa, 209
 subcylindrica, 209
Anabaenopsis, **225**, 280
 elenkinii, 225
Anacystis, 237, 279
ANASTOMOSE: to separate and come together again at another point or level; a meshwork. Fig. 544

Figure 544

Ancylonema, 20, **121**, 275
 nordenskioldii, 121
ANISOGAMETE: a sex cell which shows only slight differentiation in respect to maleness or femaleness.
Anisonema, 275
Ankistrodesmus, 7, **74**, 80, 84, 86, 87, 94, 272
 braunii, 74
 convolutus, 74

 falcatus, 74
 fractus, 74
 spiralis, 74
Ankyra, **87**, 272
 judayi, 87
Anomoeoneis, **264**, 279
 sphaerophora var. sculpta, 264
ANTAPICAL: the posterior or rear pole or region of an organism, or of a colony of cells.
ANTERIOR: the forward end; toward the top.
ANTHERIDIUM: a single cell or a series of cells in which male gametes are produced; sometimes applied to the multicellular spherule in the Characeae which is a specialized, complex branch in which antheridial cells are produced. (See Figs. 1, 2, 3, 17).
ANTHEROZOID: male sex cell; sperm.
Apatococcus, 72, 274
 lobatus, 73
APEX; APICAL: the summit, the terminus; foward end of a projection; inner part of an incision; the upper end of a filament of cells which may show basal-distal differentiation.
Aphanizomenon, **226**, 280
 flos-aquae, 226
Aphanocapsa, **237**, 238, 279
 elachista, 237
 grevillei, 237
Aphanochaetaceae, 274
Aphanochaete, **132**, 143, 274
 polychaete, 132
 repens, 132
Aphanothece, **240**, 279
 castagnei, 240
APICAL RESERVOIR: a cavity in the anterior end of flagellated cells, often adjoined by contractile vacuoles.

multi-nucleated, non-cellular plant, *e.g.*, *Vaucheria.* Fig. 301

COLLAR: a thickened ring or neck surrounding the opening in a shell or lorica through which a flagellum projects from the enclosed organism.

Collecting Algae, 8

COLONIAL MUCILAGE: a gelatinous investment or sheath which encloses one to several or many cells.

COLONY: a group of closely associated cells, a cluster, adjoined or merely aggregated and enclosed in a common mucilage; cells not arranged in a linear series to form a filament or a many-celled plant.

COLUMNAR CELLS: longitudinal cells surrounding the axial filament or its branches as in *Chara.* Fig. 2

CONCENTRIC: arranged about a common center.

CONE-SHAPED; CONICAL: a figure circular in cross section, broad at the base and tapering symmetrically to the apex.

CONFLUENT: running together or intermingling, as mucilaginous sheaths of a plant becoming intermingled and losing identity. Fig. 549

Figure 549

CONJUGATION: sexual reproduction between cells which have become joined or "yoked" together, the gametes (sex cells) becoming united through a tube or opening at the point of junction, often by an amoeboid movement. Fig. 550

Figure 550

CONJUGATION TUBES: See Conjugation.
Conochaete, **69**, 274
 comosa, 69
CONSTRICTED; CONSTRICTION: cut in or incised from a margin, usually from 2 opposite points as in the Desmids where a shallow or deep incision divides the cell into 2 semicells connected by an isthmus; indented at the cross walls between cells of a filament. Figs. 551, 555

Figure 551

CONTRACTILE VACUOLE: a small vacuole or cavity which is bounded by a membrane that pulsates, expanding and contracting, 29.
Copepods, 6
CORDATE: heart-shaped.
Coronastrum, **58**, 69, 80, 273
 aestivale, 59
Corone, **47**, 59, 271
 bohemica, 47
CORONULA: a little crown; especially applied to the cells terminating the tubular cells which invest the oogonium in the Characeae.
CORTEX; CORTICATING CELLS: cells which surround or enclose a main axis; superimposed cells.
Coscinodiscaceae, 278
Coscinodiscus, **243**, 278
 lacustris, 243
Cosmarium, 4, **104**, 275
 margaritatum, 104
 panamense, 104
Cosmocladium, **63**, 275
 tuberculatum, 63
COSTAE: ribs; linear thickenings of the cell wall.
CRENULATE: wavy with small scallops; with small crenations.

CRESCENT-SHAPED: in the form of an arc of a circle; a narrow, curved figure tapering to points from a wider or cylindrical midregion. (See Lunate).
Crucigenia, **76**, 273
 rectangularis, 76
 tetrapedia, 76
Cryocystis, 112
Cryptochrysidaceae, 276
Cryptochrysis, **31**, 196, 276
 commutata, 31
Cryptoglena, 31, 275
Cryptomonadaceae, 276
Cryptomonas, **26**, 197, 276
 erosa, 26
 splendida, 26
Cryptophyceae, 9, 276
Cryptophyta, **10**, 14, 276
Ctenocladus, 148, **149**, 274
 circinnatus, 150
Culture, 1
CUP-SHAPED: a nearly complete folded plate (as a chloroplast) which lies immediately within the cell wall, open at one position to form a cup. Fig. 552

Figure 552

CUSHION: a pad; a thickened plate; sometimes referred to a low mound of cells.
Cyanoderma, 7, **27**, 279
 bradypodis, 7, 27
Cyanophyta, 6, 9, 12, 19, 279
Cyathomonas, 276
Cyclonexis, **204**, 278
 annularis, 204
Cyclops, 72
Cyclotella, **242**, 278
 meneghiana, 243
CYLINDRICAL: a figure round in cross section, elongate with parallel margins when viewed from the side, the ends square or truncate. (See Subcylindrical).
Cylindrocapsa, **137**, 273
 geminella var. minor, 137
Cylindrocapsaceae, 273
Cylindrocystis, **55**, 60, 97, 275
 brebissonii, 56
Cylindrospermum, **225**, 280
 majus, 225
 marchicum, 225
Cylindrotheca, **260**, 279
Cymatopleura, **245**, 256, 279
 elliptica, 245
Cymbella, **255**, 279
 cistula, 256
Cymbellaceae, 279

Cystodinium, **193**, 276
 cornifax, 193

D

Dactylococcopsis, **238**, 280
 acicularis, 239
 fascicularis, 239
Dactylococcus, **72**, 79, 272
 infusionum, 72
Dactylothece, **60**, 273
 confluens, 61
Dasygloea, **217**, 280
 amorpha, 218
DAUGHTER CELLS: cells produced directly from the division of a primary or parent cell; cells produced from the same mother-cell.
DAUGHTER COLONY: a group of cells formed as a small colony within a mother-cell or mother colony.
Debarya, **123**, 274
Denticula, **246**, 260, 279
 elegans, 247
 tenuis, 247
Derepyxidaceae, 278
Derepyxis, **190**, 278
 dispar, 190
Dermatophyton, **140**, 145, 274
 radians, 140
Dermocarpa, **230**, 280
 prasina, 231
 rostrata, 231
Dermocarpaceae, 280
Desert Soil, 6
Desmatractum, **94**, 272
 bipyramidatum, 94
Desmidiaceae, 275
Desmidium, **118**, 275
 baileyi, 118
 grevillii, 118
Desmids, 4
Desmococcus, **72**, 84, 111, 141, 274
 viridis, 73
Desmokontae, 276
Desmonema, **229**, 280
 wrangelii, 229
Diachros, **175**, 184, 277
 simplex, 175
Diacronema, 278
Diacronemataceae, 278
Diatoma, **253**, 278
 vulgare, 253
Diatomaceae, 278
Diatomella, **262**, 279
Diatoms, 6, 10; representative forms, 187; description of, 241
Dicellula, **52**, 77, 106, 107, 273
 planctonica, 52
Diceras, **188**, 277
 phaseolus, 188
Dichotomococcus, **184**, 277
 elongatus, 185
Dichotomosiphon, **157**, 274
 tuberosus, 157
Dichotomosiphonaceae, 274
Dichotomosiphonales, 274
DICHOTOMOUS: dividing or branched by repeated fork

ings, usually into 2 equal portions or segments.
Dicranochaetaceae, 274
Dicranochaete, **70**, 92, 274
 reniformis, 70
Dictyochlorella, 272
Dictyochloris, **270**, 272
Dictyochloropsis, **270**, 272
Dictyococcus, **270**, 272
Dictyosphaeriaceae, 273
Dictyosphaerium, **64**, 81, 273
 pulchellum, 64
Dictyosphaeropsis, 272
Dimorphococcus, **64**, 81, 273
 cordatus, 64
 lunatus, 64
Dinema, 275
Dinobryaceae, 278
Dinobryon, 2, **187**, 194, 204, 205, 278
 sertularia, 187
Dinocapsales, 276
Dinococcaceae, 276
Dinococcales, 276
Dinoflagellates, 10, 11; cysts, 200
Dinokontae, 276
Dinophyceae, 276
Dinoxanthin, 11
DIOECIOUS: with 2 kinds of spores (megaspores and microspores) produced on separate plants; with male and female gametes produced on separate plants.
Diogenes, **106**, 112, 273
 bacillaris, 106
Dioxys, **173**, 277
 inermis, 174
Diplocolon, **229**, 280
 heppii, 229
Diploneis, **247**, 263, 279
 elliptica, 247
Diplostauron, 271
DISC; DISC-SHAPED: a flat (usually circular) figure which is decidedly less in thickness than in width; a circular plate.
Dispora, **67**, 273
 crucigenioides, 68
DISTAL: the forward or anterior end or region as opposed to the basal end.
Distigma, 275
DIVISION (Of the Plant Kingdom): defined, 3.
Docidium, **96**, 275
 undulatum, 96
DORSAL: the back or upper surface or part as opposed to the under or lower (ventral) surface.
Draparnaldia, **156**, 274
 glomerata, 157
Draparnaldiopsis, **156**, 274
 salishensis, 156
Drying Specimens, 8
Duckweed, 90
Dunaliella, **43**, 271
 salina, 43
DWARF MALE: a minute male plant, smaller than the female, as in the Oedogoniaceae, growing on or near the female sex organ (oogonium).

Figure 557

($C_6H_{10}O_5$) which causes the starch to become blue-black.

Isochrysidaceae, 278
Isochrysidales, 278
ISODIAMETRIC: a figure with all planes having an equal diameter or nearly so.
ISOGAMETE: a sex cell which shows no detectable differentiation in respect to maleness or femaleness morphologically.
ISOKONT FLAGELLATION: with flagella of equal length on an organism and usually the same morphologically.
Iwanoffia, 274

J

Jack-in-the Pulpit: See *Arisaema*.
Jenningsia, 275
JOINT: point or plane where 2 cells or 2 elements adjoin.

K

Kelp, 2, 11
Kentrosphaera, **111**, 272
   bristolae, 111
   facciolae, 111
Kephyrion, 277
Kephyriopsis, 278
Keriochlamys, **82**, 106, 273
   styriaca, 82
Kirchneriella, **66**, 88, 273
   lunaris, 66
   obtusa var. *major*, 66
Koliella, **136**, 273
Kybotion, **171**, 278
   eremita, 171
Kybotionaceae, 278
Kyliniella, 20, **129**, 164, 279
   latvica, 129

L

Lagerheimia, 107, 112, 272
Lagynion, **189**, 277
   reductum, 189
   triangularis var. *pyramidatum*, 189
LAMELLA; LAMELLATE: with layers; with plates lying against one another.
LAMINARIN: a polysaccharide carbohydrate used as a food storage product in the Phaeophyta.
LAMINATE: with layers. (See Lamella).
Lamprothamnus, 275
LATERAL CONJUGATION: sexual reproduction in which 2 cells of the same filament become interconnected by a passage way (conjugation

tube), around the cross wall permitting the union of the cell contents of the acting gametangia to unite and to form a zygospore, as in some species of *Spirogyra*.
Lauterborniella, **79**, 273
   elegantissimum, 79
Lemanea, 1, **160**, 279
   annulata, 160
Lemaneaceae, 279
Lemna, 6
   trisulca, 6, 53, 91
Lepocinclis, **34**, 275
   acuta, 34
   glabra fa. *minor*, 34
Leptosira, **149**, 274
   mediciana, 149
LEUCOSIN: a whitish food reserve characteristic to many of the Chrysophyta, especially the Heterokontae (probably a carbohydrate), 164.
Leuvenia, **176**, 277
   natans, 176
Lichens, 108
Lime (Marl Deposits), 19
LINEAR SERIES: cells or units arranged in a row consecutively.
Lithoderma, 186, 279
Lithodermataceae, 279
Lithodora, 186, 279
Lobocystis, **60**, 273
   dichotoma, 60
Lobomonas, **39**, 271
   rostrata, 39
LOBULE: a small lobe; a secondary division of a lobe.
LONGITUDINAL FURROW: a groove or sulcus extending longitudinally on the ventral side of a motile organism, especially Dinoflagellates.
LORICA: a shell-like structure of various shapes possessed by many motile organisms; an outer shell having a pore through which the flagellum extends. Fig. 558

Figure 558

Loriellaceae, 280
LUNATE: crescent-shaped, bowed.
Lutherella, **184**, 277
   obovoidea, 184
Lychnothamnus, 275
Lyngbya, **209**, 216, 280
   birgei, 209
   contorta, 209

M

Macrocystis, 2
Magnolia, 21, 146
Malleochloris, **92**, 272
   sessilis, 92
Malleodendraceae, 276
Malleodendron, **167**, 276
   gloeopus, 167
Mallomonas, **192**, 196, 278
   acaroides, 192
   caudata, 192
   pseudocoronata, 192
MANNITOL: an alcohol type of food storage, non-solid, in the Phaeophyta, 11.
Mare's Eggs, 227
MARL: calcium carbonate compounds deposited by algal physiology or by other organisms; concretions which precipitate from hard water, mixed with clay, forming bottom sediments.
Marssoniella, **239**, 280
   elegans, 239
Massartia, **199**, 276
   musei, 199
Mastogloia, **262**, 279
   danseii, 263
MEDIAN INCISION: See Incision.
Melosira, **248**, 250, 278
   granulata, 248
Menoidium, 275
Meridion, **250**, 278
   circulare, 251
Meringosphaera, **179**, 277
   spinosa, 180
Merismopedia, **234**, 236, 280
   convoluta, 234
   elegans var. *major*, 234
   glauca, 234
Merotrichia, **30**, 276
   capitata, 30
Mesostigma, **43**, 271
   viridis, 44
Mesotaeniaceae, 275
Mesotaenium, **56**, 100, 109, 275
   degreyii, 56
   macrococcum, 56
METABOLIC: plastic, changing shape in motion as in many *Euglena* species.
METABOLISM: referring to the physiological activities within a living cell.
Micractiniaceae, 273
Micractinium, **77**, 273
   pusillum, 77
   quadrisetum, 78
Micrasterias, 4, **100**, 275
   americana var. *boldtii*, 100
   foliacea, 100, 101, 116
   radiata, 100
Microchaete, **224**, 228, 280
   diplosiphon, 224
   robusta, 224
Microcoleus, **217**, 280
   lacustris, 217
   vaginatus, 217
Microcystis, **236**, 238, 280
   aeruginosa, 236
   flos-aquae, 236
MICROFAUNA: referring to all microscopic animals.

MICROFLORA: microscopic plants.
MICRON: a unit of microscopic measurement, 1/1000 of a millimeter; written mm.
Microspora, 3, 131, **138**, 273
   floccosa, 138
   loefgrenii, 138
   willeana, 138
Microsporaceae, 3, 273
Microthamnion, **152**, 274
   strictissimum, 152
Mischococcaceae, 277
Mischococcales, 276
Mischococcus, **167**, 183, 277
   confervicola, 168
Monallantus, 177, 277
MONILIFORM: arranged like a string of beads; lemon-shaped. Fig. 559

Figure 559

Monocilia, **168**, 277
   flavescens, 168
   viridis, 168
MONOECIOUS: with microspores and megaspores produced on the same plant; with male and female organs borne on the same plant.
Monomastix, **31**, 43, 276
   opisthostigma, 31
Monosiga, 278
Monosigaceae, 278
Monosigales, 278
Monostroma, **124**, 273
   latissimum, 125
MONOTYPIC: a genus with but 1 species, or a family with 1 genus, 3.
MONOXIAL: with 1 axis; with a single row of cells as a primary filament; growth occurring at the apex of one filament.
MOTHER-CELL: a cell which gives rise internally to small cells; cell wall enclosing daughter cells such as autospores, q.v.
Mougeotia, 2, 3, 119, **121**, 124, 147, 274
   elegantula, 121
   genuflexa, 121
Mougeotiopsis, **122**, 274
   calospora, 123
Mounting Specimens, ix, 8, 32

MULTIAXIAL: with more than one axis; with more than one filament of cells in the axis of a thallus; growth at

the apex of more than one series of cells.
MULTINUCLEAR; MULTINU-CLEATE: cells or units of a thallus containing more than 1 nucleus, usually many.
MULTISERIATE: a thallus or filament composed of more than 1 series of cells.
Musa, 146
Mycanthococcus, 7, **108**, 273
    antarcticus, 108
Myrmecia, **109**, 272
    aquatica, 109
Myxochloridaceae, 276
Myxochrysidaceae, 278
Myxochrysis, 278
Myxosarcina, **230**, 280
    amethystina, 230

N

Naegelliella, **206**, 278
    britannica, 207
    flagellifera, 207
Naegelliaceae, 278
Nannochloris, 106, 112, 273
Nautococcus, **109**, 272
    piriformis, 109
Navicula, **266**, 279
    digitoradiata fa. *minor*, 267
    petersenii, 267
Naviculaceae, 279
Neidium, **264**, 279
    iridis var. *subundulata*, 265
    magellanicum, 264
Nemalionales, 279
Nemalionopsis, **159**, 279
    shawi, 159
Neochloris, **269**, 272
Nephrocytium, **68**, 83, 273
    agardhianum, 68
    ecdysiscepanum, 68
    limneticum, 68
    lunatum, 68
    obesum, 68
Nephroselmidaceae, 276
Nephroselmis, **43**, 276
Netrium, **99**, 276
    digitus, 99
NEUSTONIC: free-floating and planktonic but adhering below the surface film of water.
Nitella, 6, **19**, 124, 275
    flexilis, 19
    tenuissima, 19
Nitellopsis, 275
Nitzschia, **261**, 279
    bilobata, 261
Nitzschiaceae, 279
NODE: a position on a filament or a thallus which serves as a joint from which (usually) branches arise; a differentiated region separated by internodes, *q.v.*
NODOSE: with regular, intermittent swellings or nodes in a longitudinal axis.
Nodularia, **227**, 280
    spumigena, 227
NODULE: a small swelling; a tubercle or buttonlike knob.
Nostoc, 1, 209, **227**, 280

amplissimum, 227
    caeruleum, 227
    commune, 227
    linckia, 227
    parmeloides, 227
    pruniforme, 227
Nostocaceae, 280
Nostocales, 280
Nostochopsis, **222**, 280
    lobatus, 222
Notosolenus, 275
NUCULE: specialized branch bearing the oogonium (female sex organ) in the Charophyceae.

O

OBLATE: slightly flattened sphere; almost globular. Fig. 560

Figure 560

OBLATE SPHEROID: a figure which is a flattened sphere; an almost globular figure; flattened on one side of a globular figure. See Fig. 560
OBLONG: a curved figure, elongate, with the ends broadly rounded but more sharply curved than the lateral margins. Fig. 561

Figure 561

OBOVATE: an ovate (egg-shaped) figure, broader at the anterior end of a cell. Fig. 562

Figure 562

Ochromonadaceae, 278
Ochromonadales, 278
Ochromonas, **174**, 197, 278
    minuta, 174
    verrucosa, 174
Oedocladium, **141**, 274
    hazenii, 141

Oedogoniaceae, 274
Oedogoniales, 274
Oedogonium, **130**, 141, 274
    crispum, 131
    westii, 131
Oligochaetophora, **70**, 274
    simplex, 70
Oncobyrsa, 232, 280
Onychonema (Sphaerozosma), 275
Oocardium, **126**, 275
    stratum, 126
Oocystaceae, 272
Oocystis, 4, **83**, 113, 273
    borgei, 83
    eremosphaeria, 83, 110
Oodinium, **191**, 276
    limneticum, 191
OOGONIUM: a specialized cell bearing a female gamete; an egg case often with a special pore to admit the sperm, 17.
Oonephris, 68, 273
    obesa, 68
Oophila, **88**, 272
    amblystomatis, 88
OOSPORE: a zygote with a thick, sometimes 2- or 3-layered wall; formed by the union of heterogametes (egg and sperm).
OPAQUE: not permitting the transmission of light.
Opephora, **253**, 278
    martyi, 253
Ophiocytium, **173**, 177, 179, 182, 277
    arbuscula, 173
    capitatum, 173
    cochleare, 173
    desertum, 173
    gracilipes, 173
Ophrydium, 89
Order, defined, 3
Oscillatoria, 208, **214**, 280
    rubescens, 214
    splendida, 214
Oscillatoriaceae, 280
Oscillatoriales, 280
Ourococcus, **86**, 94, 273
    bicaudatus, 86
OVAL: an elongate figure with curved margins, the poles narrowly rounded. Fig. 563

Figure 563

OVOID: shaped like an egg; a curved figure with symmetrical convex margins but broader at one pole than the other. Fig. 564

Figure 564

P

Pachycladon, **113**, 273
    umbrinus, 113
Palatinella, 277
Palmella, **24**, 54, 58, 110, 272
    miniata, 24
    mucosa, 24
Palmellaceae, 272
Palmellococcus, 89, 111, 113, 272
Palmellocystis, 65, 272
Palmellopsis, **23**, 272
    gelatinosa, 24
Palmodictyon, **59**, 127, 140, 272
    varium, 60, 127
    viride, 59, 127
Pandorina, **48**, 271
    charkowiense, 48
    morum, 48
PARAMYLUM: a whitish, starch-like food reserve produced in members of the Euglenophyta.
PARASITIC: growing on or in another living organism and taking sustenance from the host; often pathogenic.
PARIETAL: along the wall; arranged at the circumference; marginal as opposed to central or axial. Fig. 565

Figure 565

Pascheriella, 45, 271
Pascherina, **46**, 271
    tetras, 45
Paulschulzia, 272
PEAR-SHAPED: a figure which is elongate and ovate, wider at one end than the other and usually distinctly narrowed or retuse in the midregion.
PECTIN: a gelatinous carbohydrate deposited in the cell or in the cell wall of many algae.
Pectodictyon, **59**, 80, 273
    cubicum, 59
PECTOSE: See Pectin.
Pediastrum, **75**, 76, 273
    biradiatum var. *emarginatum* fa. *convexum*, 75
    boryanum, 75
    obtusum, 75
    simplex, 75
    tetras, 75
PEDICEL: the stalk of an attached cell, long or short.
Pedinellaceae, 177
Pedinomonadaceae, 272
Pedinomonas, **42**, 272
    rotunda, 42

Pedinopera, **40**, 271
  granulosa, *41*
  rugulosa, *41*
PELLICLE: the outer skin or
  membrane of a cell which
  has no true cell wall. (See
  Euglena, 10).
Penium, **97**, 275
  margaritaceum, *97*
PENNALES: rod-shaped, bacil-
  liform Diatoms; raphe pres-
  ent in many.
Peranema, 275
Peranemaceae, 275
PERFORATE: with openings; not
  continuous.
Peridiniaceae, 276
PERIDININ: a brown pigment,
  often reddish, found in the
  Dinoflagellates especially.
Peridinium, 19, **200**, 276
  wisconsinense, *200*
PERIPHERY: the outer bound-
  ary; an outer region.
PERIPLAST: See Pellicle.
Perone, **171**, 277
  dimorpha, *172*
Peronia, **259**, 279
  erinacea, *259*
Peroniella, **173**, 182, 277
  hyalothecae, *173*
Petalomonas, 275
Phacotaceae, 271
Phacotus, **37**, 271
  lenticularis, *37*
Phacus, **33**, 275
  curvicauda, *33*
  triquetra, *33*
Phaeaster, 174, **183**, 277
  pascheri, *183*
Phaeodermatium, **168**, 201, 277
  rivulare, *168*
Phaeogloea, 278
Phaeophyta, **11**, 13, 279
Phaeoplaca, **206**, 277
  thallosa, *206*
Phaeoplacaceae, 277
Phaeoplacales, 277
Phaeosphaera, 206
Phaeothamniaceae, 279
Phaeothamnion, **186**, 278
  confervicola, *186*
Phalansteriaceae, 278
Phalensterium, 278
Phormidium, **216**, 217, 280
  ambiguum, *216*
  favosum, *216*
  inundatum, *216*
PHYCOCYANIN: a blue phyco-
  bilin pigment found in the
  Cyanophyta and the Rho-
  dophyta, 9
PHYCOERYTHRIN: a red phyco-
  bilin pigment found in the
  Cyanophyta and Rhodo-
  phyta, 9
Phycopeltis, **20**, 138, 148, 274
  arundinacea, *20*
PHYCOPYRRHIN: a brownish
  pigment found in some
  Dinoflagellata, 11.
Phyllobium, **90**, 272
  sphagnicola, *90*
Phyllosiphon, 7, **90**, 157, 277
  (Phyllosiphonaceae,
  Vaucheriales)
  arisari, *90*
Phylum, defined 3, 9, 12

Phymatodocis, **118**, 275
  nordstedtiana, *118*
Physolinum, **146**, 274
  monilia, *146*
Pigmentation, 9, *sq.*
PIGMENT-SPOT: See Eye-Spot.
Pinnularia, **266**, 279
  borealis, *266*
  viridis, *266*
Pithophora, **150**, 154, 274
  mooreana, *150*
  oedogonia, *150*
PLACODERM DESMIDS: those
  Desmids which have the
  cell wall in 2 sections that
  adjoin in the midregion; the
  wall pitted with pores and
  often furnished with granules, spines or verrucae.
  (See Saccoderm Desmids).
Placosphaera, 61, 272
PLANE (END WALLS): cross
  walls of cells in filamentous
  algae which are smooth and
  even (not infolded, q.v.).
  Fig. 566

Figure 566

PLANKTON: the organisms
  which drift or which cannot
  swim against currents.
Plankton Net, 7
Planktosphaeria, **66**, 84, 272
  gelatinosa, *65*
PLASTID: a body or organelle
  within the cell, sometimes
  containing pigments, as
  chloroplasts, *e.g.*
PLATE: a section of the cell wall
  as in the Armored Dino-
  flagellata; a disclike arrangement of cells in forming a colony. Fig. 567

Figure 567

Platychloris, 271
Platydorina, **45**, 271
  caudatum, *45*
Platymonas, **41**, 271
  elliptica, *41*
Plectonema, **212**, 280
  wollei, *212*
Pleodorina, **50**, 271

californica, *50*
  illinoisensis, *50*
Pleurocapsa, **232**, 280
  minor, *233*
Pleurocapsaceae, 280
Pleurochloridaceae, 276
Pleurochloridella, 276
Pleurochloridellaceae, 276
Pleurococcus, 72, 84, 111, 141
Pleurodiscus, **120**, 274
  purpureus, *120*
Pleurogaster, **178**, 180, 277
  lunaris, *178*
Pleurosigma, 274
Pleurotaenium, **96**, 275
  nodosum, *96*
  trabecula, *96*
Podohedra, 273
POLAR; POLE: the apices of
  cells or of a thallus, often
  showing basal-distal differ-
  entiation.
Polyblepharidaceae, 271
Polyblepharides, **44**
  fragariiformis, *44*
Polychaetophora, **69**, 274
  lamellosa, *70*
Polyedriopsis, **113**, 114, 115, 272
  quadrispina, *114*
  spinulosa, *113*
POLYGONAL: a figure or a cell
  with many sides.
POLYHEDRAL: a figure or cell
  with more than 4 sides.
Polytoma, 19, **34**, 271
  granuliferum, *34*
  obtusum, *34*
Polytomella, 271
Pond Scum, 1
Pond Silk, 2
Population, 2
Porphyridium, **23**, 167, 279
  cruentum, *23*
  magnificum, *23*
Porphyrosiphon, **215**, 280
  notarisii, *215*
POSTERIOR: toward the rear;
  the pole or region at or
  toward the base of a cell
  which is differentiated in
  polarity, or of a thallus.
Potassium Iodide, ix
Prasinocladus, **127**, 272
  lubricus, *127*
Prasinophyceae, 31
Prasiola, 119, **125**, 273
  crispa, *125*
Prasiolaceae, 273
Preparation, ix
Preserving Algae, ix, 8
PROCESS: an extension; horn,
  arm or protuberance.
  Fig. 568

Figure 568

Prokaryotic, Prokaryota, 9, 12
Prorocentraceae, 276
Protochrysis, **158**, 165, 197, 276
  phaeophycearum, *158*

Protococcus, 72, 84, 111, 274
  viridis, *73*
Protoderma, **144**, 274
  viride, *144*
PROTONEMA: an early or embryonic stage in the development of a larger or mature plant, either threadlike
  and filamentous or platelike; a primary growth
  stage.
PROTOPLAST: the living protoplasm within a cell wall or
  periplast; living material
  contained within a plasma
  membrane or a pellicle.
Protosiphon, **91**, 274
  botryoides, *91*
Protosiphonaceae, 274
Protozoa, 2, 10
Prunus, 2
Prymnesiaceae, 278
Prymnesiales, 278
PSAMMON: the association of
  organisms existing immediately beneath a surface
  layer or sand or loose soil,
  6.
Pseudendoclonium, 129, **152**,
  274
  submarinum, *152*
Pseudochaete, **142**, 274
  crassisetum, *142*
  gracilis, *142*
PSEUDOCILIA: fine, fibrillar
  hairs (false flagella), the axial canals (fibrils) being different from those of true
  flagella, 54, 61.
Pseudokephyrion, 278
PSEUDOPARENCHYMATOUS: a
  cushion or mound of cells,
  sometimes composed of
  closely adherent filaments,
  superficially resembling a
  multicellular thallus or
  frond. Fig. 569

Figure 569

Pseudopedinella, 277
PSEUDOPODIUM: a protoplasmic extension of an amoeboid organism, used in locomotion; a "false foot,"
  10.
PSEUDORAPHE: a narrow, clear
  area in the axis of a Diatom
  valve which superficially resembles a true raphe, *q.v.*
Pseudoschizomeris, **128**, 135,
  273
  caudata, *128*
Pseudostaurastrum, **177**, 277
  enorme, *177*
  hastatum, *177*
  muticum, *177*
Pseudotetraedron, **176**, 277
  neglectum, *176*

Pseudotetraspora, **23**, 65, 273
  gainii, 23
PSEUDOVACUOLE: a pocket of
  gas in some blue-green algal cells, appearing as a
  vacuole; a false vacuole,
  usually light refractive, 9.
Pseudulvella, **144**, 274
  americana, 145
Pteromonas, **35**, 271
  aculeata, 35
Pulchrasphaera, **270**, 272
PUNCTATE; PUNCTA: with
  minute, pinpoint pores in
  the cell wall, not necessarily extending completely
  through the wall, some-
  times in rows or patterns.
  Fig. 570

Figure 570

PYRAMIDAL: in the shape of a
  pyramid; a pointed, 4-sided
  figure with a broad base.
  Fig. 571

Figure 571

Pyramimonaceae, 271
Pyramimonas (Pyramidomonas),
  12, **44**, 45, 271
  tetrarhynchus, 44
PYRENOID: a proteinaceous
  body, usually within a
  chloroplast, around which
  starch or paramylum col-
  lects in a cell; may be on
  the surface of a chloroplast
  or rarely free in the cytoplasm; a glistening granule,
  10.
PYRIFORM: See Pear-shaped.
  Fig. 572

Figure 572

Pyrobotrys, 47, 271
Pyrrhophyta, **10**, 13, 14, 19, 276

Q

Quadrate: 4-sided, with a
  general outline showing
  4 sides.
Quadrichloris, **45**, 271
  carterioides, 45
Quadrigula, **68**, 273
  chodatii, 68

R

Raciborskia, **191**, 276
  bicornis, 191
Raciborskiella, **46**, 271
  uroglenoides, 46
RADIATE: extending outward in
  several planes from a common center; extending arms
  or rays, or filaments in one
  plane from a common
  center.
Radiococcaceae, 273
Radiococcus, **65**, 273
  nimbatus, 65
Radiofilum, **134**, 273
  conjunctivum, 135
  flavescens, 135
Radiosphaera, 272
Ragweed, 28
RAPHE: a longitudinal fissure in
  the valve wall of some Diatoms; the Pennales may
  have such a fissure in one
  or both valves; lacking in
  the Centrales.
Raphidiopsis, **218**, 280
  curvata, 218
  mediterranea, 219
Raphidium, 74, 87
Raphidonema, 7, **136**, 273
  nivale, 136
Rayssiella, 83, **84**, 273
  hemispherica, 84
RECTANGULAR: a figure with
  4 sides, with 2 parallel
  sides longer than the other
  2.
RECTILINEAR: arranged in
  straight rows in two directions.
RED ALGAE: See Rhodophyta.
RED SNOW: snow (and ice) at
  high altitudes reddened by
  the presence of cryoplank-
  ters which contain a reddish carotinoid pigment
  haematochrome.
RED TIDE: a profuse development of microscopic algae
  which are reddish in color
  or which have pseudovacuoles which refract light
  (some Dinoflagellata; *Trichodesmium*), coloring extensive areas of the ocean
  especially over the conti-
  nental shelf, or of freshwater bodies.

RENIFORM: kidney-shaped;
  bean shaped. Fig. 573

Figure 573

REPLICATE: infolded cell cross
  walls, as in some species
  of *Spirogyra;* not a plane or
  straight cross wall. Fig. 574

Figure 574

RESERVOIR: a cavity in the
  anterior end of a flagellated
  cell from which the organs
  of locomotion arise, 11, 29.
RETICULATE: netted; arranged
  to form a network; with
  openings. Fig. 575

Figure 575

Rhabdoderma, **239**, 280
  linearis, 240
Rhabdomonas, 275
Rhizochloridaceae, 276
Rhizochloridales, 276
Rhizochrysidaceae, 277
Rhizochrysidales, 277
Rizochrysis, **193**, 205, 277
  limnetica, 193
Rhizoclonium, **130**, 147, 164, 274
  hieroglyphicum, 130
  hookeri, 130
RHIZOID; RHIZOIDAL: irregu-
  larly branched, rootlike
  extensions of cells or of
  protoplasm, used by some
  organisms for locomotion;
  extensions of some cells or
  of thalli for anchorage.
RHIZOPODAL: with pseudo-
  podia; with protoplasmic,
  rootlike extensions, 10.
Rhizosolenia, **249**, 278
  eriensis, 249
Rhizosoleniaceae, 278
Rhodochorton, **163**, 279
Rhodochytrium, **28**, 90, 166, 272
  spilanthidis, 28
Rhododendron, 21
Rhodomelaceae, 279
Rhodomonas, **25**, 276
  lacustris, 25
Rhodophyta, 11, 13, 19, 279
Rhoicosphenia, **258**, 279
  curvata, 258

RHOMBOID: a figure with 4
  oblique angles; referring to
  cells which are somewhat
  angular and nearly isodia-
  metric.
Rhopalodia, **260**, 279
Riccia, 6
  fluitans, 6
Ricciocarpus, 6
  natans, 6
Rivularia, **220**, 280
Rivulariaceae, 280
Romeria, **215**, 280
  elegans var. *nivicola*, 215
Rosa cinnamonea, 2
Rotifera, 6
Roya, **100**, 275
  obtusa, 100
Ruttnera, 278
Ruttneriaceae, 278

S

SACCATE: like a sack; balloonlike cell, or a colony of
  cells; a bulbous plant body.
  Fig. 576

Figure 576

SACCODERM DESMIDS: des-
  mids which have a plane,
  smooth (usually) cell wall in
  one piece and without
  pores, seldom ornamented.
  (See Placoderm Desmids).
Sacconema, **219**, 280
  rupestre, 219
Sacheria, 279
Salpingoeca, 278
Salpingoecaceae, 278
SARCINA ARRANGEMENT:
  cells arranged in the form
  of a cube or block.
SCALARIFORM CONJUGATION:
  ladderlike; sexual reproduction by conjugation tubes
  formed between cells of
  2 filaments, forming a ladderlike figure.
Scenedesmaceae, 273
Scenedesmus, 4, **52**, 77, 79, 273
  bijuga var. *alternans*, 52
  incrassatulus var. *mononae*,
  52
  opoliensis, 52
  quadricauda, 52
Scherfellia, 30, **41**, 271
  cornuta, 41
Schizochlamys, **61**, 272
  gelatinosa, 61
Schizodictyon, 55, 272
Schizogoniaceae, 126, 273

SCROBICULATE; SCROBICULA-
  TION: with saucerlike
  depressions in a plane surface, *e.g.,* cell wall, sometimes deep. Fig. 577

Figure 577

SEMICELL: a cell-half, as in the
  Placoderm Desmids in
  which the cell wall is in
  2 parts and the cell contents paired, usually with a
  constriction in the midregion, and with an isthmus
  connecting the 2 semicells.
Semiparasitic, 7
SEPTUM: a cross partition,
  either complete or incomplete through a cell cavity,
  transverse or longitudinal,
  as in some Diatoms.
SERRATE: toothed or jagged
  margin.
SESSILE: sedentary; attached
  without a stalk or pedicel.
SETA: a hair, or fine spine, arising either from the cell wall
  or from within the cell and
  extending through a pore.
  Fig. 578
Sewage oxidation, 1
SHEATH: Covering or envelope,
  usually of mucilage, either
  soft or firm; the covering
  external to a group or colony of cells, sometimes
  layered (lamellate).

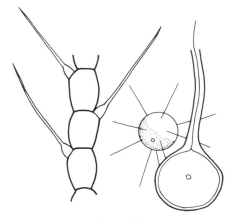

Figure 578

SICKLE-SHAPED: acutely
  curved, crescent-shaped,
  but more sharply curved
  than an arc of a circle.
  Fig. 579

Figure 579

SIPHON; SIPHONOUS: a tube; a
  thallus without cross partitions, usually multinucleate.
SKEIN: a weblike expanse; a
  thin membranous sheet of
  cells or of a sheath.
SNOW ALGAE: See Red Snow.

SPHERULE: a specialized
  branch which forms a globular structure involving
  sperm-producing cells in
  the Characeae, sometimes
  referred to as an antheridium, 17.
SPICULE: a scale or needle in
  the walls of certain cells,
  usually siliceous.
SPINDLE-SHAPED: See Fusi-
  form.
SPINE: a sharply pointed projection from a cell wall, either
  needlelike or tapering.
SPORANGIUM: a cell, either an
  unspecialized vegetative
  cell or a morphologically

differentiated cell which
  produces spores (motile or
  non-motile asexual reproductive elements); the case
  which forms around the zygospore in the Zygnem-
  atales.
Staining, x
STARCH: a solid carbohydrate
  $(C_6H_{10}O_5)$n formed as a food
  reserve, especially in green
  algae and in higher plants.
STARCH TEST: See Iodine Test.
STAR-SHAPED: See Stellate.
STAUROS: a wide and conspicuous central nodule under
  the valve wall of Diatoms,
  best seen in valve view.
STELLATE: star-shaped; with radiating projections from a
  common center, either in
  one or in several planes.
STINGING CELLS: organelles
  (trichocysts) either immediately within a cell membrane, or scattered, able to
  throw out threads (protec-
  tive?), 10, 11.
STIPE: a stalk or pedicel, slender or stout, 11, 29, 31.
STRATIFIED: with layers. (See
  Lamellae.)
STRIAE: lines or etchings
  (especially in Diatom frustules); sometimes com-
  posed of rows of fine dots
  (punctae).

Strombomonas, 32, 275
Stylococcaceae, 277
Stylodinium, **192**, 276
  globosum, 192
Stylosphaeridium, **93**, 272
  stipitatum, 93
SUBAERIAL: See Aerial.
SUBAPICAL: near the anterior
  end but not at the apex of a
  cell or of a thallus.
SUBCYLINDRICAL: nearly cylindric, with very slightly convex lateral margins.
SUBFLAGELLAR: below the flagella; points or granules
  where flagella arise.
SUBQUADRANGULAR: nearly
  square; somewhat rectangular; sometimes square with
  rounded corners.
SUBSPHERICAL: near spherical.
  (See: Oblate Spheroid).
SULCUS: a furrow or groove,
  especially the longitudinal
  furrow on the ventral side
  of a Dinoflagellate. (See
  Peridinium, Fig. 394).
Surirella, **246**, 256, 279
  oblonga, 246
  ovalis, 246
Surirellaceae, 279
SUTURE: a groove; a winglike
  margin; a rim about the
  zygospore in many Zygnemataceae; the groove between the plates of the Armored Dinoflagellata.
Symbiotism, 165, 166
Symploca, **216**, 280
  muscorum, 216
Syncrypta, **203**, 278
  volvox, 203
Synechococcus, **238**, 280
  aeruginosus, 238
Synechocystis, **233**, 280
  aquatilis, 233
Synedra, **254**, 278
Synura, 2, **203**, 278
  uvella, 203
Synuraceae, 278
Synuropsis, 278

T

Tabellaria, **252**, 279
  fenestrata, 252
Taxon, defined, 3
Tea (*Thea*), 21, 146
TEGUMENT: a skin; a firm outer
  covering.
Teilingia, **116**, 275
  granulata, 116
Temnogametum, **123**, 275
Terpsinoe, **247**, 278
  americana, 247
TEST: an outer shell; a firm
  outer covering, either
  transparent or opaque, with
  a space between the en-
  closed protoplast.
Tetmemorus, **95**, 275
  laevis, 95
Tetrablepharis, 271
Tetrachloris, 45, 271

Tetracladus, **53**, 272
  mirabilis, 53
Tetracyclus, **252**, 279
  lacustris, 252
Tetracystis, **270**, 272
Tetradesmus, **79**, 273
  smithii, 79
  wisconsinense, 79
Tetradinium, **190**, 276
  simplex, 191
Tetraedriella, **181**, 277
  gigas, 181
Tetraedron, 4, **114**, 115, 177, 272
  caudatum, 114
  incus, 114
  minimum, 114
TETRAGONAL: with 4 angles
  which are in 2 opposite
  planes.
Tetragonidium, **188**, 276
  verrucatum, 188
Tetragoniella, **181**, 277
TETRAHAEDRON: See Tetrago-
  nal.
Tetrallantos, **67**, 80, 273
  lagerheimii, 67
Tetrapedia, **238**, 280
  reinschiana, 238
Tetraselmiaceae, 271
Tetraselmis, **41**, 271
Tetraspora, **54**, 60, 61, 127, 272
  cylindrica, 54
  gelatinosa, 54
Tetrasporaceae, 272
Tetrasporales, 268, 272
Tetrasporopsis, **206**, 277
  perforata, 206
Tetrastrum, **76**, 273
  heteracanthum, 76
Thallochrysidiaceae, 277
THALLUS; THALLOID: a plant
  body without differentiation
  into true roots, stem and
  leaves.
Thalpophila, **223**, 280
  imperialis, 223
Thamniochaete, **145**, 274
  huberi, 145
Thea: See Tea, 21, 146
THECA; THECATE: a skin; cell
  wall composed of plates as
  in the Armored Dino-
  flagellata.
Thoracomonas, **37**, 271
  feldmannii, 37
Thorea, **160**, 279
  ramosissima, 161
Thoreaceae, 279
Thylakoids, 9
Tolypella, **17**, 275
  intricata, 17
Tolypothrix, **228**, 280
  distorta, 228
Tomaculum, **67**, 273
  catenatum, 67
TOW; TOW SAMPLE: a collection made by casting or
  towing a plankton net.
Trachelomonas, 19, 28, **32**, 275
  ampulla, 32
  conica, 32
Trachychloron, **181**, 277
  biconnicum, 181
TRANSVERSE FURROW: a gir-
  dle or sulcus extending
  around the cell transversely, especially in the Dinoflagellata. Fig. 580

Figure 580

TRAPEZIFORM; TRAPEZOID: a
  figure which has 2 parallel
  sides; shaped somewhat
  like a trapezoid. Fig. 581

Figure 581

Trebouxia, **108**, 272
  cladoniae, 108
Trentepohlia, 7, 12, **22**, 148, 151,
  274
  aurea, 22
  jolithus, 22
Trentepohliaceae, 274
Trentepohliales, 274
Trentonia, 276
Treubaria, **115**, 273
  crassispina, 115
Tribonema, **169**, 277
  bombycinum, 169
  bombycinum var. *tenue*, 169
  utriculosum, 169
Tribonemataceae, 277
Tribonematales, 277
Triceratium, 278
TRICHOCYST: See Stinging
  Cells.
Trichodesmium, **215**, 216, 280
  erythraceum, 215
  lacustre, 215
TRICHOGYNE: a necklike exten-
  sion of the carpogonium,
  the female reproductive
  organ in the Rhodophyta.
TRICHOME: a fine hair; a bristle;
  the thread of cells in the
  Cyanophyta exclusive of
  the sheath.
Trichophilus, 7, **140**, 274
  welcheri, 7, **140**
Trichosarcina, **129**, 152, 273
  polymorpha, 129
Triplastrum, **97**, 275
  indicum, 97
Triploceras, **95**, 275
  gracile, 95
Trochiscia, **108**, 273
  granulata, 108
  obtusa, 108
  reticulata, 108
Tropidoscyphus, 275
TRUE BRANCH: a branch
  formed by the lateral division of a cell in the main
  filament or principal trichome (as in some blue-
  green algae).
TRUNCATE: cut off abruptly at
  the apex; flat or plane at
  the end as opposed to be-

ing rounded or pointed.
  Fig. 582

Figure 582

Trypanochloridaceae, 277
Trypanochloris, 277
TUBERCLE: a buttonlike knob or
  protuberance.
TUBULAR; TUBULAR THALLUS:
  an elongate, strandlike or
  threadlike growth; a hollow
  thallus without cross walls;
  a plant body in the form of
  a tube.
TUMID: swollen or convex; inflated, especially on lateral
  margins of a cell or thallus.
  Fig. 583

Figure 583

Tuomeya, **161**, 279
  fluviatilis, 161
Turtle, 7, 151
TYCHOPLANKTON: the unat-
  tached, free-floating organisms near shore and/or intermingled with vegetation
  in shallow water. (See
  Euplankton).
Typha, 162

U

Ulothrix, **135**, 137, 273
  aequalis, 135
  cylindricum, 135
  zonata, 135
Ulotrichaceae, 273
Ulotrichales, 273
Ulvaceae, 273
Ulvales, 273
Ulvella, **145**, 274
  involvens, 145
UNDULATE: regularly wavy at
  the margin or over the surface.

UNILATERAL: arising on one side only.

UNISERIATE: cells arranged in a single series or row.

Urceolus, 275

Urococcus, 27, 110, 276
  insignis, 27

Uroglena, **202**, 278
  volvox, 202

Uroglenopsis, 202, **203**, 278
  americana, 204

Uronema, **137**, 273
  elongatum, 137

UTRICLE: a saclike or tubular sheath, usually a firm, mucilaginous cover. Fig 584

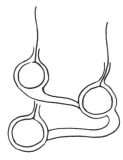

Figure 584

Utricularia, 8

Uva, **47**, 271
  gracilis, 48

V

Vacuolaria, **29**, 276
  virescens, 29

VALVE: one of the 2 parts of a Diatom cell wall, the flat surfaces of the frustule, one (the epivalve) larger than the lower (hypovalve).

VALVE VIEW: the aspect of a Diatom cell when seen from above or below so that the valve is in view as opposed to the side or girdle view.

Vaucheria, 19, **157**, 277
  geminata, 157
  sessilis, 157

Vaucheriaceae, 277

Vaucheriales, 277

Vaucheriopsis, 277

VEGETATIVE: non-reproductive cells, or state; reproduction by cell division or fragmentation without the formation of special reproductive elements.

VENTRAL: the under or lower side of an organism or thallus as opposed to the top, back or dorsal surface.

VERMIFORM: elongate and threadlike; wormlike.

VERRUCA: a warty or spiny, thick projection from the cell wall; with knots or tubercles. Fig. 585

Figure 585

VERRUCOSE: roughened; with irregular thickenings on the surface. Fig. 586

Figure 586

VESICLE: a sac or balloonlike cell or thallus; a globular pocket.

Vitamins, 1

Volvocaceae, 271

Volvocales, 271

Volvox, **48**, 271
  tertius, 48

Volvulina, 271

W

WATER BLOOM: a profuse, conspicuous growth of planktonic algae which clouds or colors the water, often forming floating scums and causing biological disturbances, 1, 48.

WATER NET: See *Hydrodictyon*.

WATER SUPPLIES: Pollution, 1

Westella, **81**, 273
  botryoides, 81

WHORL: several lateral parts (branches or cells) arising at one level from an axis and from different sides.

Wislouchiella, **36**, 271
  planctonica, 37

Wollea, **225**, 280
  saccata, 226

X

Xanthidium, **105**, 275
  cristatum var. *uncinatum*, 105

Xanthophyceae, 10, 164, 276

XANTHOPHYLL: a yellow pigment of several kinds associated with chlorophyll ($C_{46}H_{50}O_2$), 9 , 10.

Xenococcus, **231**, 280
  schousbei, 231

Y

Yellow-green algae, 10

Z

Zoochlorella, **88**, 272, 273
  conductrix, 89
  parasitica, 89

ZOOSPORE: an asexual, motile spore, equipped with 1 or more flagella, and usually with an eye-spot.

Zygnema, 3, 118, 119, **120**, 275
  pectinatum, 120

Zygnemataceae, 3, 274

Zygnematales, 3, 274

Zygnemopsis, **119**, 275
  decussata, 119
  desmidioides, 119

Zygogonium, **120**, 275
  ericetorum, 120

ZYGOSPORE: a thick-walled resting spore formed as a result of union of iso- or anisogametes; a zygote with a thick wall in a dormant state.